BIOMASS GASIFICATION

BIOMASS GASIFICATION

Principles and Technology

Edited by T.B. Reed

Solar Energy Research Institute
Golden, Colorado

NOYES DATA CORPORATION

Park Ridge, New Jersey, U.S.A.

1981

Library of Congress Catalog Card Number: 81-9667
ISBN: 0-8155-0852-2
Printed in the United States

Published in the United States of America by
Noyes Data Corporation
Noyes Building, Park Ridge, New Jersey 07656

Library of Congress Cataloging in Publication Data

Reed, Thomas B.
Biomass gasification.

(Energy technology review ; no. 67)
Bibliography: p.
Includes index.
1. Biomass energy. I. Title. II. Series.
TP360.R42 665.7'76 81-9667
ISBN 0-8155-0852-2 AACR2

Foreword

This detailed review of biomass gasification principles and technology was written to aid in determining the areas of gasification which are ready for commercialization and those areas in which further research and development will be most productive. The book presents relevant scientific background information, surveys the current status of gasification activities, and examines various questions concerning the uses of the product gases.

Biomass (any material derived from growing organisms) has the potential for being an energy source with few significant environmental drawbacks and some important environmental benefits. Direct use of biomass as a fuel offers a limited field of application because of problems of distribution, combustion, and emissions. Gaseous fuels, however, have been used for more than a century because they are clean burning and easy to distribute. In addition, the gases can be converted to other (liquid) fuels or can serve as feedstocks for strategic chemicals. Thus, gasification could continue to supply the "convenience" gaseous and liquid fuels that the nation has come to depend on during the age of low-cost fossil fuels.

The book is structured to serve as a handbook on topics pertinent to gasification, as well as to provide reviews of past and present activities which will be of use to both the generalist and the specialist.

The information in the book is from: *A Survey of Biomass Gasification. Volume I—Synopsis and Executive Summary* (SERI/TR-33-239-V.1, July 1979), *Volume II—Principles of Gasification* (SERI/TR-33-239-V.2, July 1979), and *Volume III—Current Technology and Research* (SERI/TR-33-239-V.3, April 1980) edited by T.B. Reed of the Solar Energy Research Institute (SERI), prepared for the U.S. Department of Energy.

The table of contents is organized in such a way as to serve as a subject index and to provide easy access to the material contained in the book.

In order to keep the price of this book to a reasonable level, it has been reproduced by photo-offset directly from the original reports and the cost savings passed on to the reader. Due to this method of publishing, certain portions of the reports may be less legible than desired.

Acknowledgements

This survey was compiled by a number of SERI staff members and consultants under the direction of T.B. Reed. Although many authors contributed to the survey and are listed in the Contents and Subject Index, many others had less formal input and are herewith thanked for their efforts.

Contents and Subject Index

PART I: SUMMARY

PART II: GASIFICATION PRINCIPLES

Part I

Summary

The information in Part I is from *A Survey of Biomass Gasification. Volume I–Synopsis and Executive Summary* (SERI/TR-33-239), edited by T.B. Reed of the Solar Energy Research Institute, prepared for the U.S. Department of Energy, July 1979.

Executive Summary

The production of energy from biomass (any material derived from growing organisms) is now seen by many to be a leading near-term solar energy technology. Already, 1% to 2% of U.S. energy is generated by combustion of biomass, and this established technology is being commercialized wherever possible and with as much speed as possible. However, solid fuels have limited applications in modern industrial society and many environmental problems as well.

Fortunately, biomass can be gasified by a number of existing or developing processes. Air gasification (burning with a limited amount of air) is already being commercialized, but much engineering and scientific work remains before oxygen gasification (burning with limited oxygen) or pyrolytic processes (breaking down of matter, usually by heat) for gasification are ready for commercialization. SERI believes that gasification will be the leading edge of thermal biomass development for at least a decade. Therefore, before beginning specific projects a survey was made of existing knowledge and present work in this area and in adjoining technologies (fuel synthesis, gas cleanup) whose development will enable gasification to have maximum impact.

The survey has a number of important goals:

- to examine the properties and potential of the biomass resource relevant to gasification (Chapters 1 to 4);
- to summarize the basic science of biomass gasification (Chapters 5 to 7);
- to look at the present state of research, development, and commercialization of gasifiers (Chapters 8 to 10);
- to examine processes associated with gasification for gas cleanup and synthesis of other fuels from biomass-gas (Chapters 11 to 13);
- to determine means by which gasification technology can be introduced more rapidly (Chapter 14); and
- to identify the areas where research and development will be needed in an intensified gasification development program (Chapter 15).

The survey fills over 400 pages and assembles in one place a wide range of technical and institutional information as an aid to engineers and decisionmakers in this field. The background and conclusions that are believed to be of interest to policymakers and the larger nontechnical audience involved in energy policy are highlighted in this summary. Those interested in greater technical depth are referred to the main body of the survey.

INTRODUCTION (Chapter 1)

Gaseous fuels have many advantages over solid fuels. Gases can be burned more efficiently and with less emissions; the gas flame is more easily controlled for sensitive industrial processes such as glassmaking and drying; gases can be distributed easily for domestic and industrial use; gases can be used to operate engines for power generation and transport; modern gas/oil burners can be retrofitted easily to use gas generated from biomass residues or coal but not solid fuels; some gases can be used for chemical synthesis of liquid fuels and chemicals such as methanol, gasoline, or ammonia. Solid fuels can be gasified efficiently in central plants, the cleaned gas can be distributed in pipelines, and the ashes and pollutants can be disposed of efficiently. This type of fuel distribution is necessary to the continued existence of our large cities, where local burning of solid fuels would entail enormous distribution and emission problems.

The gasification of coal and biomass began in about 1800 and the superior properties of gaseous fuels relative to solid fuels caused this technology to develop so fast that by about 1850 gas light for streets was commonplace. Before the construction of natural gas pipelines in the United States between 1935 and 1960, there were about 1,200 municipal "gasworks" serving larger towns and cities. During the petroleum shortages of World War II in Europe, almost a million small gasifiers were used to run cars, trucks and buses, using primarily wood as fuel. Although coal has been the preferred fuel for larger gasifiers in the past, technical and environmental changes are likely to give biomass a larger role in gasification in the future.

Gasification of solid fuels is accomplished in high-temperature processes similar to combustion that convert the fuel to a gas with minimal loss (typically 10 to 30%) of the energy of the solid fuel. The methods used for gasification can be divided into the four categories shown in Figure S-1. Air gasification is the simplest process but gives a gas of low energy content that must be "close-coupled" to its immediate use for heat or power. Air gasification is already being commercialized. Oxygen gasification gives a gas of higher energy content that can be distributed in industrial pipelines or used for chemical synthesis of a variety of fuels and chemicals such as methanol, ammonia, methane, and gasoline. Commercial prototypes have been operated successfully. Pyrolysis also can yield gas of medium energy but in addition yields oils and chars that have a utility of their own. Pyrolytic processes are still in the development stage.* Fast pyrolysis can yield a gas especially rich in unsaturated hydrocarbons that can form the basis of gasoline or alcohol synthesis. The energy contents of various gases are listed in Table S-1 along with their uses. [We have used the terms "low energy gas" (LEG) etc., as more descriptive than "low Btu gas" (LBG) etc., and as compatible with international usage and the SI system.]

THE POTENTIAL BIOMASS RESOURCE BASE (Chapter 2)

The importance of biomass conversion technologies depends on the quantity of biomass that can be made available for conversion to gas. The existing resource base is comprised of agricultural residues, manures, wood and bark mill residues, logging residues, noncommercial (cull) trees in the forests, and the organic fraction of municipal solid wastes. The quantities potentially available are summarized in Table S-2, which shows an enormous total potential of about 15 quads. Not all of this resource can be collected, and the amount used will depend on energy costs, competition from other fuel and solar energy sources, environmental and ecological factors, etc.

In addition to these forms of existing biomass, there are several other large reservoirs of biomass energy that are even more difficult to quantify. A number of "biomass mines," consisting of past residues, have accumulated over the years. These include municipal wastes, sometimes even now digesting to give methane; food processing plant residues; and bark piles. Though only available on a one-time basis, the biomass mines are a potentially low-cost and environmentally attractive energy source.

A second unexploited category of biomass is that available through land improvement. Many acres of land have been laid waste by man and can support only the growth of such plant species as scrub, mesquite, and chapparal. Harvesting of these plants for their biomass energy and conversion of this energy to fuels could pay for the cost of improving the land.

Finally, there is the large potential of "energy plantations," in which land or even oceans and lakes could be used to raise biomass for energy purposes. Again, the economics of these processes, and energy needs, will determine the degree to which they are developed.

---

*Hydrogen can be used under pressure to give higher energy gases or liquids, but hydrogasification of biomass is still in its infancy.

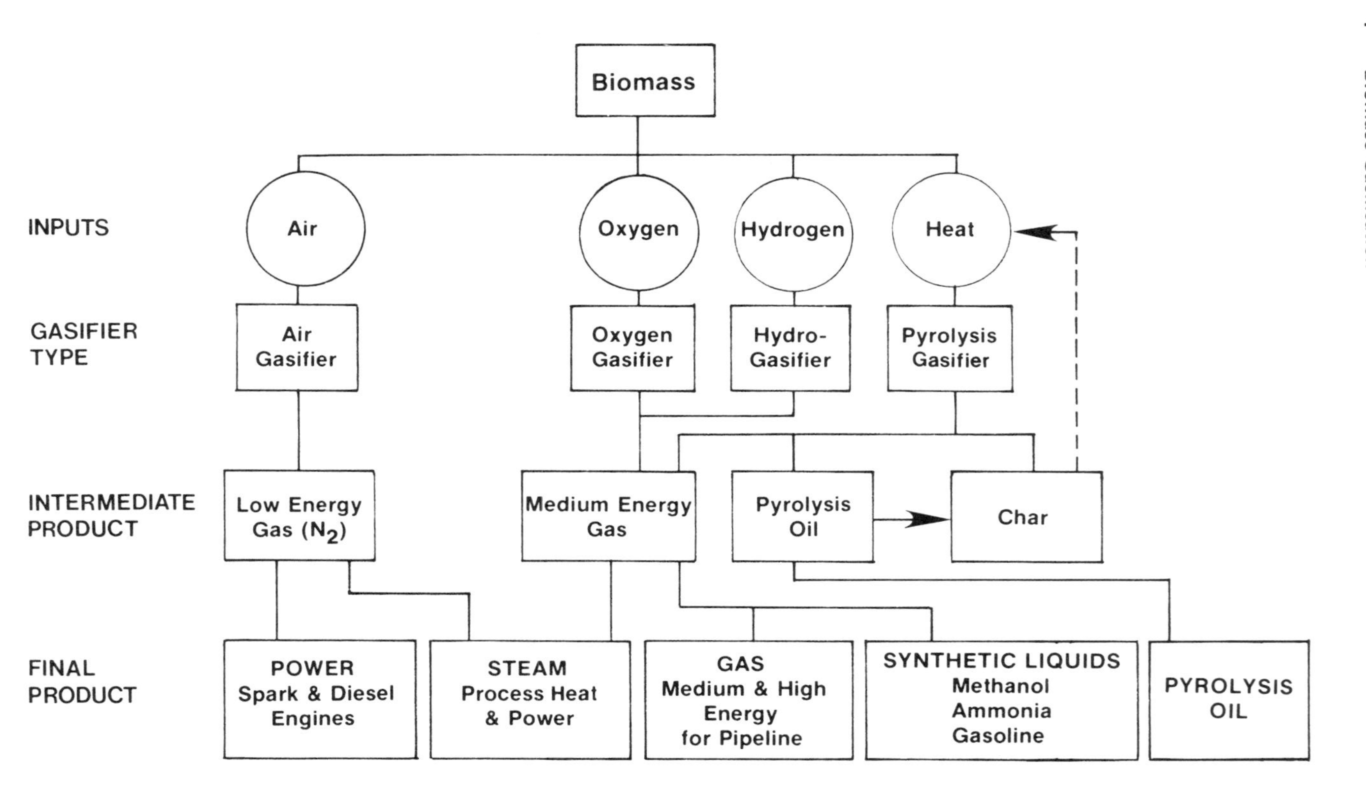

Figure S-1. Gasification Processes and Their Products

Table S-1. ENERGY CONTENT OF FUEL GASES AND THEIR USES

Name	Source	Energy Range (Btu/SCF)	Use
Low Energy Gas (LEG) [Producer Gas, Low Btu Gas]	Blast Furnace, Water Gas Process	80-100	On-site industrial heat and power, process heat
Low Energy Gas (LEG) [Generator Gas]	Air Gasification	150-200	Close-coupled to gas/oil boilers Operation of diesel and spark engines Crop drying
Medium Energy Gas (MEG) [Town Gas; Syngas]	Oxygen Gasification Pyrolysis Gasification	300-500	Regional industrial pipelines Synthesis of fuels and ammonia
Biogas	Anaerobic Digestion	600-700	Process heat, pipeline (with scrubbing)
High Energy Gas (HEG) [Natural Gas]	Oil/Gas Wells	1000	Long distance pipelines for general heat, power, and city use
Synthetic Natural Gas (SNG)	Further Processing of MEG and Biogas	1000	Long distance pipelines for general heat, power, and city use

Table S-2. SUMMARY OF THE ANNUAL ENERGY POTENTIAL OF EXISTING SOURCES OF BIOMASS

Resource	10^6 Dry Tons/Year	Quads/Year
Crop Residues	278.0	4.15
Animal Manures	26.5	0.33
Unused Mill Residues[a]	24.1	0.41
Logging Residues	83.2	1.41
Municipal Solid Wastes	130.0	1.63
Standing Forests[b]	384.0	6.51
TOTALS:	925.8	14.44

[a] Does not include unused bark from wood pulp mills
[b] Surplus, noncommercial components

PROPERTIES OF BIOMASS RELEVANT TO GASIFICATION (Chapter 3)

Biomass is easier to gasify than coal because it has a much higher volatile content (typically 70% to 90%) and because it contains its own oxygen and water, two elements important in forming gaseous molecules from high-carbon feedstocks. With a few exceptions, biomass has less than 2% ash (while coal is typically 5% to 20%), and the typical biomass sulfur content is less than 0.1% as compared to 2% to 4% in coal. Biomass materials have carbon contents considerably lower than coals and the hydrogen/carbon ratio is typically 1.5; for coal it is close to 1.0.

These advantages of biomass for gasification are offset in part by a high moisture content, generally requiring drying before gasification, and by a lack of large concentrations of biomass, thus favoring small gasifiers with higher costs. However, very large quantities of biomass associated with many biomass processing plants (wood, lumber, food) are likely to be important in making these industries energy self-sufficient. Municipal solid waste also occurs in large quantities in the cities.

Biomass has three principal components—cellulose, hemicellulose, and lignin—and both the structure and the gasification of the infinite varieties of biomass can be understood in terms of the behavior of these components. In addition, minor amounts of extractables—hydrocarbons, tannins, oils, and resins—can add to the fuel and chemicals derived from biomass.

Because of the many forms in which biomass occurs, it is difficult to make general statements about the thermal properties of biomass relevant to gasification. The heat of combustion of pure cellulose is 7,250 Btu/lb and that of pure lignin is 11,500 Btu/lb, so that the heat of combustion of the various mixtures of cellulose and lignin in different forms of biomass ranges from about 7500 Btu/lb to 9500 Btu/lb, a much smaller range than for coals. Thermal conductivities are very low for biomass materials, ranging from 0.01 Btu/h-ft^2 (°F/ft) to 0.1 Btu/h ft^2 (°F/ft) depending on form, and this is important in the behavior of biomass during gasification. Other properties important in understanding the gasification process are the heat capacity and the diffusivities, particularly of transition charcoal forms. Though it is known that the porosity of charcoal greatly improves the kinetics of gasification, very few data are available on porosity, heat capacity, and diffusivity. Work is in progress at SERI to learn more about some of these properties.

BENEFICIATION OF BIOMASS (Chapter 4)

In many cases the energy content of biomass is unavailable because the biomass form is unsuitable for conversion. Often gasification processes require beneficiation of the biomass (improving its properties so that energy can be recovered more economically) before it can be used, and it is important to know the energy costs of each step.

Biomass often has a high moisture content, and some gasification processes require dry feedstock. Though in theory this water can be vaporized with an applied heat of 1000 Btu/lb water vaporized, in practice it requires 1500 to 2500 Btu/lb, depending on the efficiency of the drier. Fortunately, low-grade heat such as stack heat can be used for this purpose. Commercial equipment is widely available for both wood and agricultural biomass.

Often the physical form of available biomass is wrong for gasification because fixed-bed gasifiers require relatively large, solid pieces to allow room for gas passage, while fluidized and suspended gasification may require powders or dusts. Commercial equipment is available for reducing larger wood pieces to a size of half an inch; the energy needed to do this is less than 1% of the amount of energy contained in the wood. An interesting combination of size reduction and drying is accomplished in the "hot dog," a device used by forest industries to dry chips with waste stack heat.

Reducing particle size below half an inch becomes increasingly costly in energy. A new process, ECO-FUEL II, which uses a mild chemical attack on the biomass during milling, reduces required milling energy by an order of magnitude to make particles of about 200 μm. These very small particles, if available at a low cost, may make fast pyrolysis, with its high production of olefins, economically attractive.

Biomass has many properties that make it superior to coal as a fuel, but its bulk density is very low, thus increasing shipping and collection costs and reducing conversion rates in gasifiers and combustion units. Densification is a new technology that overcomes these disadvantages and makes essentially "instant coal" from biomass residues such as sawdust, bark, and straw. The biomass is dried to about 20% moisture content and then, under high pressure, it is pressed or extruded to form pellets, briquettes, or logs. These have a specific gravity of 1 to 1.3 depending on the process, as compared to a specific gravity of 0.4 to 0.6 for wood and even less for other biomass forms. There is synergism between pelletizing and gasification: pellets are a superior feedstock for gasification, and gas fuel (from pellets) is an efficient way of drying pellets. The energy required for making pellets is 1% to 2% of the amount of energy in the dry biomass. Wet biomass must be dried, but this drying energy is largely recovered in the more efficient final gasification or combustion of the pellets.

PYROLYSIS (Chapter 5)

Pyrolysis is the breakdown of biomass by heat at temperatures of 200 C to 600 C to yield a medium energy gas, a complex pyrolysis oil, and char. All biomass gasification and combustion processes involve pyrolysis as a necessary first step: in combustion, subsequent oxidation of the products leads to total heat release; in gasification the products are used directly or are converted to other fuel forms.

There are two kinds of pyrolysis: slow and fast. At slow heating rates or with large pieces of biomass, pyrolysis leads to a high proportion of charcoal that must then be gasified. At the most rapid heating rates, cellulose is largely converted to a gas containing a high proportion of olefins that are valuable as a chemical feedstock; char production is minimal.

Although not yet proven quantitatively, it is commonly accepted that the pyrolysis of the many complex forms of biomass can be understood as the sum of the breakdown of its three components: cellulose, hemicellulose, and lignin. This is borne out qualitatively by comparison of laboratory analyses of the pyrolysis of components with those of whole biomass.

Pyrolysis is studied in the laboratory using several types of thermal analysis instruments. Thermogravimetric analysis (TGA) yields data on the weight loss of biomass as a function of either time (isothermal TGA) or temperature (dynamic TGA). These measurements yield the proximate analysis of the biomass sample, giving the percentages of moisture, volatiles, char, and ash. TGA data are useful in determining the rates of pyrolysis and are qualitatively relevant to pyrolysis in gasifiers though fixed bed gasifiers probably pyrolyze at slower rates than are convenient in the laboratory while fast pyrolysis is beyond the range of ordinary laboratory instruments. The decomposition rate of cellulose is usually fitted by a classical kinetic equation of the form:

$$dV/dt = V\ A\ \exp(-E/RT)$$

where V is the remaining volatile component at temperature T, A is an adjustable constant, R is the gas constant, and E is the activation energy. This equation can also predict the decomposition of hemicellulose and lignin but with less accuracy.

Another very useful technique in understanding pyrolysis is the semiquantitative technique of differential thermal analysis (DTA) that has been supplanted recently by the quantitative differential scanning calorimetry (DSC). Both techniques measure the heat input to the sample at a constant heating rate and determine whether various stages of the pyrolysis are endothermic (requiring heat) or exothermic (producing heat).

At fast heating rates leading primarily to gas, pyrolysis seems to be endothermic across the entire temperature range. Thus, the faster pyrolysis techniques require a moderate heat input at pyrolysis temperatures. That slower pyrolysis leads to more char formation and is exothermic at higher temperatures is consistent with the observation that pyrolysis can be "autothermic," and a pyrolytic gasifier, if properly arranged and insulated, requires no net heat input for partial gasification.

The gases and liquids evolved during pyrolysis are commonly measured with mass spectrometry, infrared spectrophotometry, or gas and liquid chromatography. Analysis suggests that at the temperature of pyrolysis the primary products are not affected by the presence of air, steam, or hydrogen, and that pressure is not an important variable except as it influences the escape of primary products. A great deal of work has been done on the chemical mechanisms involved in the breakdown of cellulose, with less known about lignin, wood, and hemicelluloses. More work is required on the effect of particle size and heating rates on both primary and secondary pyrolysis of the products.

An emerging field that is relevant to gasification is "fast pyrolysis," the very rapid heating of finely divided biomass resulting in maximal gas yields. A number of investigations, some aimed at converting solid municipal waste to energy forms, have determined the composition of the products resulting from various heating techniques. In addition, some investigators are examining the subsequent "gas phase pyrolysis" of the oils produced from the solid, a process which is likely to become very important if gas is the only product desired. Furthermore, this vapor cracking can yield other products, primarily olefins, of much greater value than the products obtained in conventional solid pyrolysis. These products are valuable precursors to gasoline or alcohol.

THERMODYNAMICS OF GAS-CHAR REACTIONS (Chapter 6)

Pyrolysis at temperatures of 200 C to 600 C is a nonequilibrium process. However, in gasification pyrolysis is generally followed by an oxygen, air, or steam conversion of the resulting oils, tar, and char to CO, H_2, or methane, and under some conditions the combined reactions closely approach equilibrium. Thermodynamic calculations, while not necessarily enabling accurate predictions of gas compositions in gasification, are at least restrictive in that they set the boundaries to what is possible in gasification.

As a part of the survey of the current state of knowledge of gasification, we used a computer program to predict the equilibrium gas compositions to be expected under a wide variety of conditions encountered in gasification. This allows rapid comparison with experimental results and often suggests useful modifications to processes.

A useful parameter in understanding the various gasification and combustion processes is the adiabatic flame (reaction) temperature (AFT), the temperature that would be reached by the products of the reaction if equilibrium were achieved. This temperature is shown in Fig. S-2 as a function of the equivalence ratio (ER), the ratio of the actual oxygen content of the air supplied to the oxygen required for complete combustion. Thus, for an equivalence ratio of 1.0, the flame temperature of biomass when burned in pure oxygen is about 2800 C, while for combustion in air it is 2050 C, close to that observed in wood combustion.

Gasification with air or oxygen occurs at an equivalence ratio of 0.25 to 0.3. In this region the reaction temperature is only 700 C to 1100 C in air and about 100 C higher in oxygen.

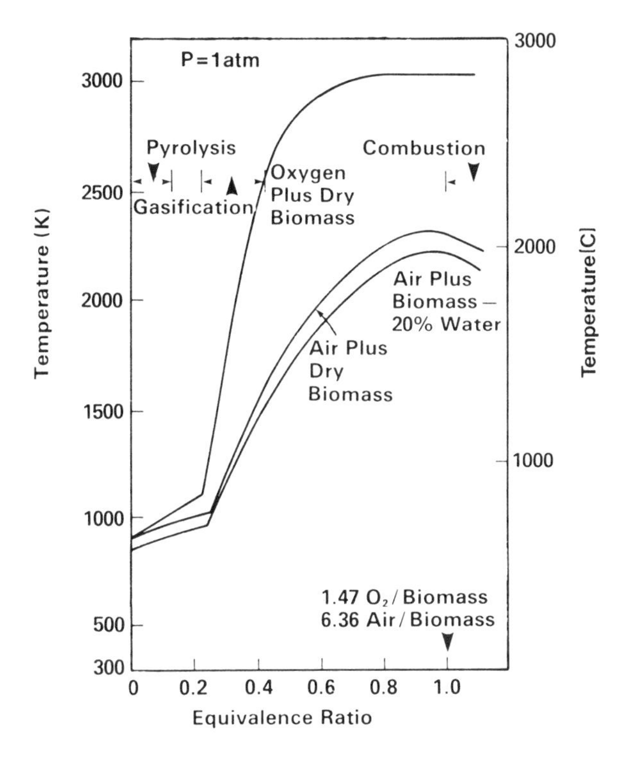

Figure S-2. Biomass Adiabatic Reaction Temperatures

The equilibrium gases produced during these processes are shown in Fig. S-3 where it is also shown that at the lowest equivalence ratios some methane is formed while CO and H_2 are the predominant fuel gases. At increasing equivalence ratios, char is converted to gas up to an equivalence ratio of 0.25. As the equivalence ratio approaches 1.0 for complete combustion, fuel gases are converted to the combustion products. This can also be seen in Fig. S-4, which shows the energy content of char and gas as the ER increases. The heating values (heat of combustion per unit volume) of the gases produced in oxygen and air gasification are shown as a function of ER in Fig. S-5.

A phenomenon occurring in the gasification region called "flame temperature stabilization" is an important factor in comprehending the operation of gasifiers. A series of reactions involving carbon, hydrogen, and oxygen are highly endothermic above about 500 C; in these reactions the initial combustion products H_2O and CO_2 are reduced to form the fuel gases H_2 and CO. Though kinetically slow at temperatures below 800 C (see section on kinetics), these reactions become very fast above 1200 C. Thus as long as any H_2O or CO_2 is present in the gas-char mixture, temperature increases will be suppressed and fuel gas will be produced. For this reason gasification equipment is relatively simple and does not have to be made of the highly temperature resistant materials used in combustion equipment. Furthermore, this buffering of the flame temperature also gives relatively stable gas compositions.

Finally, an equivalence ratio of zero corresponds to no oxidation and pure pyrolysis. Figure S-2 shows the surprising result that even without any oxygen or energy addition, biomass could reach a temperature of about 900 K (627 C) if a kinetic route to equilibrium could be found. Biomass pyrolysis can be regarded as a means of bringing biomass to equilibrium with a minimum of energy loss, time, and equipment. Unfortunately, this equilibrium includes formation of about 30% char; so, a second task in gasification is

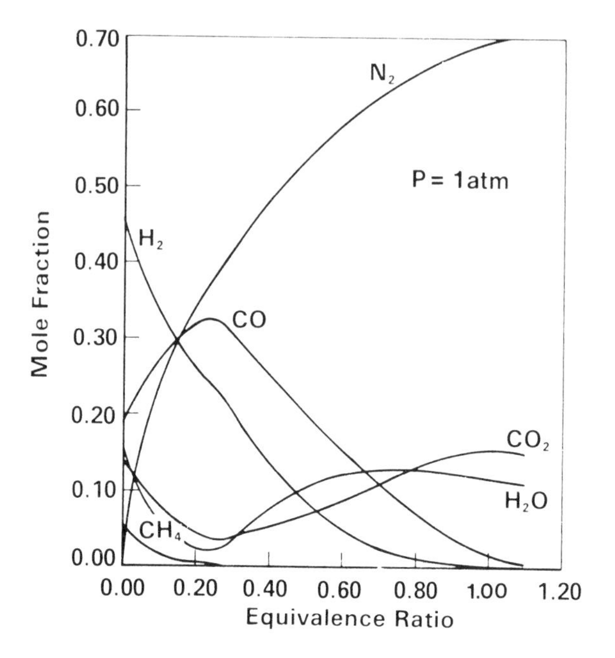

Figure S-3. Equilibrium Composition for Adiabatic Air/Biomass Reaction

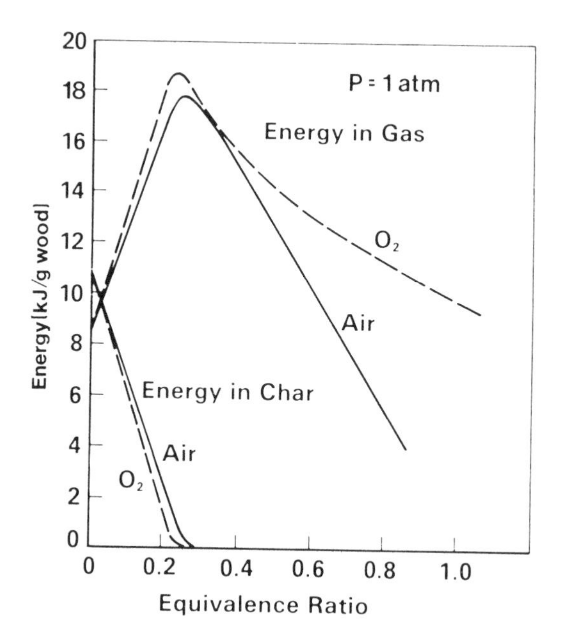

Figure S-4. Gas and Char Energy Content for Oxygen and Air/Biomass Equilibrium

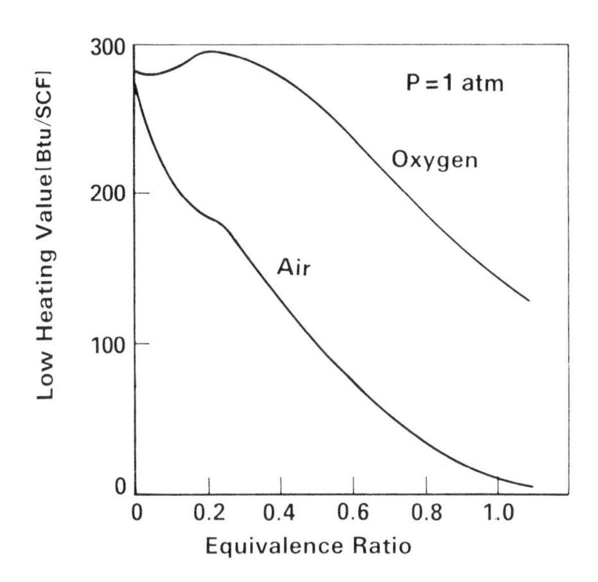

Figure S-5. Low Heating Value for Dry Equilibrium Gas for Air/O_2 - Biomass Reaction

conversion of any unwanted char to gas. This is accomplished most easily by using the char to reduce H_2O to H_2, but this, in turn, complicates the process. (Biomass often contains as much as 50% excess water that can thus be put to good use here.) The effect of water addition on the conversion of char has been examined at various temperatures and pressures.

In updraft gasifiers the initial reaction is in the hot zone (where equilibrium may be approached) but subsequent reactions occur at successively lower temperatures so that it is not expected that these equilibrium calculations will have much relevance to the final gas composition, though they are important in understanding the reactions at the grate. In downdraft gasifiers combustion occurs first and then the gases are drawn through the hot charcoal, thus having a good chance to reach a quasi-equilibrium. Finally, in fluidized bed gasifiers a number of variations of temperature can be used to produce specific intermediate equilibrium states, thus giving better control over gas composition.

KINETICS OF CHAR GASIFICATION REACTIONS (Chapter 7)

Although equilibrium favors the formation of fuel gases in any system where there is an excess of char, the rate of conversion of char to gas depends in a rather complex fashion on the kinetics of the reactions. Without catalysts, very little reaction occurs below about 800 C, but at higher temperatures the reactions become very rapid and equilibrium considerations dominate. The degree of reaction is influenced by the particle size; the physical properties of the char, especially its porosity and lifetime; and the methods of contacting gas with char in fixed-bed, fluidized-bed, or suspended flow gasification.

Fortunately, a great deal of work done in the gasification of coal is also applicable to biomass. However, very little of this work has been applied to biomass, a task for the coming years.

In the reaction of char to form gas, the following steps occur in series and each can, under certain conditions, limit the reaction rate:

- diffusion of reactants across the boundary layer at the external char surface;
- diffusion of gas through the pores of the solid;
- adsorption, surface reaction, and desorption of gas on the pore wall;
- diffusion of products out of the pore; and
- diffusion of products across the boundary layer.

The overall reaction rate (i.e., the effective reaction rate in a practical situation) is composed of two factors: the rate of heat and mass transfer between the bulk gas surrounding the char particle and the particle, and the true kinetics of reaction at the char surface or in the pores. A very useful parameter in evaluating the relative importance of these two factors is the "effectiveness factor," a measure of the effect of pore volume and surface on reaction rate.

The external heat and mass transfer are described by well-known equations in terms of the heat and mass transfer coefficients which, in turn, depend on diffusion coefficients, thermal conductivities, reactant concentrations, and other gas properties. At sufficiently high temperatures, these coefficients do not change rapidly with temperature.

The mass transfer coefficient behaves like a diffusion coefficient. If an Arrhenius behavior is assigned to the mass transfer coefficient, at sufficiently high temperatures the effective activation energy is very low, only about 4 kcal/mole. There is also an activation energy required for heat transfer, and as a practical consequence at high temperatures the particle temperature can be significantly lower (endothermic reaction) or higher (exothermic reaction) than the surrounding gas temperature.

At lower temperatures, the gasification reactions occur principally within the char particle, requiring the reactants to diffuse into the pores to the reacting surface. The average rate of diffusion within the pores relative to the rate of diffusion to the particle surface is given by the effectiveness factor. Effectiveness factors are estimated for biomass chars and, at low temperatures with small particles, external heat and mass transfer are not limiting. At temperatures over 1100 C for gasification reactions and at lower temperatures for combustion reactions, the effectiveness factor approaches zero and external heat and mass transfer are limiting. The porosities of chars produced from biomass materials are such that comparable gasification rates are obtained at temperatures 100 C to 200 C lower than those required for coal.

Particle size also determines the degree to which mass and heat transfer are limiting. For the small particles encountered in suspended or fluidized-bed gasification, external transfer is never important below about 1100 C. However, for fixed-bed operation and large particles, transfer becomes limiting at lower temperatures. Adaptation to biomass of the heat and mass transfer equations developed in coal gasification is an important task in gasification research.

Much theoretical and experimental work has been done to determine the mechanism of the reaction of CO_2 and steam with chars. Such mechanistic studies are necessary to elucidate these gasification reactions for biomass. These studies should be coupled with experimental work on the reactivities of the various forms of char that arise during pyrolysis and that change as the char is consumed. Data show that chars from biomass are much more reactive than those from coal. Several investigators have determined the effect of catalysis on char gasification and found mixed results, ranging from the anticatalytic effects of many minerals to a tripled reaction rate catalyzed by K_2CO_3.

An interesting field now being explored is hydrogasification of coal. The rapid heating of char in a hydrogen atmosphere enhances hydrocarbon yields. Few studies of the kinetics of biomass hydrogasification have been done, but this should be a fruitful field of research.

A SURVEY OF GASIFIER TYPES (Chapter 8)

The central problem in gasification is to convert all of the elements comprising solid biomass into gases containing the highest possible energy. Yet no combination of the constituent elements of dry biomass leads directly to gas only. For instance, an equilibration of dry biomass at 1000 C would give:

$$CH_{1.4}O_{0.6} \rightleftarrows 0.7\ H_2 + 0.6\ CO + 0.4\ C\ \text{(solid)}$$

in which $CH_{1.4}O_{0.6}$ is a representative formula for biomass and the solid char formed contains a significant part of the biomass energy. Gasification at lower temperatures avoids equilibrium and produces a high proportion of oil in addition to char. Conversion of these chars and oils to gases can be done by four basic types of gasification: air gasification, oxygen gasification, hydrogasification, and pyrolytic processes comprising generally more complex cycles.

Air Gasification

The simplest form of gasification is air gasification, in which the excess char formed by pyrolysis is burned with a limited amount of air at an equivalence ratio of about 0.25, requiring 1.6 g air per gram of biomass.

The simplest air gasifier is the updraft gasifier shown in Fig. S-6. Air is drawn up through a fixed bed of biomass on a grate. At the lowest and hottest level on the grate, combustion and char gasification occur; as the gases rise they reach the successively lower temperature pyrolysis and drying zones and exit the gasifier at low temperatures, saturated with pyrolysis oils and water. Ideally, this gas is burned immediately in a boiler, the so-called "close-coupled" operation. The temperature of the output gas must be kept high enough to prevent condensation of oils before combustion, yet low enough to prevent the oils from coking. A number of these units are now in operation in the United States.

Oil production is largely eliminated in downdraft gasifiers (Fig. S-7), where air is introduced between the char zone and the pyrolysis zone. Heat from the char zone pyrolyzes the biomass above; the tars and oils pass down through a bed of hot charcoal where they are cracked and reduced, mostly to H_2 and CO fuel gas. Several million of these gasifiers were used in Europe during World War II to operate cars and trucks.

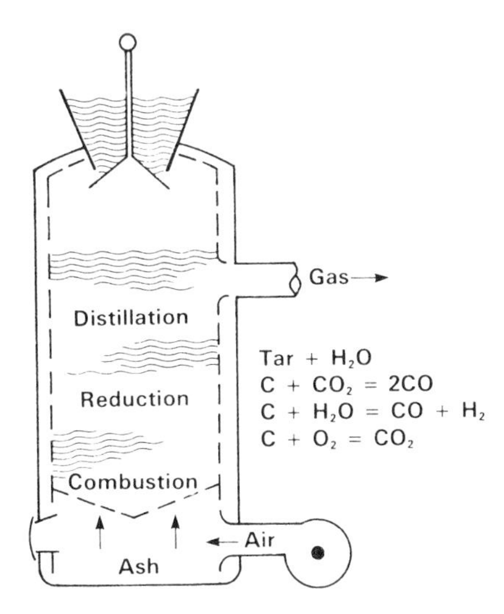

Figure S-6. Schematic Diagram of Updraft Gasifier

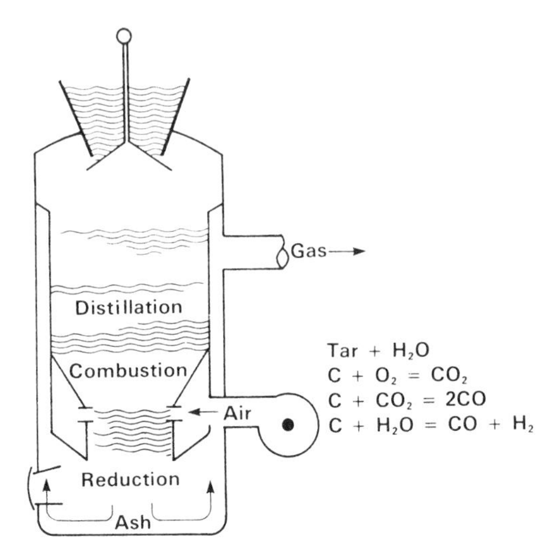

Figure S-7. Schematic Diagram of Downdraft Gasifier

Fixed-bed gasifiers require biomass of a relatively uniform size larger than several centimetres in the smallest dimension, so that gas passages are provided in and around the particles. A wider range of particle size and higher throughput can be achieved with fluidized-bed gasifiers, in which a sufficiently large flow of gas is maintained to provide a fluidized bed. Fluidized-bed gasifiers often contain a solid heat transfer agent such as a catalyst or sand and generally require a recycling of the product gas to maintain fluidization. It is claimed that these gasifiers minimize oil production and maximize char consumption, but they are in the early stages of development.

Air gasifiers are simple, cheap, and reliable and have operated almost continuously for decades at a time. Their chief drawback is that the gas produced is low in energy and would be uneconomical to distribute; it must be used on-site for process heat to operate engines and for power generation.

Oxygen Gasification

The production of low energy gas is not a problem in oxygen gasifiers, in which the product is undiluted by nitrogen from air and could be distributed in an industrial pipeline network, as town gas was distributed in the United States until 1940. In addition, the medium energy gas is a necessary precursor to the manufacture of methanol, ammonia, methane, or gasoline.

Updraft oxygen gasification has been demonstrated with municipal solid waste (MSW - Purox process). A small downdraft oxygen gasifier has been operated on a SERI contract. The chief disadvantage of oxygen gasification is that it requires an oxygen plant or nearby source of oxygen and thus increases the cost of gasification.

Hydrogasification

Research is just beginning on the effects of added H_2 (or CO) on gasification, with emphasis on enhanced direct methane production.

Pyrolysis Gasification

Oxygen and air gasifiers consume char directly by increasing the oxygen content of the biomass to permit gas formation. In pyrolytic processes gas, oil, and char all are formed simultaneously in a reactor. Subsequently the char and oil are converted in a separate reactor to heat and additional gas. The subsequent process recirculates hot solids or hot gases as a heat exchange medium for additional conversion of the char and oil to gas. A high moisture content in the biomass, a liability in air and oxygen gasification, contributes hydrogen in pyrolytic processes.

The four types of gasifiers mentioned in Fig. S-1 can be grouped into a large number of subdivisions according to various characteristics:

- Fuel type: including biomass, solid municipal waste, peat, coal;
- Fuel size: chunks, shreds, pellets, powder;
- Fuel gas contact: updraft (counterflow), downdraft (co-flow), fluidized bed, suspended particle;
- Ash form: dry ash for grate temperature below about 1100 C; slagging for temperatures above 1300 C, depending on feed;
- Pressure: Although no pressurized biomass gasifiers now exist, there are a number of advantages to building gasifiers operating at 10 to 100 atm; and
- Catalyst use.

Of the many types of gasifiers, those for which examples are given in the main body of the report are listed in Table S-3.

DIRECTORY OF GASIFIER MANUFACTURERS (Chapter 9)

Questionnaires were sent to the manufacturers and researchers listed in Table S-3, who are currently working on gasifiers; the results are given as a directory listing the various characteristics of existing gasifiers by manufacturer.

SURVEY OF GASIFIER RESEARCH (Chapter 10)

Where scientific and engineering studies are in progress for gasification processes, the processes are summarized. Gas compositions, salient features, and the present status of many of the projects listed in Table S-3, among others, are given in more detail. Some of the projects are primarily research, developing information useful to the gasification community; others are in the development stage, characterizing a particular gasifier in engineering terms and determining and solving operational problems. Others have been built on a commercial scale and are being use-tested. References in the literature are provided where available. The listing is not complete, relying heavily on current studies supported by DOE. Additions and corrections are welcome.

ECONOMICS OF GASIFICATION FOR EXISTING GAS/OIL SYSTEMS (Chapter 11)

A particularly attractive feature of gasification is that it permits continued use of existing gas/oil equipment. This retrofit capability has caused a great deal of interest in air gasification and a number of companies have been formed to manufacture and sell air gasifiers.

In comparing the cost of retrofitting existing equipment to new installations, it is estimated that the purchase of an air gasifier in the size range from 5 MBtu/h to 100 MBtu/h, for attachment to existing boilers, will cost about two-thirds of the cost of a new solid fuel installation, as shown in Fig. S-8. Furthermore, the simplicity of gas-burning boilers suggests that a gasifier combined with a new gas boiler will be comparable in price to installing a new, solid-fueled boiler. The gasifier combination offers lower emissions and higher turndown ratios than the solid-fueled boiler, and the option to burn gas or oil.

Table S-3. SURVEY OF GASIFIER RESEARCH, DEVELOPMENT, AND MANUFACTURE[a,b]

Organization	Gasifier Type: Input	Gasifier Type: Contact Mode	Fuel Products	Operating Units	Size Btu/h
Air Gasification of Biomass					
Alberta Industrial Dev. Edmonton, Alb., Can.	A	Fl	LEG	1	30M
Applied Engineering Co., Orangeburge, SC 29115	A	U	LEG	1	5M
Battelle-Northwest Richland, WA 99352	A	U	LEG	1-D	—
Century Research, Inc. Gardena, CA 90247	A	U	LEG	1	80M
Davy Powergas, Inc. Houston, TX 77036	A	U	LEG-Syngas	20	—
Deere & Co. Moline, IL 61265	A	D	LEG	1	100kW
Eco-Research Ltd. Willodale, Ont. N2N 558	A	Fl	LEG	1	16M
Forest Fuels, Inc. Keene, NH 03431	A	U	LEG	4	1.5-30M
Foster Wheeler Energy Corp. Livingston, NH 07309	A	U	LEG	1	—
Fuel Conversion Project Yuba City, CA 95991	A	D	LEG	1	2M
Halcyon Assoc. Inc. East Andover, NY 03231	A	U	LEG	4	6-50M
Industrial Development & Procurement, Inc. Carle Place, NY 11514	A	D	LEG	Many	100-750kW
Pulp & Paper Research Inst.,[c] Pointe Claire, Quebec H9R 3J9	A	D	LEG	—	—
Agricultural Engr. Dept. Purdue University W. Lafayette, IN 47907	A	D	LEG	1	0.25M
Dept. of Chem. Engr. Texas Tech University Lubbock, TX 79409	A	Fl	LEG	1	0.4M
Dept. of Chem. Engr. Texas Tech University Lubbock, TX 79409	A	U	LEG	1	—
Vermont Wood Energy Corp. Stowe, VT 05672	A	D	LEG	1	0.08M
Dept. of Ag. Engr. Univ. of Calif. Davis, CA 95616	A	D	LEG	1	64,000
Dept. of Ag. Engr. Univ. of Calif. Davis, CA 95616	A	D	LEG	1	6M
Westwood Polygas (Moore)	A	U	LEG	1	
Bio-Solar Research & Development Corp. Eugene, OR 97401	A	U	LEG	1	- -

[a]Table notation defined at end of table.
[b]Unless noted otherwise, the gasifiers listed here produce dry ash (T < 1100 C) and operate at 1 atm pressure. (Coal gasifiers and future biomass gasifiers may operate at much higher pressures.)
[c]Operates at 1-3 atm pressure.

Table S-3. SURVEY OF GASIFIER RESEARCH, DEVELOPMENT, AND MANUFACTURE (continued)

Organization	Gasifier Type Input	Gasifier Type Contact Mode	Fuel Products	Operating Units	Size Btu/h
Oxygen Gasification of Biomass					
Environmental En. Eng. Morgantown, WV	O	D	MEG	1P	0.5
IGT-Renugas	O,S	Fl	MEG		
Pyrolysis Gasification of Biomass					
Wright-Malta Ballston Spa, NY[a]	PG	O	MEG (C)	1R, 1P	4
Coors/U. of MO	P	Fl		1P	
U. of Arkansas	P	O	MEG (C)	IR	
A & G Coop Jonesboro, AR	P	O	MEG (C)	1C	
ERCO Cambridge, MA	P	Fl	PO, C	1P, (1C)	16, (20)
ENERCO Langham, PA	P		MEG, PO, C	1P, 1C	
Garrett Energy Research	MH		MEG	1P	
Tech Air Corporation Atlanta, GA 30341	P	U	MEG, PO, C	4P, 1C	33
M. Antal Princeton Univ. NS	PG	O	MEG, C	1 R	--
M. Rensfelt Sweden	PG	O	MEG, C	1 R	
Texas Tech Lubbock, TX	PG	Fl	MEG	1 P	
Battelle-Columbus Columbus, OH					
Air Gasification Solid Municipal Waste (CSMW)					
Andco-Torrax[b] Buffalo, NY	A	U	LEG	4C	100M
Battelle NW Richmond, VA 99352					
Oxygen Gasification of SMW					
Union Carbide (Linde) Tonowanda, NY[b]	O	U	MEG	1	100M
Catorican Murray Hills, NS	O	U			9M
Pyrolysis Gasification of SMW					
Monsanto, Landgard, Enviro-chem.	P, C	K	LEG, O, C	1 D	20 (375)
Envirotech, Concord, CA	P	MH	LEG	1 P	
Occidental Res. Corp El Cajon, CA	P	Fl	PO, C, MEG	1 C	

[a]Operates at 10 atm pressure.

[b]These gasifiers produce slagging (T > 1300 C) instead of dry ash.

Table S-3. SURVEY OF GASIFIER RESEARCH, DEVELOPMENT, AND MANUFACTURE (concluded)

Organization	Gasifier Type Input	Gasifier Type Contact Mode	Fuel Products	Operating Units	Size Btu/h
Garrett En. Res. & Eng. Hanford, CA	P	MH	MEG	1P	
Michiga Tech, Houghton, MI	P	ML	MEG		
U. of W. Va-Wheelebrator Morgantown, WV	P, G, C	Fl	MEG	1P	
Pyrox Japan	P, G, C	Fl	MEG	1C	
Nichols Engineering	P		MEG, C		
ERCO Cambridge, MA	P	Fl	MEG	1P	16
Rockwell International Canoga Park, CA	P	MS	MEG, C	1P	16
M. J. Antal Princeton, NS	P	O	MEG, C	2R	--

TABLE NOTATION: (by columns)

Input: A = air gasifier; O = oxygen gasifier; P = pyrolysis process; PG = pyrolysis gasifier; S = steam; C = char combustion

Contact Mode: U = updraft; D = downdraft; O = other (sloping bed, moving grate); Fl = fluidized bed; S = suspended flow; MS = molten salt; MH = multiple hearth

Fuel Products: LEG = low energy gas (~150-200 Btu/SCF) produced in air gasification; MEG = medium energy gas produced in oxygen and pyrolysis gasification (350-500 Btu/SCF); PO = pyrolysis oil, typically 12,000 Btu/lb; C = char, typically 12,000 Btu/lb

Operating Units: R = research; P = pilot; C = commercial size; CI = commercial installation; D = demonstration

Size: Gasifiers are rated in a variety of units. Listed here are Btu/h derived from feedstock throughput on the basis of biomass containing 16 MBtu/ton or 8000 Btu/lb, SMW with 9 MBtu/ton. ( ) indicate planned or under construction.

In order to compare gas costs of various technologies, SERI has adapted the cost analysis method developed at the Electric Power Research Institute (EPRI). This method was used to estimate the costs of gas produced in two gasifiers. The resulting costs are shown in Table S-4 for biomass costing \$20/dry ton. Since gasifiers are low in capital costs, the conversion and operating costs (first year) are \$0.17/MBtu to \$0.26/MBtu. At \$20/dry ton, total costs for gas are to \$2.58/MBtu to \$4/MBtu. However, many manufacturers have biomass residues available at a cost considerably lower than \$20/ton.

Table S-4. DETAILED COST BREAKDOWN FOR \$20/TON FUEL (\$/MBtu)

	Gasifier "A" (15 MBtu/h)		Gasifier "B" (85 MBtu/h)	
	1978 Cost	Levelized Cost	1978 Cost	Levelized Cost
Operating Costs	\$0.11	\$0.15	\$0.13	\$0.19
Capital Costs	0.06	0.09	0.13	0.19
Fuel Cost	2.55	3.75	2.32	3.40
TOTAL COSTS	\$2.72	\$3.99	\$2.58	\$3.78

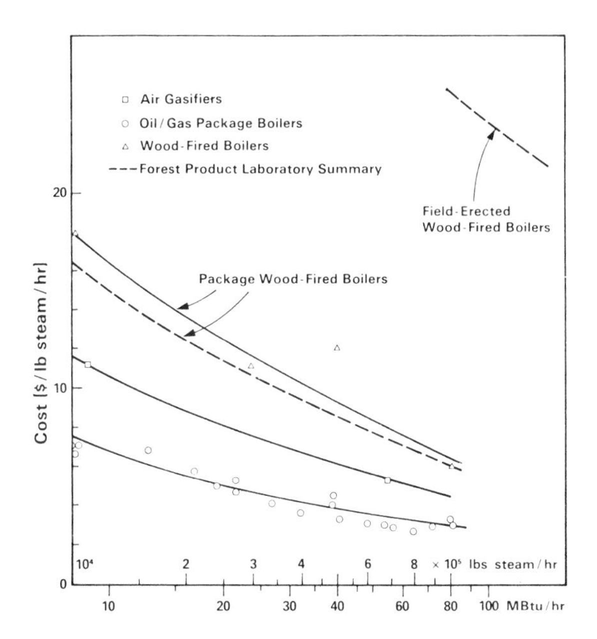

Figure S-8. Cost Comparisons Between Retrofitting Existing Equipment and New Installations

GAS CONDITIONING PROCESSES (Chapter 12)

Any working gasifier is only a part of a system involving solid feed delivery, gas conditioning, and final use. Conditioning the gas can be as costly and difficult as gasification itself. The Mittelhauser Corporation has made a thorough study of the existing methods and the costs of gas scrubbing, one form of gas conditioning.

If the gas from a gasifier is to be used directly for heat (close-coupled operation) there is probably no need for conditioning. In all other cases, however, oils, tars, and hydrocarbons contained in the gas may prohibit its distribution in a pipeline or its use as a chemical feedstock. To condition the gas for its final use, it is necessary to employ a range of available commercial equipment.

The raw gas typically contains as much as 5% (by weight) of oxygenated oils and tar vapor. These can be removed by scrubbing with a spray of the oil itself or with water in a variety of scrubber designs, followed by a mist eliminator or an electrostatic precipitator, depending on the final application. If the gas is to be used primarily for heat, this treatment is generally sufficient.

If the gas is to be used for chemical synthesis of methanol, ammonia, gasoline, or natural gas, further conditioning is required because of the presence of hydrocarbons that can affect the catalyst and possibly of sulfur (though biomass is relatively low in sulfur). Also, the carbon/hydrogen ratio of the gas must be adjusted to the proper value for chemical synthesis. The processes of hydrogenation, re-forming, and cryogenic separation to accomplish these ends are discussed.

The design of gas conditioning plants is studied and commercial practice is illustrated. Examples of costs for hydrogenation, re-forming, and cryogenic separation plants are developed. For instance, the capital cost of gas cleanup for methanol manufacture is

$127/daily ton. Although it is probable that improvements can be made in both gasification itself and in cleanup, this is a very sizable fraction of processing cost and must not be overlooked.

PRODUCTION OF LIQUID FUELS AND CHEMICALS FROM BIOMASS GASIFICATION (Chapter 13)

Gasification is already becoming important for the production of manufactured gases to replace natural gas and oil. Ultimately of equal importance may be the production of liquid fuels and chemicals, from what is known as "synthesis gas," often called "syngas," a mixture of CO and H_2. Commercial processes for using this gas already exist and are summarized in Table S-5. Here it is evident that a wide variety of useful products can be made, provided that syngas can be produced from biomass. This chapter, prepared by Science Applications Inc., provides an understanding of syngas technology and some examples of the costs of making synthetic fuels and chemicals.

Also shown in Table S-5 are the percentages of the heating value of syngas lost in conversion to the products shown and the "equilibrium" temperature for the conversion reaction. Conversion must normally be made at temperatures below this value and therefore will require catalysts and often high pressure. There also is an energy loss involved in conversion, though the penalty is justified by the higher value of the product.

The most important of the syngas reactions in the United States today is the production of methanol. Currently about a billion gallons per year are made from natural gas, primarily for the plastics industry. The reaction utilizes a CuO-ZnO catalyst at a pressure of 100 atm at about 300 C. All of the syngas conversions are exothermic, and reactors must be specially designed to carry this heat away; on the other hand, this heat is available at a relatively high temperature and can be used for compression and power generation.

Several other methanol catalysts are also available, and a new, liquid phase methanol synthesis process is being developed that removes the reaction heat more efficiently. Projections show a cost advantage of about 15% over present processes. Present processes based on natural gas have efficiencies of 50% to 70%. Biomass processes are projected to have overall efficiencies in the 30% to 50% range.

A number of studies have been made of the cost of methanol production from wood, refuse, gas, and coal in the past five years. The results of these studies, brought to a common basis for comparison, are presented in Table S-6. Here production costs from wood are projected to be $0.50 to $1.35/gal based on feedstock costs from $20 to $48/dry ton. Methanol costs from refuse are projected to be $0.72 to $0.42/gal based on a $6 to $14/ton credit for waste disposal.

An interesting new concept in the manufacture of methanol is that of the hybrid biomass-methane plant. Syngas produced from biomass is hydrogen-poor, and increasing the hydrogen content requires additional processing. Syngas from re-forming natural gas is hydrogen-rich. Therefore there would be considerable advantage in using a biomass-methane feedstock anywhere that isolated gas wells can be used. Depending on the gasification process, it is expected that the yield would be increased two to five times over that achievable with the biomass alone, and processing costs would be reduced.

The hybrid system has advantages for retrofitting existing natural gas methanol plants, with biomass replacing as much as 30% of the natural gas feedstock, possibly reducing methanol cost. For long-term development, methane could be derived from anaerobic digestion of biomass, municipal solid waste, sewage sludges, or peat. Another variation envisions augmenting methanol production with hydrogen from electrolysis of water or thermochemical closed cycles driven by solar energy. The oxygen from electrolysis could be used in the gasifier.

Although methanol synthesis is the most highly developed alcohol production process at present, catalysts containing alkali or alkaline earth oxides with acid metals (chromates, manganates, molybdates) have been used to produce a mixture of alcohols with 42%

Table S-5. SELECTED GAS CONVERSION SYNTHESES

Reaction	Approximate T°C at Which $\Delta F = 0$[a]	ΔH[a] (kcal/mol product)	Percent of Heating Value of Syngas Lost[c]
Methanol: $CO + 2H_2 = CH_3OH$	140	-10.3[b]	15.2[b]
Ethanol: $2\ CO + 4H_2 = C_2H_5OH + H_2O$	300	-11.8[b]	17.4[b]
Methane: $CO + 3H_2 = CH_4 + H_2O$	690	-12.3	18.2
Nonane: $9CO + 19H_2 = C_9H_2O + 9H_2O$	410	-12.0	17.8
Decane: $10CO + 19H_2 = C_{10}H_{22} + 10H_2O$	410	-12.0	17.8
Alkane + CH_2: $R\text{-}R' + CO + 2H_2 = RCH_2R' + H_2O$	380	-12.0	17.8
Ethylene: $2CO + 4H_2 = C_2H_4 + 2H_2O$	380	-8.4	12.4

[a] All species in standard gas states unless otherwise noted.
[b] Alcohol in liquid state.
[c] Syngas heating value is approximately 67.8 kcal/mol.

Table S-6. SUMMARY COMPARISON OF PROJECTED METHANOL PRODUCTION COSTS ($ 1980)[a]

Source	Plant Size (Ton MeOH /day)	Feedstock		Reforming Oxidation or Gasification Process	Methanol Synthesis Process	Capital Cost[a] (Million $)	Annual Operation & Maintenance Cost (Million $)	Feedstock Cost	Unit Production Cost		
		Type	Throughout per/day						($/gal) MeOH)	($/ Ton MeOH)	($/ MBtu)
Badger Plants, Inc.	58,300	Coal	63,000	Slagging Gasifier	Lurgi low pressure	3,800	593	$31/ton	0.23	69	3.7
Ralph M. Parsons	245	Refuse 25.8% moisture	1,500 tons	Purox (Union Carbide)	Low pressure	126	16	$-14/ton[b]	0.72	217	10
Mathematical Sciences Northwest	275	Refuse 25% moisture	1,500 tons	Purox (UC)	ICI low pressure	31	3.1	$-6.4/ton[b]	0.42	127	6.5
Reed, T.	300	Wood (dried)	900	not reported	Available commercial process	45	5.0	30.3	0.58	173	8.9
Intergroup Consulting Economists (Canada)	1,000	Wood 35% moisture	2,380	Purox	Available commercial process	223	16	37	0.76	229	11.8
Mackay and R. Sutherland (Canada)	1,000	Wood (dried)	3,160	not reported	ICI medium pressure	223	13.8	46	0.96	290	15
MITRE	1,340	Wood 50% moisture	3,400	Purox	ICI low pressure	130	21	45	0.66	199	10
MITRE	335	Wood 50% moisture	850	Purox	ICI low pressure	46	8.9	45	0.84	253	13
Raphael Katzen Associates	500	Wood waste 50% moisture	1,500	Moore-Canada	Vulcan Cincinnati intermediate pressure	90	7	48	1.35	404	20.7
Raphael Katzen Associates	2,000	Wood waste 50% moisture	6,000	Moore-Canada	Vulcan Cin. I. P.	237	N/A	48	1.02	304.0	15.6
SRI	666	Wood 50% moisture	1,000	Oxygen blow gasification	not specified	100.8	9.0	19.1	0.51	154	7.96
SRI	1990	Wood 50% moisture	3,000	Oxygen blow gasification	not specified	268.7	29.4	19.1, 38.2	0.50, 0.62	150, 185	7.77, 9.53

[a]Costs were extrapolated to 1980 dollars by using the Chemical Engineering Cost Index with appropriate extrapolation.
[b]Negative numbers mean that the methanol producer receives money by taking the feedstock (refuse in this case). This money comes from the refuse and drop charges.

methanol, 38% higher alcohols, and 15% aldehydes and acetals. Higher alcohols have a higher energy content than methanol and high octane properties, and investigations of these catalysts should be a part of any alcohol fuel program.

Hydrocarbon fuels have been made from synthesis gas since the 1920s by the Fischer Tropsch process and were an important route to synthetic fuels used by Germany during World War II. They have been produced in South Africa since the early 1950s, and capacity there is now being increased fivefold. The Fischer Tropsch process suffers from having a very wide variety of products, including olefins, alcohols, and waxes. The principal components of the catalyst are cobalt and iron. Nitrided and carburized iron catalysts improve yields of middle distillates and reduce yields of waxes and olefins. Synthesis occurs at about 250 C at 20 atm. Recent work at Exxon is directed toward sulfur resistant catalysts. Since biomass contains little sulfur, use of biomass for Fischer Tropsch processing could offer considerable savings.

Recently, the Mobil Corporation has announced a new process for converting methanol to gasoline using molecular sieves. If the C_3 and C_4 olefins are alkylated with the isobutane produced in the reaction, the process gives over 90% yields of high octane gasoline from methanol. Conversion is projected to cost $0.06/gal of gasoline and requires 2.4 gallons of methanol per gallon of gasoline produced. Gasoline from methanol requires 23% more energy than is contained in the methanol feedstock. Since methanol can be burned in spark engines with 26% to 45% higher efficiency than gasoline, this is a severe energy penalty. The cost of producing gasoline from wood by the Mobil process has been estimated to range from $1.89 to $2.51/gal.

Ammonia has been called a "fuel for biomass," because modern farming achieves efficient production of biomass with ammonia fertilization. Furthermore, ammonia is produced in a series of reactions from synthesis gas in plants basically similar to those used for methanol production. Thus it is natural to include the possibility of product ammonia in any biomass gasification scheme, and a methanol/ammonia plant small enough for operation on farm residues at a farmer's cooperative would go a long way toward making the American farmer independent of fossil fuel inputs.

Typically, ammonia is made at pressures to 200 atm using $FeO-Fe_2O_3$ catalysts and small additions of other metallic oxides. Recent studies of the synthesis of ammonia from wood show a mass conversion efficiency of 1.7 to 2.0 tons of biomass required per ton of ammonia produced. For wood costing $20 to $45/dry ton, ammonia would cost $120 to $300/ton.

Since these costs are competitive with ammonia produced by current industrial processes, production of ammonia may well be the first chemical use of biomass derived synthesis gas. With current technology, methanol is the best liquid fuel that can be produced thermally from biomass feedstocks. In the long term, new technologies may play a significant role in improving the economics of all the gasification processes for producing alcohols, gasoline, methane, H_2, and chemicals.

INSTITUTIONAL SUPPORT OF BIOMASS GASIFICATION AND RELATED ACTIVITIES (Chapter 14)

A questionnaire asking for opinions on possible roles for government assistance was sent by Pyros, Inc., to a number of manufacturers, researchers, and members of government and private institutional groups interested in biomass utilization and gasification in particular. Twenty responses were received and are summarized.

RECOMMENDATIONS FOR FUTURE GASIFICATION RESEARCH AND DEVELOPMENT (Chapter 15)

This survey has been written to outline the value of gasification, the technical base on which future work can proceed, and the activities now underway. Various people reading this information will draw different conclusions. The conclusions on which work at SERI will be based are given here. It is recommended that the national program be guided in this direction also. None of these conclusions is immutable and comment is invited as to their validity.

- Both coal and biomass gasification shall be developed rapidly, because these two technologies will be required soon to supplement fuel supplies as oil and gas become increasingly costly or unavailable. Gasification can provide not only the gas needed for clean heat and power in our cities, but also the basis for synthesis of liquid fuels, SNG, ammonia, and olefins.
- Air gasifiers may find a place in domestic and commercial heating, but they certainly will be used in process heating and producing power for the biomass industries. Although research in progress may improve air gasification, immediate commercialization is recommended at the present level of development.
- Large-scale oxygen gasifiers may play a prominent role in the conversion of municipal waste. If small oxygen gasifiers and plants could be developed (50 tons/day), they could play a crucial role in energy self-sufficient farms, manufacturing ammonia and methanol or gasoline from residues at the farmers' cooperative level to eliminate the heavy dependence on fossil fuels that makes our farms vulnerable to inflating fuel costs and uncertain supply. Development of a 50 ton/day to 100 ton/day pressurized oxygen gasifier to operate on farm or forest residues is recommended. From preliminary operation of a downdraft gasifier on oxygen, and from the thermodynamics presented in the survey, it is believed that it will be possible to design an oxygen gasifier that produces clean synthesis gas in one step, eliminating the need for costly gas conditioning. In this regard it is recommended that support be provided for research on energy efficient methods to separate oxygen from air.
- Pyrolytic gasifiers are not as well developed as oxygen gasifiers, but the majority of the research supported by EPA and DOE has been in this area. Continuing research and pilot work are recommended on many of these systems because they promise higher efficiencies and lower costs than oxygen gasification in production of medium energy gas. However, because it is not clear to what degree medium energy gas will be distributed in the United States, full-scale development of pyrolytic gasifiers must wait on decisions concerning the gas infrastructure in the United States. These decisions hinge on the costs of converting gas to methane for distribution versus distribution of lower energy and lower cost gas. One possible development would be the use of medium energy gas in captive installations and industrial parks but conversion of coal to methane for domestic distribution.
- Top priority development is recommended for fast pyrolysis processes that give a high yield of olefins which can be converted directly to gasoline or alcohols. This seems to be the one new development in gasification since World War II. Evaluation of various feedstocks and particle size options is recommended at the bench level, combined with bench and engineering studies of process designs giving the very high heat transfer and short residence times necessary to produce these products. Evaluation of processes for reducing particle size at reasonable costs, since this seems to be a necessary adjunct to fast pyrolysis, is also recommended.
- Finally, a continuing effort to determine the molecular details of pyrolysis under carefully controlled but realistic laboratory conditions is recommended to provide a firm foundation for understanding and thus improving all gasification processes.

A number of systems studies also should be performed as adjuncts to the technical program.

- The scale of gasification plants should be studied immediately and, where appropriate, programs should be initiated to overcome scale limitations. In particular, coal is likely to supply gas heat for our cities, where large plants can clean the gas sufficiently and make methane for distribution. Because biomass is much cleaner it can be used on a smaller scale, a fact which is compatible with its wider distribution. If biomass residues must be processed at the 1,000 ton/day level or greater to be economically viable, very little biomass will be used as an energy source in this country. If it can be processed economically at the 100 ton/day level, it can be used more widely.
- A systems study of biomass energy refineries is recommended to be used in conjunction with farming and forestry operations, taking residues and converting them to the ammonia and fuel required to operate the farm and forestry operation, and shipping any surplus energy to the cities in the form of gaseous or liquid fuels.

For the longer term, and for biomass conversion plants of larger scale, economic analyses should be performed to identify suitable hybrid schemes. These include:

- production of methanol using a combination of biomass (low hydrogen/carbon ratio) and natural gas (high hydrogen/carbon ratio);

- joint electrolytic/gasification systems in which waste generates hydrogen and oxygen electrolytically, the oxygen is consumed in gasification and the hydrogen increases the hydrogen/carbon ratio; and
- solar fast pyrolysis, in which the high intensity heat is supplied by solar collectors.

Part II

Gasification Principles

The information in Part II is from *A Survey of Biomass Gasification. Volume II–Principles of Gasification* (SERI/TR-33-239), edited by T.B. Reed of the Solar Energy Research Institute, prepared for the U.S. Department of Energy, July 1979.

Chapter 1

Introduction

T.B. Reed and D. Jantzen
SERI

1.1 HISTORY OF BIOMASS GASIFICATION

If fire is a cornerstone of civilization, the use of gaseous and liquid fuels has become the foundation of the modern age of technology. Many processes we now use would be impossible without these refined fuels, and all processes would be less efficient, less convenient, and more polluting. Although civilization might survive the exhaustion of fossil gas and liquid fuels, modern technology will be crippled unless we find a substitute. The gasification technology described here provides the basis for a continuing supply of both liquid and gaseous fuels.

It is difficult for modern man to conceive of a world without gaseous fuels, but gas was not discovered in the laboratory until the end of the 18th century and did not come into commercial and domestic use until 1830. By 1850 large parts of London had gas lights and there was a flourishing gas industry manufacturing gas from coal and biomass.

The early "gasworks" used iron retorts to heat the fuel, pyrolyzing it to gas, oils, and coke or charcoal. Later improvements were the use of fireclay and then silica retorts to achieve higher pyrolysis temperatures. The plants operated with a thermal efficiency which converted 70% to 80% of the energy in the fuel to salable products, producing a gas containing 500 Btu/SCF.

Another widely used process was the "blue water-gas process." The solid fuel was heated to very high temperatures with a blast of air (the "blow"), which formed a low energy gas (100 Btu/SCF) called "producer gas" for use as fuel for manufacturing processes. When sufficiently hot, the air was cut off and steam was blown in from the opposite end of the vessel (the "run"). This produced a higher energy gas (300 Btu/SCF). This "blue water-gas" (blue because it burned with a blue flame) could be converted to "carburetted water-gas" by using the high off-gas temperature to crack oils, yielding a gas with 500 Btu/SCF.

Using these processes, the gas industry grew rapidly and by the time of World War II there were 1,200 plants in the United States producing and selling gas. With the coming of the "big inch" and other pipelines in the 1930s natural gas gradually replaced manufactured gas, and these plants have almost all closed down. Now, with the increased cost of natural gas, gas producers are again being installed. A Wellman Incandescent gas plant operating on coal has recently been installed in York, Pa.

Gas has many advantages over solid fuels. Gas can be distributed easily; its combustion can be controlled to give high efficiency; it can be burned automatically; and it burns with low emissions, making "smokeless cities" possible. It burns with a higher temperature needed in many industrial processes and no local storage is necessary. It is ideal for cooking and heating in homes and is a necessity for many modern manufacturing processes. A given amount of energy is worth two to four times as much energy in the form of gas as it would be in the form of a solid fuel.

In addition, gas can be used to operate spark and diesel engines or turbines to generate power. The use of "producer gas" to run an engine was first tried around 1881. By the 1920s portable gas producers were being used to run trucks and tractors in Europe. These gas generators operated on either wood or charcoal and produced a gas with a rather high tar content. While it was possible to run engines on this gas, it was not convenient, and solid fuels for automotive use did not achieve wide acceptability. There was continued activity aimed at improving gas generators by individual inventors and a few companies until World War II. Commercial installations to run both stationary and mobile engines continued at a low level.

The beginning of World War II and the scarcity of liquid fuels in Europe intensified the search for domestically available fuels and resulted in a great surge of activity in designing and installing gas generators. In Sweden, approximately 75,000 vehicles (40% of the automotive fleet) were converted to generator gas operation within two years. Gas generators were also used on tractors, boats, motorcycles, and even on railway shunting engines. Techniques were developed for converting both diesel and spark ignition engines to generator gas operation. These engines operated reliably, although there was a derating of power output to approximately 75% of the gasoline rating, and considerable additional maintenance of filters, coolers, and the generator itself was required of the operator. It required 20 lb of wood to replace 1 gal. of gasoline (Generator Gas 1979).

The end of the war brought renewed supplies of liquid fossil fuels and a rapid reconversion of vehicles to diesel and gasoline. Since the war a few generators have been in operation, primarily in underdeveloped countries. The Swedish government has also maintained low level research and development programs for gasifiers, with the intention of maintaining military and economic preparedness in the event of a fossil fuel embargo. There has been limited experience with operation of a gas turbine on generator gas, but the indications are that no significant problems are expected. Commercial applications of gas turbines fueled with producer gas have not been attempted to date.

With the increase in oil prices following the formation of OPEC, there has been a renewed interest in all forms of gasification. A number of research projects are underway, aimed at producing fuel gas for pipeline use (see Part III), and more than a score of manufacturers and research groups are developing air gasifiers for retrofitting existing boilers and power generation (Retrofit 1979).

1.2 TYPES OF GASIFICATION PROCESSES

Biomass can be converted to a number of useful products through the processes shown in Fig. 1-1. (Various terminologies are used, often loosely, to describe these processes. We will use the following terminology in this survey.)

Pyrolysis is the destructive decomposition of biomass using mainly heat to produce char, pyrolysis oil, and medium Btu gas. "Pyrolysis" is the name of an important stage in all gasification and combustion processes for both coal and biomass. However, it is also the name of a process which produces gas, char, and oil simultaneously. Therefore, its meaning must be inferred from context.

Pyrolysis Gasification. Pyrolysis processes historically have been operated primarily to yield char and oil products, with the gas burned to operate the process. However, some processes burn the oil and char to recover their heat in the form of higher yields of medium energy gas. The gas produced typically contains 300-500 Btu/SCF.

Air Gasification. If biomass is burned with a limited supply of air it produces a low energy gas containing primarily H_2 and CO, but diluted with nitrogen, typically containing 150-200 Btu/SCF. This gas is suitable for operation of boilers or engines but is too dilute to be transported in pipelines.

Oxygen Gasification. If biomass is burned with a limited supply of oxygen it will yield a medium energy gas equivalent to the "town gas" of the 1930s, suitable for limited pipeline distribution, and containing 300 Btu/SCF. This gas can be used for industrial process heat or as synthesis-gas to make methanol, gasoline, ammonia, methane, or hydrogen.

Hydrogasification. Biomass has a low ratio of hydrogen to carbon compared to most liquid and gaseous fuels. In principle, biomass can be converted to gaseous or liquid fuels under pressure with hydrogen.

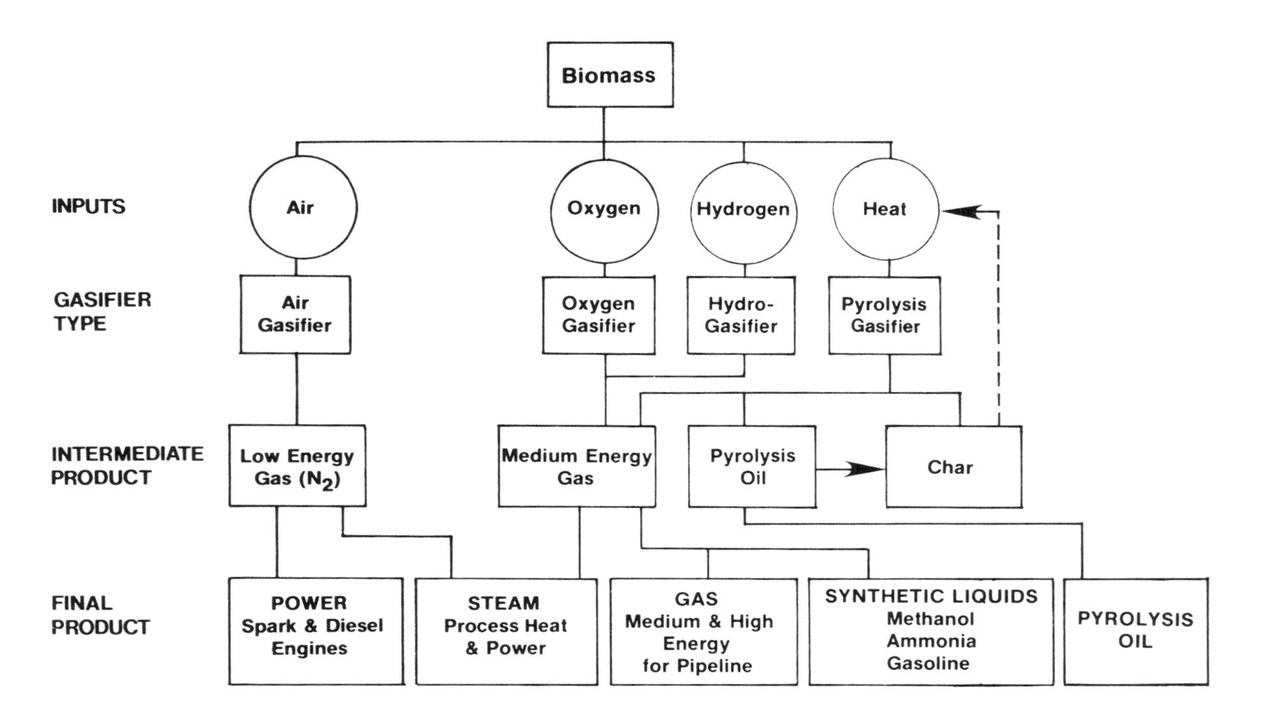

Figure 1-1. Gasification Processes and Their Products

1.3 TYPES OF GASIFIERS

In addition to the general types of processes just defined, there are a number of types of gasifiers which are classified by different process variables. They are briefly defined here and the reader is referred to Part III, Chapter 8, for a more complete discussion.

1.3.1 Method of Gas/Fuel Contact

Fixed Bed Gasifiers are used for bulky fuels such as wood chips, pellets, or corn cobs. They include updraft gasifiers (also called counterflow) in which air or oxygen is passed up through the reacting bed while the fuel passes down, producing a gas high in oil and tar; downdraft gasifiers (coflow), in which both fuel and air or oxygen pass downward through the hot bed, thus removing most of the tars from the product gas; and crossdraft, sloping grate, and other gasifier configurations.

Fluidized Bed Gasifiers typically use a wider range of fuel sizes, and the fuel is suspended in an upward flowing column of air. In addition to the biomass there is often a high percentage of an inert material, such as sand, which aids in the heat transfer to the fuel particles.

Suspended flow gasifiers use finely divided particles for very rapid gasification.

1.3.2 Ash Form

Dry ash gasifiers maintain grate temperatures below about 1100 C so that the ash can be removed as a fine powder. Slagging gasifiers maintain the grate temperatures above 1300 C so that the ash can be poured out as a liquid. Intermediate temperatures lead to ash with the consistency of molasses, which prevents further operation.

1.3.3 Gas Pressure

Atmospheric pressure gasifiers are the most easily constructed and operated. Suction gasifiers actually operate slightly below atmospheric pressure and are used mostly to power engines. Pressurized gasifiers typically operate at pressures of 10-100 atm, yielding a gas that can be put in pipelines or used immediately to operate turbines or as feedstocks for chemical synthesis.

1.4 ENERGY CONTENT OF FUEL GASES

The gases produced through gasification have a wide range of energy content and corresponding applications. These are summarized in Table 1-1. Note that natural gas has the highest energy content and can be used for any of the other applications. Its high energy content is important in long distance shipping but has little effect on process use. Use of gases with energy content below about 200 Btu/SCF may result in some loss of performance in engines or boilers.

1.5 THE RELATIVE MERITS OF BIOMASS AND COAL

Although coal was used in the larger producer gas installations described in Section 1.1, in many other cases wood or biomass were used because of ease of gasification and availability. We list here a number of factors which will influence the relative degree of development of coal and biomass for energy and fuels in the coming years.

1.5.1 Availability

- Coal is available in many places in high concentrations; other areas are located at great distances from the mines and involve higher costs for coal than for biomass.

Table 1-1. ENERGY CONTENT OF FUEL GASES AND THEIR USES

Name	Source	Energy Range (Btu/SCF)	Use
Low Energy Gas (LEG) (Producer Gas, Low Btu Gas)	Blast Furnace, Water Gas Process	80-100	On-site industrial heat and power, process heat
Low Energy Gas LEG (Generator Gas)	Air Gasification	150-200	Close-coupled to gas/oil boilers Operation of diesel and spark engines Crop drying
Medium Energy Gas (MEG) (Town Gas, Syngas)	Oxygen Gasification Pyrolysis Gasification	300-500	Regional industrial pipelines Synthesis of fuels and ammonia
Biogas	Anaerobic Digestion	600-700	Process heat, pipeline (with scrubbing)
High Energy Gas (HEG) (Natural Gas)	Oil/Gas Wells	1000	Long distance pipelines for general heat, power, and city use
Synthetic Natural Gas (SNG)	Further Processing of MEG and Biogas	1000	Long distance pipelines for general heat, power, and city use

- Biomass is widely available in smaller quantities and this favors dispersed use.
- Biomass may become available in larger quantities if energy plantations are developed (See Chapter 2).
- Biomass is renewable and will always be available in reasonable quantities (probably from 10 to 40 quads in the United States).

1.5.2 Technical Factors

- Biomass has a low energy density and occurs in a wide variety of forms, many unsuitable for combustion or gasification without pretreatment (drying or densification).
- Biomass is easier to burn or gasify because of its low pyrolysis temperatures and high concentration of volatiles.
- Biomass contains oxygen and water, which may be advantageous in gasification.

1.5.3 Environmental Factors

- Coal is high in sulfur content; biomass sulfur content is an order of magnitude lower.
- Coal has a high content of ash with no value; biomass ash content is lower and has value as fertilizer or for synthesizing chemicals. Coal conversion mobilizes toxic trace metals, and coal tars are highly carcinogenic.
- Coal mining is generally destructive of the land; proper biomass residue use or proper harvesting improves the land, but overcutting can also be very destructive.
- Coal combustion increases the CO_2 concentration in the atmosphere; steady-state biomass use does not increase CO_2 levels.

1.5.4 Economic Factors

- Coal, like gas and oil, has lower direct costs than biomass; however, consideration of environmental costs makes coal more comparable in cost to biomass; we do not now have methods for quantifying these costs.
- Biomass residues (such as solid municipal waste) can have negative or low cost, but collection and processing increases these costs.
- Small-scale use of biomass is favored by lower shipping costs and less difficult conversion.
- Large-scale use of coal is favored by the economies of scale required to offset the high cost of environmental control.

1.5.5 Conclusions

The combustion and gasification of both coal and biomass are feasible and necessary. Each energy resource will find its proper niche in the years to come, as dictated by the interplay of the considerations just discussed.

1.6 PURPOSE OF THIS SURVEY

The purpose of this survey is:

- to examine the properties and potential of the biomass resource relevant to gasification (Chapters 2 to 4);
- to summarize the basic science of biomass gasification (Chapters 5 to 7);

- to survey the present state of research, development and commercialization of gasifiers (Chapters 8 to 10);
- to examine processes associated with gasification for gas cleanup and synthesis of other fuels from biomass-gas (Chapters 11 to 13);
- to determine governmental means by which gasification technology can be introduced more rapidly (Chapter 14); and
- to identify the areas where research and development will be needed in an intensified gasification development program (Chapter 15).

It is believed that this survey accomplishes these tasks and will serve as a foundation for gasification research technology expansion over the next decades.

1.7 UNITS

In writing this survey the following dilemma must be faced: If English units are used the survey will be understood in the United States, Canada, and England but will be less comprehensible in the rest of the world. If SI units are used it will be more understandable in the world at large and, possibly in several decades, worldwide. If both kinds of units are used the tables and figures will be greatly complicated. Therefore, in each chapter the units now in common use for the subject matter were used. For conversion the reader is referred to any of dozens of sets of conversion tables but included here are a few particularly applicable conversion factors.

Table 1-2. CONVERSION FACTORS

1 Btu/SCF[a]	=	34.54/N - m^3 [a]
1 Btu	=	1054 J
1 acre	=	4047 m^2
1 cal	=	4.184 J
1 hp	=	746 W
1 atm	=	1.013×10^5 Pa
1 ft^3	=	0.0283 m^3

[a]The volume of a gas in a standard cubic foot (SCF) is the volume that gas would occupy at atmospheric pressure and 60 F. The volume in normal cubic metres (N - m^3) is the volume it would occupy at 1 atm and 0 C.

1.8 REFERENCES

Generator Gas. 1979. Golden, CO: The Solar Energy Research Institute; SERI/SP-33-140; Jan.

Retrofit 79: Proceedings of a Workshop on Air Gasification. 1979. Golden, CO: The Solar Energy Research Institute; SERI/TP-49-183.

Chapter 2

The Potential Biomass Resource Base

R. Inman
SERI

The ultimate applicability of all biomass conversion technologies, including biomass gasification, is restricted by the quantity of feedstocks that can be made available for conversion. A meaningful impact on the nation's energy supply could not be made, regardless of the number of potential applications or the developments achieved in conversion technologies, if the feedstock supply were inadequate. Hence the utility of biomass gasification is, ultimately, resource-limited.

The biomass resource base in the United States is immediately distinguished from other solar energy resources by its high degree of diversity. The corollary to this characteristic is that, while not all biomass or its components are equally suited to gasification, its diversity is translatable into versatility and hence affords the opportunity to produce diverse energy end-products and to develop diverse energy applications. A second distinguishing factor of this resource is its juxtaposition, and in some cases its supraposition, to the resource base used for food and fiber products. The special relationships between these feedstock sources, including in some cases direct competition for their use, weighs heavily upon the economics of energy applications.

The existing resource base is comprised of agricultural crop residues, manures from confined livestock and poultry operations, wood and bark mill residues from primary wood product manufacturing plants, bark residues from the wood pulp industry, logging residues from timber harvesting operations, noncommercial components of standing forests, and the organic fraction of municipal solid wastes. In addition to the existing base, it is believed that future biomass supplies could be supplemented by feedstock produced on energy farms. Overall, it would appear that there is a resource base of significant size and that this base will, in all probability, be expanded in future years as timber harvests increase and as energy farming needs and technologies develop. Each component of the resource base is characterized in this chapter.

2.1 AVAILABLE RESOURCES

2.1.1 Crop Residues

Crop residues consist of plant remains left in the field following harvest or harvested material discarded during the preparation of produce for packing and shipping. Approximately 320 million dry ton equivalents (DTE) of this potential energy feedstock are generated each year (Inman and Alich 1976), and it is estimated that about 278 million DTE are "available" (i.e., are already collected or could be collected with existing machinery [Table 2-1]). Almost half of this resource category consists of straw from the cultivation of small grain (wheat, rye, barley and rice) and grass seed crops, and more than one third of stover (the dried stalks and leaves) from corn and sorghum production. Only about 2.5% of the available resource is collected during the course of normal operations. Deterrents to the use of crop residues as an energy feedstock include: (1) their seasonality, (2) their high cost of collection and transport, and (3) their current ecological value in situ. Almost three fourths of the resource base is returned to the soil (plowed under) each year (Table 2-2). Some, largely corn stover, is pastured to livestock following harvest, and small portions are sold (straw, sugar beet pulp, and cotton gin

Table 2-1. ANNUAL AVAILABILITY OF CROP RESIDUES (1971-1973) [a]

Residue Category	Million Dry Tons
Corn and sorghum (field)	96.6
Small grains and grasses (field)	131.8
Other crops (field)	42.3
Collected residues	7.3
Total	278.0

[a]From Anderson 1972.

Table 2-2. DISPOSITION OF CROP RESIDUES (1971-1973) [a]

Disposition	Million Dry Tons
Sold for profit	11.3
Fed to livestock	52.3
Used as fuel	1.7
Disposed of at cost	6.8
Returned to soil	205.9
Total	278.0

[a]From Anderson 1972.

trash), used as a fuel (bagasse), or disposed of at cost (burned in the field). The great majority of the crop residue resource would be amenable to use as a gasification feedstock, should sustainable soil conservation practices permit.

2.1.2 Animal Manures

Animal manures are only marginally attractive as a gasification feedstock but could be used after drying. This resource, however, is relatively small (Table 2-3) and in all probability will eventually be used in its entirety as a substrate for methane production through anaerobic digestion, as a soil amendment, or as a recycled livestock feed.

Table 2-3. DISPOSITION OF ANIMAL MANURES FROM CONFINED ANIMAL OPERATIONS (1971-1973) [a]

Disposition	Million Dry Tons
Sold for profit	3.9
Fed to livestock	0.2
Used as fuel	0.02
Disposed of at cost	4.5
Returned to soil	17.9
Total	26.5

[a]From Anderson 1972.

2.1.3 Mill Residues

Wood and bark are preferred gasification feedstocks. One potential source of this feedstock is the residue from sawmills, plywood plants, and other primary wood manufacturing mills. These residues occur in a large variety of forms (slabs, edgings, sawdust, planer shavings, sander dust, ends, veneer trimmings, defective products, etc.). Over 86 million dry ton equivalents (DTE) per year are generated (Howlett and Gamache 1977), but less than one third of this resource is available for extended use as an energy feedstock (Table 2-4). Current uses for this material are dominated by the use of the coarse wood fraction for pulp manufacture (Howlett and Gamache 1977) and the direct combustion of the remaining fractions for process steam and/or electric power generation is increasing rapidly. It is widely believed that the entire mill residue resource soon will be consumed by the forest products industry itself for pulp and fuel.

The figures presented in Table 2-4 do not include bark residues from pulp mills, which have been estimated to total about four million DTE per year (Inman and Alich 1976). Moreover, large piles of this material have been allowed to accumulate at certain pulp mills, forming veritable "biomass mines." The use of bark by the pulping industry to produce steam and electric power is also increasing because energy requirements of this segment of the wood industry dwarf those of primary wood manufacturing plants.

Table 2-4. WOOD AND BARK MILL RESIDUES: GENERATION AND DISPOSITION ANNUALLY BY REGION (1970)[a]

	Million Dry Tons		
Region	Total Generated	Residues Used	Residues Unused
Northeast	6.6	4.3	2.3
North Central	6.4	4.3	2.1
Southeast	11.4	6.9	4.5
South Central	16.7	12.1	4.6
Pacific Northwest	27.8	23.6	4.2
Pacific Southwest	8.8	5.5	3.3
Northern Rockies	6.6	4.5	2.1
Southern Rockies	1.8	0.8	1.0
Totals	86.1	62.0	24.1

[a]From Howlett and Gamache 1977.

2.1.4 Logging Residues

Portions of harvested or felled trees left in the woods following logging operations total over 83 million DTE annually (Howlett and Gamache 1977). The total resource is split almost evenly between hardwood and softwood residues (Table 2-5), but there are tremendous regional variations in this distribution. Virtually none of this resource is currently used as an energy feedstock due to the high cost of collection and the lack of appropriate collection machinery. It is widely expected, however, that changes in conventional energy economics will bring this resource into use within the mid-term.

2.1.5 Standing Forests

By far the largest existing resource is the surplus and noncommercial components of the standing forests. The total annual productivity of these components has been estimated to be almost 400 million DTE (Salo and Henry 1979) (Table 2-6). The harvest of this resource for energy production in all likelihood would be closely associated with both commercial timber harvest and timber stand improvement practices. Environmental concerns also will have to be served. Some of this resource could conceivably be managed as a renewable energy feedstock source.

2.1.6 Municipal Solid Wastes (MSW)

The organic component of MSW totals approximately 130 million DTE annually (Anderson 1972) This represents a generation rate of 3.5 lb per person per day, an amount which

Table 2-5. ANNUAL GENERATION OF LOGGING RESIDUES BY REGION AND TIMBER CATEGORY (1970) [a]

Region	Million Dry Tons		
	Softwood	Hardwood	Total
New England	1.94	2.03	3.97
Middle Atlantic	0.52	4.81	5.33
Lake States	0.55	3.12	3.67
Central States	0.07	4.47	4.54
Southern Atlantic	3.22	8.60	11.82
East Gulf	3.05	2.57	5.62
Central Gulf	4.13	6.45	10.58
West Gulf	5.02	4.68	9.70
Pacific Northwest	17.52	0.84	18.36
Pacific Southwest	4.31	0.63	4.94
Northern Rockies	3.60	Trace	3.60
Southern Rockies	0.98	0.10	1.08
Totals	44.91	38.29	83.21

[a]From Howlett and Gamache 1977.

Table 2-6. THE ANNUAL ENERGY RESOURCE REPRESENTED BY UNUSED STANDING FOREST PRODUCTIVITY (1976) [a]

Region	Million Dry Tons			
	Surplus Growth	Mortality	Noncommercial[b] Timber	Total
Northeast	34.47	14.00	14.06	62.53
Northern Plains	0.94	0.76	1.00	2.70
Corn Belt	3.24	1.94	5.24	10.42
Southeast	37.06	9.65	11.18	57.89
Appalachian	40.29	9.35	13.41	63.05
Southern Plains	6.47	1.41	6.41	14.29
Delta States	23.18	6.18	8.71	38.07
Lake States	19.82	11.24	4.41	35.47
Pacific	0.00	20.29	20.18	40.47
Mountain	18.53	11.00	29.53	59.06
Totals	184.00	85.82	114.13	383.95

[a]From Inman and Alich et al. 1976.

[b]Includes noncommercial timber growth on commercial forest land and all timber growth on noncommercial forest land producing less than 20 ft^3 per acre-year of commercial timber.

may even increase in the future. Most of this material is currently disposed of in landfills at a significant cost. Gasification of this refuse would appear to be an ideal "disposal" method.

2.1.7 Summary of Available Resources

As shown in Table 2-7, the existing resource base totals almost 15 fuel-quad equivalents. Only a portion of this resource base, however, could ever be expected to be applied to energy production. Economic and environmental concerns will influence the application of the two major resource components, standing forests and crop residues. Use of MSW probably will serve adequately only in large metropolitan areas where sufficient disposal credits can be realized. Wood and bark residues are largely captive resources of the forest products industry.

Table 2-7. SUMMARY OF THE ANNUAL ENERGY POTENTIAL OF EXISTING SOURCES OF BIOMASS

Resource	10^6 Dry Tons/Year	Quads/Year
Crop residues	278.0	4.15
Animal manures	26.5	.33
Unused mill residues[a]	24.1	.41
Logging residues	83.2	1.41
Municipal solid wastes	130.0	1.63
Standing forests	384.0	6.51
Totals	925.8	14.44

[a]Does not include unused bark from wood pulp mills.

2.2 POTENTIAL BIOMASS RESOURCES

The presently available resources listed above provide sufficient incentive to develop biomass collection, combustion, and gasification systems. However, biomass production is the principal method of solar energy collection, and in the future we will need to expand our biomass base by more efficient utilization of present resources and development of new species and land for energy production. The following major categories, while more difficult to quantify than existing residues, are likely ways for enlarging the biomass energy base.

2.2.1 Biomass Mines

In addition to the continuing production of residues inventoried in this chapter, there are "biomass mines" composed of accumulations of residues from past years and including bark piles, the dumps of food processing industries, and the municipal landfills of cities. At present no estimate is available of the recoverable energy in these forms, but if it were assumed that 10% of the 6 quads/yr of municipal, crop, and mill residues dumped over the last 20 years could be recovered, we estimate that there might be 12 quads available in this form. In addition, removal of these wastes would be environmentally attractive. We recommend that a good assessment of this energy base be made.

2.2.2 Land Improvement Residues

Another category of biomass is that available through land improvement. Many millions of U.S. acres of land have been laid waste by man and presently support species of low value such as scrub, mesquite, and chapparal. Harvesting these plants for their biomass energy could pay the cost of improving this land.

2.2.3 Energy Farming

In the future, energy farming may supplement energy feedstock supplies. It has been estimated that from four to eight fuel-quad-equivalents of biomass could be produced should the need arise (Inman et al. 1977), presuming that research were directed to develop this agronomic technology to the point at which biomass yields were sufficient to make cash crop energy farming an economically competitive venture.

At present, the potential biomass resource base would not restrict the development of biomass gasification as an energy conversion technology. The extent to which this resource will actually be used as a gasification feedstock will depend upon a large number of factors whose interactions cannot be predicted accurately at this time.

2.3 REFERENCES

Anderson, L. L. 1972. Energy Potential from Organic Wastes: A Review of the Quantities and Sources. Washington, D.C.: U.S. Department of the Interior. Bureau of Mines. Bureau of Mines Information Circular 8549.

Howlett, K.; Gamache, A. 1977. "Silvicultural Biomass Farms." Volume VI of Forest and Mill Residues as Potential Sources of Biomass. Final Report. McLean, VA: The MITRE Corporation/Metrek Division; ERDA Contract No. E (49-18) 2081; MTR7347.

Inman, R. E.; Alich, J., et al. 1976. An Evaluation of the Use of Agricultural Residues as an Energy Feedstock, Volume I. Final Report. NSF Grant No. NSF/RANN/SE/GI/18615/FR/76/3.

Inman, R. E. et al. 1977. Silvicultural Biomass Farms. Volumes I-VI. Final Report. McLean, Va: The MITRE Corporation/Metrek Division; ERDA Contract No. E (49-18) 2081; MTR 7347.

Salo, D. J.; Henry, J. F. 1979. Wood-Based Biomass Resources in the United States. McLean, VA: The MITRE Corporation/Metrek Division.

Chapter 3

Properties of Biomass Relevant to Gasification

M. Graboski and R. Bain
Colorado School of Mines

An understanding of the structure and properties of biomass materials is necessary in order to evaluate their utility as chemical feedstocks. This section summarizes available information on a variety of such properties including chemical analysis, heats of combustion and formation, physical structure, heat capacities, and transport properties of biomass feedstocks and chars. Much of the information reported is for wood materials; however, where data were available for other forms of biomass such as municipal solid waste and feedlot waste, they were included.

3.1 BULK CHEMICAL ANALYSIS OF BIOMASS

In evaluating gasification feedstocks, it is generally useful to have proximate and ultimate analyses, heats of combustion, and sometimes ash analyses. These provide information on volatility of the feedstock, elemental analysis, and heat content. The elemental analysis is particularly important in evaluating the feedstock in terms of potential pollution.

Table 3-1 lists the standard methods for evaluating carbonaceous feedstocks.

A number of instruments have been developed for determining elemental composition, most often, in biomass conversion, for carbon, hydrogen, nitrogen, sulfur, and oxygen. Chlorine normally is not determined by such analyzers. Most of these systems employ a catalytic combustion or pyrolysis step to decompose the sample to carbon dioxide, water, hydrogen sulfide, and nitrogen, which are then determined quantitatively by gas chromatography using flame ionization (FID) or thermal conductivity (TC) detectors. Oxygen is usually determined by catalytic conversion to carbon monoxide over a platinized carbon catalyst followed by GC analysis. A short list of some representative instruments is given in Table 3-2.

3.1.1 Proximate Analyses

The proximate analysis classifies the fuel in terms of its moisture (M), volatile matter (VM), ash, and (by difference) fixed carbon content. In the test procedure, the volatile material is driven off in an inert atmosphere at high temperatures (950 C) using a slow heating rate. The pyrolysis yield is representative of that for slow pyrolysis processes; fast pyrolysis techniques employing very rapid heating rates normally yield more volatile matter. The moisture determined by the proximate method represents physically bound water only; water released by chemical reactions during pyrolysis is classified with the volatiles. The ash content is determined by combustion of the volatile and fixed carbon fractions. The resulting ash fraction is not representative of the original ash, more appropriately termed mineral matter, due to the oxidation process employed in its determination. In the most exact analysis, small corrections to the ash weight are necessary to correct it to a mineral matter basis. The fixed-carbon content of an as-received sample is calculated by material balance. Thus:

$$FC = 1 - M - ASH - VM. \quad (3\text{-}1)$$

Table 3-1. ASTM STANDARDS METHODS FOR GASIFICATION FEEDSTOCKS

Method	Test No.[a]	Repeatability (wt %)[b]	Reproducibility (wt %)[b]
Proximate Analysis			
Moisture	D-3175-73		
Less than 5%		0.2	0.3
More than 5%		0.3	0.5
Volatile Matter	D-3175-77		
High Temp. Coke		0.2	0.4
Bituminous Coal		0.5	1.0
Lignite		1.0	2.0
Ash	D-3174-73	0.5	1.0
Ultimate Analysis			
C	D-3178-73	0.3	—
H	D-3178-73	0.07	—
O	None		
N	D-3177-75	0.05	—
S	D-2361-66 [1978]		
less than 2%		0.05	0.10
more than 2%		0.10	0.20
Gross Heating Value	D-3286-77	50 Btu/lb	100 Btu/lb
Ash Analysis	D-295-69 [1974] D-3682-78 D-3683-78		
SiO_2		1.0	2.0
Fe_2O_3		0.3	0.7
CaO		0.2	0.4
K_2O		0.1	0.3
Na_2O		0.1	0.3
MgO		0.3	0.5
P_2O_5		0.05	0.15

[a]The two digit number following the second dash is the year the method was approved. The date in brackets is the year the test was reapproved without change.
[b]Taken fron Instit. of Gas Technology 1978.

The fixed carbon is considered to be a polynuclear aromatic hydrocarbon residue resulting from condensation reactions which occur in the pyrolysis step.

Table 3-2. ELEMENTAL ANALYZER EQUIPMENT

Instrument	Oxidant	Capability	Detection
Carlo Erba 1104	oxygen	C, H, N, O	FID & TC
Chemical Data Systems (CDS 1200)	oxygen	C, H, N, O, S and functional groups	FID & TC
Hewlett-Packard HP-185	MnO_2 added	C, H, N	FID & TC
Perkin Elmer 240	oxygen	C, H, N, O, S	TC

The most useful basis for reporting proximate analysis is the dry basis. In this instance the compositions are normalized to a moisture-free basis (denoted by *):

$$VM^* + FC^* + ASH^* = 1 , \qquad (3\text{-}2)$$

and, for example,

$$VM^* = VM/(1 - M).$$

The moisture is reported as grams of moisture per gram of dry feedstock. Typical proximate analyses for solid fuels are given in Table 3-3, from which it is evident that common biomass materials are more readily devolatilized (pyrolyzed) than lignite and bituminous coals, yielding considerably less fixed-carbon residue. This is due to the much more aromatic structure of the coals which is produced by the geological coalification process. The higher volatile content of biomass materials makes them potentially useful feedstocks for pyrolysis processes. In general, the ash content of biomass materials is considerably lower than for coals. This is due to the fact that the bulk of the coal ash was deposited in coal beds by processes such as siltation and did not come from the parent carbonaceous material. An exception is municipal solid waste, which contains a high mineral content due to nonvolatile trash components such as metals and glass.

Table 3-3 also gives proximate analyses of wood chars derived from low-temperature carbonization. The volatile content, while reduced, is still a significant portion of the resulting chars.

Table 3-3. PROXIMATE ANALYSIS DATA FOR SELECTED SOLID FUELS AND BIOMASS MATERIALS
(Dry Basis, Weight Percent)

	Volatile Matter (VM*)	Fixed Carbon (FC*)	Ash*	Reference
Coals				
Pittsburgh seam coal	33.9	55.8	10.3	Bituminous Coal Research 1974
Wyoming Elkol coal	44.4	51.4	4.2	Bituminous Coal Research 1974
Lignite	43.0	46.6	10.4	Bituminous Coal Research 1974
Oven Dry Woods				
Western hemlock	84.8	15.0	0.2	Howlett and Gamache 1977
Douglas fir	86.2	13.7	0.1	Howlett and Gamache 1977
White fir	84.4	15.1	0.5	Howlett and Gamache 1977
Ponderosa pine	87.0	12.8	0.2	Howlett and Gamache 1977
Redwood	83.5	16.1	0.4	Howlett and Gamache 1977
Cedar	77.0	21.0	2.0	Howlett and Gamache 1977
Oven Dry Barks				
Western hemlock	74.3	24.0	1.7	Howlett and Gamache 1977
Douglas fir	70.6	27.2	2.2	Howlett and Gamache 1977
White fir	73.4	24.0	2.6	Howlett and Gamache 1977
Ponderosa pine	73.4	25.9	0.7	Howlett and Gamache 1977
Redwood	71.3	27.9	0.8	Howlett and Gamache 1977
Cedar	86.7	13.1	0.2	Howlett and Gamache 1977
Mill Woodwaste Samples				
-4 Mesh redwood shavings	76.2	23.5	0.3	Boley and Landers 1969
-4 Mesh Alabama oakchips	74.7	21.9	3.3	Boley and Landers 1969
Municipal Refuse and Major Components				
National average waste	65.9	9.1	25.0	Klass and Ghosh 1973
Newspaper (9.4% of average waste)	86.3	12.2	1.5	Klass and Ghosh 1973
Paper boxes (23.4%)	81.7	12.9	5.4	Klass and Ghosh 1973
Magazine paper (6.8%)	69.2	7.3	23.4	Klass and Ghosh 1973
Brown paper (5.6%)	89.1	9.8	1.1	Klass and Ghosh 1973
Pyrolysis Chars				
Redwood (790 F to 1020 F)	30.0	67.7	2.3	Howlett and Gamache 1977
Redwood (800 F to 1725 F)	23.9	72.0	4.1	Howlett and Gamache 1977
Oak (820 F to 1185 F)	25.8	59.3	14.9	Howlett and Gamache 1977
Oak (1060 F)	27.1	55.6	17.3	Howlett and Gamache 1977

Table 3-4. ULTIMATE ANALYSIS DATA FOR SELECTED SOLID FUELS AND BIOMASS MATERIALS
(Dry Basis, Weight Percent)

Material	C	H	N	S	O	Ash	Higher Heating Value (Btu/lb)	Reference
Pittsburgh seam coal	75.5	5.0	1.2	3.1	4.9	10.3	13,650	Tillman 1978
West Kentucky No. 11 coal	74.4	5.1	1.5	3.8	7.9	7.3	13,460	Bituminous Coal Research 1974
Utah coal	77.9	6.0	1.5	0.6	9.9	4.1	14,170	Tillman 1978
Wyoming Elkol coal	71.5	5.3	1.2	0.9	16.9	4.2	12,710	Bituminous Coal Research 1974
Lignite	64.0	4.2	0.9	1.3	19.2	10.4	10,712	Bituminous Coal Research 1974
Charcoal	80.3	3.1	0.2	0.0	11.3	3.4	13,370	Tillman 1978
Douglas fir	52.3	6.3	0.1	0.0	40.5	0.8	9,050	Tillman 1978
Douglas fir bark	56.2	5.9	0.0	0.0	36.7	1.2	9,500	Tillman 1978
Pine bark	52.3	5.8	0.2	0.0	38.8	2.9	8,780	Tillman 1978
Western hemlock	50.4	5.8	0.1	0.1	41.4	2.2	8,620	Tillman 1978
Redwood	53.5	5.9	0.1	0.0	40.3	0.2	9,040	Tillman 1978
Beech	51.6	6.3	0.0	0.0	41.5	0.6	8,760	Tillman 1978
Hickory	49.7	6.5	0.0	0.0	43.1	0.7	8,670	Tillman 1978
Maple	50.6	6.0	0.3	0.00	41.7	1.4	8,580	Tillman 1978
Poplar	51.6	6.3	0.0	0.0	41.5	0.6	8,920	Tillman 1978
Rice hulls	38.5	5.7	0.5	0.0	39.8	15.5	6,610	Tillman 1978
Rice straw	39.2	5.1	0.6	0.1	35.8	19.2	6,540	Tillman 1978
Sawdust pellets	47.2	6.5	0.0	0.0	45.4	1.0	8,814	Wen et al. 1974
Paper	43.4	5.8	0.3	0.2	44.3	6.0	7,572	Bowerman 1969
Redwood wastewood	53.4	6.0	0.1	39.9	0.1	0.6	9,163	Boley and Landers 1969
Alabama oak woodwaste	49.5	5.7	0.2	0.0	41.3	3.3	8,266	Boley and Landers 1969
Animal waste	42.7	5.5	2.4	0.3	31.3	17.8	7,380	Tillman 1978
Municipal solid waste	47.6	6.0	1.2	0.3	32.9	12.0	8,546	Sanner et al. 1970

3.1.2 Ultimate Analyses

Ultimate analyses generally report C, H, N, S and (by difference) O in the solid fuel. Table 3-1 lists the appropriate ASTM tests for these elements while Table 3-2 lists several manufacturers of modern elemental analyzers. Care must be exercised in using ultimate analyses for fuels containing high moisture content because moisture is indicated in the ultimate analysis as additional hydrogen and oxygen.

In order to avoid confusion and give a good representation of the fuel itself, ultimate analyses should be performed and reported on a dry basis; when this is done all hydrogen determined is truly a constituent of the fuel. For certain biomass materials like municipal solids and animal waste, the determination of chlorine is important because it represents a possible pollutant and corrosive agent in gasification and combustion systems.

Typical ultimate analyses for a variety of feedstocks are presented in Table 3-4.

All biomass materials have carbon contents considerably lower than coals; the atomic carbon to hydrogen ratio is much higher in coals than in biomass materials. For coal, the H/C ratio is unity, while for biomass the ratio is typically 1.5. The bound oxygen content of biomass materials is considerably higher, due to the ether, acid, and alcohol groups in the cellulose, hemicellulose, and lignin fractions of biomass, as will be discussed later in this section. The nitrogen and sulfur contents in coal are considerably higher than those in biomass. Thus, in direct biomass combustion, pollutants resulting from bound nitrogen and sulfur in the fuel generally are present in small enough quantities to meet EPA standards, although the high chlorine contents that are found in animal wastes can pose a severe pollution problem.

The relative "quality" of the volatile matter can be estimated using the ultimate analysis and simple stoichiometry. If it is assumed that the fixed carbon contains only carbon, then all hydrogen and oxygen plus a portion of the carbon are associated with the volatile material. Table 3-5 presents a typical calculation for the volatile fraction of lignite and Douglas fir bark.

Table 3-5. **ELEMENTAL ANALYSIS OF VOLATILES LIBERATED BY PYROLYSIS FOR TWO SELECTED FUELS**

Fuel	Wt % in Volatiles, Dry Basis			Molar Ratio Volatile		
	C	H	O	C	H	O
Lignite	17.4	4.22	19.17	1	2.91	0.83
Douglas fir bark	23.4	5.9	36.7	1	3.03	1.17

The C/H/O ratios of these volatile fractions are very similar despite the difference in feedstock. In the pyrolysis process, at relatively high temperatures,

$$\text{Volatiles} \rightarrow CH_4 \quad (3\text{-}3)$$

$$\text{Volatiles} \rightarrow CO + CO_2 \quad (3\text{-}4)$$

$$\text{Volatiles} \rightarrow H_2O. \quad (3\text{-}5)$$

If we assume that CO is produced exclusively we can calculate the product analysis from pyrolysis.

Therefore, assuming:

$$C + 4H \rightarrow CH_4 \quad (3\text{-}6)$$

$$2H + O \rightarrow H_2O \quad (3\text{-}7)$$

$$C + O \rightarrow CO, \quad (3\text{-}8)$$

let X be the moles of carbon converted to methane, Y the oxygen converted to water, and Z the carbon to CO.

The material balance equations yield:

$$X = \frac{2 + (H/C) - 2\,(O/C)}{6} \quad (3\text{-}9)$$

$$Z = 1 - X \quad (3\text{-}10)$$

$$Y = \frac{O}{C} - Z. \quad (3\text{-}11)$$

In the calculation for methane it should be pointed out that as long as water-gas shift reaction equilibrium is attained, it makes no difference whether the nonhydrocarbon products are CO and H_2O or a mixture of CO, CO_2, H_2, and H_2O.

Table 3-6 presents such an analysis on a dry basis of 100 lb of fuel.

Table 3-6. EVALUATION OF FEEDSTOCKS FOR PYROLYSIS BY MATERIAL BALANCE CALCULATION

Feedstock	SCF Gas / 100 lb Dry Feed	Mole Fractions CH_4	CO	H_2O	lb C in CH_4 / 100 lb C in feed
Lignite	754	0.395	0.334	0.271	14.7
Douglas fir bark	1196	0.277	0.341	0.382	18.7

The gas derived from lignite is higher in quality than that from the fir bark due to the bark's greater potential to form water. The quantity of gas produced is greater for the fir bark due to the greater quantity of volatiles present. The most important factor is the fraction of carbon converted to methane. The woody material shows a greater potential to form methane on a carbon feed basis, indicating that it is a higher quality feedstock for pyrolysis. This may be attributed to the higher degree of aromaticity exhibited in coals.

Table 3-7 presents ultimate analysis for typical pyrolysis chars derived from biomass feedstocks. Except for the municipal solid waste char, all contain considerable quantities of voltatile constituents, including H and O, due to the low processing temperature.

The C/H and C/O ratios are greater in all chars than in the fresh feed materials. The high-temperature municipal waste char has been almost completely devolatilized, as is evidenced by the low H and O contents.

3.1.3 Moisture Content of Fuels

Woody fuels and municipal solid waste samples are available with various moisture contents. The moisture is important in determining drying costs and as-received heat contents of the fuels.

Table 3-8 presents approximate ranges of moisture for typical biomass fuels. The effect of moisture on the recoverable heat is dramatic due to the heat requirements for vaporizing the moisture plus superheating the vapor.

Table. 3-7. ULTIMATE ANALYSIS DATA FOR SELECTED PYROLYSIS CHARS
(Dry Basis, Weight Percent)

Material	C	H	N	S	O	Ash	Higher Heating Value (Btu/lb)	Reference
Fir bark char	49.9	4.0	0.1	0.1	24.5	21.4	8,260	Pober and Bauer 1977
Rice hull char	36.0	2.6	0.4	0.1	11.7	49.2	6,100	Pober and Bauer 1977
Grass straw char	51.0	3.7	0.5	0.8	19.7	24.3	8,300	Pober and Bauer 1977
Animal waste char[a]	34.5	2.2	1.9	0.9	7.9	48.8	5,450	Pober and Bauer 1977
Municipal solid waste char (high temperature)	54.9	0.8	1.1	0.2	1.8	41.2	8,020	Sanner et al. 1970
Redwood charcoal (790 F to 1020 F)	75.6	3.3	0.2	0.2	18.4	2.3	12,400	Boley and Landers 1969
Redwood charcoal (860 F to 1725 F)	78.8	3.5	0.2	0.2	13.2	4.1	13,100	Boley and Landers 1969
Oak charcoal (820 F to 1185 F)	67.7	2.4	0.4	0.2	14.4	14.9	10,660	Boley and Landers 1969
Oak charcoal (1060 F)	64.6	2.1	0.4	0.1	15.5	17.3	9,910	Boley and Landers 1969

[a]Contains 3.7% Cl lumped with oxygen.

Table 3-8. APPROXIMATE MOISTURE CONTENTS OF TYPICAL BIOMASS FUELS

Biomass Fuel	Moisture Content (wt %)
Bark	25-75
Coarse wood residue	30-60
Shavings	16-40
Sawdust	25-40
Sander dust	2-8
Municipal refuse	20
Air dry feedlot waste	12

3.1.4 Heating Values

The heating value of carbon feedstocks is determined by the ASTM method listed in Table 3-1. The experimental method employs an adiabatic bomb calorimeter which measures the enthalpy change between reactants and products at 25 C. The heating value obtained is termed the higher heating value because the water of combustion is present in the liquid state at the completion of the experimental determination.

The heating value may be reported on two bases. These are the gross or higher heating value and the net or lower heating value. The higher heating value (HHV) represents the heat of combustion relative to liquid water as the product. The lower heating value (LHV) is based on gaseous water. The difference in the heating value is the latent heat of the water of combustion. Heating values often are reported on both wet and dry fuel bases. The conversion between bases is simple in the case of the higher heating value, involving only normalizing out the moisture (M). This is true because the moisture present in the raw fuel is in the same state before and after combustion.

$$HHV^* = \frac{HHV}{(1 - M)} \quad . \tag{3-12}$$

Lower (net) heating values depend on the moisture content in a more complicated fashion. Since both the product water and moisture are present as vapor after combustion, a portion of the heat of combustion is used to evaporate the moisture. Therefore, using the latent heat of water, $\lambda = 980$ Btu/lb,

$$HHV^* = \frac{LHV + M\lambda}{(1 - M)} \quad . \tag{3-13}$$

To convert between higher (gross) and lower (net) heating values, the amount of water produced by combustion reactions, but not including moisture, must be known. If this is called W, lb water/lb fuel, then the heating values are related by:

$$HHV = LHV + W\lambda . \tag{3-14}$$

All heats reported in this chapter are higher (gross) heating values on a dry basis.

Table 3-4 reports higher heating values on a dry basis for a variety of biomass fuels.

Typically, the heating values for coals are much greater than for biomass materials, ranging from 10 MBtu/lb to 14 MBtu/lb and 5 MBtu/lb to 9 MBtu/lb, respectively. This is principally due to the higher carbon content of the coals. Table 3-7 gives higher heating values for biomass chars. The values are low due to the high ash content of the chars; however, on a dry, ash-free basis, the heating values are similar to those of the coals.

A common method for estimating heating values of solid fuels is the Dulong-Bertholot equation (Spiers 1962) which permits the heating value to be estimated from the ultimate analysis. Table 3-9 presents a comparison of calculated and experimental gross heating values for biomass fuels and chars. For the fresh biomass feeds, the method consistently underpredicts the heating value. For the 14 feedstocks listed in Table 3-9, the average error in heating values is -6.8% or -500 Btu/lb. The method is least accurate for the samples with the highest oxygen content. In the case of the chars, the method is much more accurate, yielding an average error of 3.1% or 220 Btu/lb. The bias error for the five chars is only 1.2%, indicating that the equation is more applicable to the chars than to the fresh biomass.

A second method for estimating heating values is that of Tillman (1978). As shown in Table 3-9, the results for Tillman's equation, which uses only the carbon content, are much more accurate for the biomass materials than the Dulong-Berthelot equation. The average error is roughly 180 Btu/lb for the fresh feedstocks. Further, the predictions show no statistical bias. For the chars, however, the errors are roughly double those of the Dulong-Berthelot equation.

Table 3-9. COMPARISON OF EXPERIMENTAL AND CALCULATED HIGHER (GROSS) VALUES USING PUBLISHED GHV CORRELATIONS

Material	Experimental HHV* (Btu/lb)	Dulong - Berthelot[a] Calc. (Btu/lb)	Dulong - Berthelot[a] Error (%)	Tillman[b] Calc. (Btu/lb)	Tillman[b] Error (%)	IGT[c] Calc. (Btu/lb)	IGT[c] Error (%)
Fresh biomass:							
Douglas fir	9052	8499	-6.1	9114	+0.7	9152	1.1
Douglas fir bark	9500	9124	-4.0	9848	-3.5	9694	2.1
Pine bark	8780	8312	-5.3	9114	+3.8	8947	1.9
Western hemlock	8620	7840	-10.7	8757	+1.6	8536	-1.0
Redwood	9040	8441	-6.6	9340	+3.3	9115	0.8
Beech	8906	8311	-5.1	8990	+2.6	8990	0.9
Hickory	8610	8036	-7.3	8620	-0.6	8746	1.6
Maple	8671	7974	-7.1	8802	+2.6	8684	0.2
Poplar	8920	8311	-6.8	8990	+0.8	8990	0.8
Rice hulls	6610	8128	-7.3	6520	-1.4	6707	1.5
Rice straw	6540	6160	-5.8	6652	+1.7	6648	1.7
Sawdust pellets	8814	7503	-14.9	8156	-7.8	8270	-6.2
Animal waste	7380	7131	-3.4	7310	-1.0	7542	2.2
Municipal solid waste (MSW)	8546	8128	-4.9	8231	-3.7	8642	-1.1
Paper	7572	6582	-13.1	7441	-1.7	7329	-3.2
Absolute Avg. Error			7.2		2.5		1.7
Bias Error			-7.2		-0.2		+0.4
Chars:							
Fir bark	8260	7961	-3.6	8663	+4.9	8184	-0.9
Rice hulls	6100	6026	-1.2	6050	-0.8	6058	-0.7
Grass straw	8300	8309	+0.1	8870	+6.7	8403	1.2
Animal waste	5450	5722	+5.9	5768	+5.8	5830	7.0
MSW	8020	8399	+4.7	9603	+19.7	8088	0.8
Absolute Avg. Error			3.1		7.6		2.1
Bias Error			+1.2		7.3		+1.5

[a]Dulong-Berthelot Equation: HHV, Btu/lb = $146.76\ C + 621\ H - \frac{N + O - 1}{8} + 39.96\ S$
[b]Tillman Equation: HHV, Btu/lb = $188\ C - 718$.
[c]IGT Equation: HHV, Btu/lb = $146.58\ C + 568.78\ H + 29.45 - 6.58\ A - 51.53\ (O + N)$.

Nomenclature: All values are weight percent, dry basis

A = Ash
C = Carbon
H = Hydrogen
N = Nitrogen
O = Oxygen
S = Sulfur

% Error = 100 [Calc. HHV - Exptl. HHV]/[Exptl. HHV]

Absolute Average Error = $\frac{\Sigma_{n}^{N} |\%\ Error|}{N}$

Bias Error = $\frac{\Sigma_{n}^{N} \%\ Error}{N}$

N = number of data points.

A third method of estimating gross heating values has been developed at IGT (Inst. of Gas Technology 1978) using the experimental heating values and ultimate analyses of more than 700 coal samples. When this heating value correlation is used to estimate the higher heating values of fresh biomass materials, the average error that results is approximately 130 Btu/lb with a small positive statistical bias of approximately 26 Btu/lb. When used to predict biomass char heating values, the IGT correlation error is smaller than the errors for both the Dulong-Bertholot and Tillman correlations.

Of these three correlations, the IGT method seems to give the best estimates of biomass and biomass char heating values. The experimental error in the ASTM heating value is $\pm$ 100 Btu/lb while the IGT method yields an average error for chars and fresh biomass of about 150 Btu/lb. Experimental values should be used in cases where the elemental analysis is much different from materials previously tested.

3.1.5 Heats of Formation

In thermodynamic calculations, the heat of formation of the feedstocks is required. Heats of formation may be calculated rigorously from the heats of combustion, assuming that the only materials oxidized are C, H, N, and S, by posing the following reactions:

$$\text{Fuel} + O_2 \longrightarrow \begin{cases} CO_2, \; -94{,}052 \text{ cal/mole} & (3\text{-}15) \\ H_2O(L), \; -68{,}317 \text{ cal/mole} & (3\text{-}16) \\ NO_2, \; +7{,}960 \text{ cal/mole} & (3\text{-}17) \\ SO_2, \; -70{,}940 \text{ cal/mole} & (3\text{-}18) \end{cases}$$

The heat of formation of the fuel may be calculated as follows, assuming no chemical heat involving ash reactions:

$$H_f \; (25\ C) = (HHV^* + 0.018 \sum_{products} [H_{fi} n_{fi}])/(1 - ASH), \qquad (3\text{-}19)$$

in Btu/lb, dry, ash-free basis.

In this equation, n_{fi} is the moles of species i formed per 100 lb of dry biomass on combustion (i can be CO_2, $H_2O(l)$, NO_2, SO_2) while H_{fi} is the heat of formation of i at 25 C in cal/mole. The factor 0.018 puts the formation enthalpy on a Btu per pound of biomass basis. The HHV is treated as a positive number. The heat of formation is normalized to a dry, ash-free basis for purposes of comparison. Table 3-10 presents heat of formation for a variety of feedstocks. The data show a definite trend in terms of the rank (degree of aromatization) of the materials involved. Biomass is very low in rank since its structure consists of only single aromatic rings (benzene derivatives). Fuels of higher rank—peat, lignite, bituminous, and anthracite coals — have structures containing progressively larger aromatic clusters. Typical bituminous coal structures contain from four to six condensed aromatic rings. The fuel of highest rank is graphite. The coals tend to have low heats of formation which increase in the exothermic sense as the rank decreases. Most woody materials exhibit a constant heat of formation in the range of -2200 Btu/lb. Materials such as straw and rice hulls have higher heats of formation, on the order of -2700 Btu/lb. The biomass chars generally exhibit heats of formation intermediate between coals and fresh biomass materials. Figure 3-1 shows how the heats of formation depend on the H/C ratio of the feedstock. It is evident that the biomass chars, although similar in ultimate analysis to coals, do not correlate with the coals in terms of H/C ratio. This is probably due to the coal's greater degree of aromatization, which is a result of the coalification process.

Heats of combustion for biomass materials can be calculated using the heat of formation data based on the following empirical correlation for biomass materials:

Table 3-10. HEATS OF FORMATION FOR TYPICAL FUELS AND BIOMASS MATERIALS
(Basis: Dry, Ash-Free Solid)

Material	H_f(77 F) (Btu/lb)	H/C, Mole Ratio
Charcoal	+ 142	
Pittsburgh seam coal	- 209	
Western Kentucky No. 11 coal	- 323	
Utah coal	- 540	
Wyoming Elkol	- 648	
Lignite	-1062	
Douglas fir	-2219	1.45
Douglas fir bark	-2081	1.26
Pine bark	-2227	1.33
Western hemlock	-2106	1.38
Redwood	-2139	1.33
Beech	-2480	1.45
Hickory	-2344	1.57
Maple	-2203	1.43
Poplar	-2229	1.45
Rice hulls	-2747	1.78
Rice straw	-2628	1.56
Sawdust pellets	-1860	1.65
Animal waste	-2449	1.55
Municipal solid waste	-2112	1.51
Fir bark char	-1580	0.96
Rice hull char	-1136	0.87
Grass straw char	-1581	0.87
Animal waste char	-1536	0.76
Municipal solid waste char	-213.8	0.18

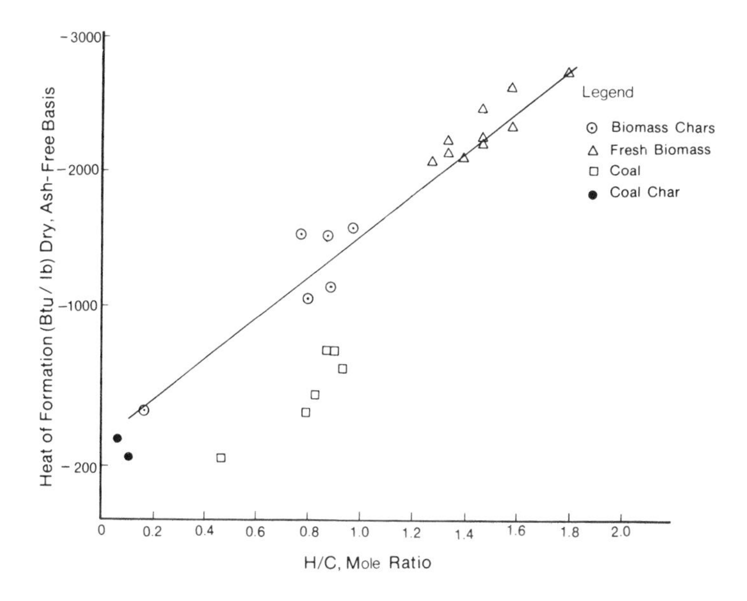

Figure 3-1. Heats of Formation of Carbonaceous Fuels

$$H_f\ (77\ F) = -1437\ H/C - 149 \tag{3-20}$$

with H/C as mole ratio, Btu/lb, dry ash-free basis.

For natural biomass materials and their chars, the following equation results, based on the ultimate analysis and the pertinent combustion reactions:

$$HHV^* = (141C + 615H - 10.2N + 39.95S) - (1 - ASH)\left(\frac{17{,}244H}{C}\right) + 149. \tag{3-21}$$

The HHV* is the gross heating value on a dry basis, Btu/lb, and the analytical data are expressed in weight percent. This equation cannot be expected to function for manmade materials such as plastics or for noncellulose-derived materials like leather. Table 3-11 shows that this equation predicts heating values more accurately than the previously tested methods, yielding errors of only $\pm$ 100 Btu/lb of material, which is within experimental error. The equation is similar in form to the IGT equation.

Table 3-11. COMPARISON OF CALCULATED AND EXPERIMENTAL HEATS OF COMBUSTION USING HEAT OF FORMATION EQUATION

	Experimental HHV* (Btu/lb)	Calculated HHV* (Btu/lb)	% Error	
Douglas fir	9052	9039	-0.12	
Douglas fir bark	9500	9617	1.23	
Pine bark	8780	8938	1.80	
Western hemlock	8620	8590	-0.35	
Redwood	9040	9124	+0.93	
Beech	8096	8906	1.67	
Hickory	8610	8610	-0.69	
Maple	8671	8671	+1.06	
Poplar	8920	8906	-0.15	
Rice hulls	6610	6646	+0.54	
Rice straw	6540	6728	2.88	
Sawdust pellets	8814	8154	-7.49	
Animal waste	7380	7442	0.85	1.63%
Municipal solid waste	8546	8357	-2.21	120 Btu/lb
Paper	7572	7385	-2.47	
Fir bark char	8260	8295	+0.43	
Rice hull char	6100	5967	-2.19	1.44%
Grass straw char	8300	8434	1.61	102 Btu/lb
Animal waste char	5450	5595	+2.66	
Municipal solid waste char	8020	7994	-0.32	

3.1.6 Ash

Table 3-12 shows that the ash content of most woods is on the order of 1%. The ash is composed principally of CaO, K_2O, Na_2O, MgO, SiO_2, Fe_2O_3, P_2O_5, SO_3 and Cl (Wise 1946). The first five oxides generally comprise the bulk of the ash although P_2O_5 is present in some ashes in concentrations as high as 20%. Calcium oxide generally represents half of the total ash, and the potassium oxide content is on the average 20%. Trace metal analysis also indicates the presence of aluminum, lead, zinc, copper, titanium, tin, nickel, and thallium.

3.2 CHEMICAL COMPOSITION OF WOODS

In characterizing and correlating reactivity data for pyrolysis and gasification, it is necessary to have some idea of the chemical structure of the reactant material. Woods can be analyzed in terms of fractions of differing reactivity by solvent extraction techniques. This section provides some of the relevant information on the structure and composition of these reactive fractions which will be useful in later discussions of gasification kinetics (Chapter 7) and pyrolysis (Chapter 5).

Woods can be separated into three fractions: extractables, cell wall components, and ash. The extractables, generally present in amounts of 4% to 20%, consist of materials derived from the living cell. The cell wall components, representing the bulk of wood, are principally the lignin fraction and the total carbohydrate fraction (cellulose and hemicellulose) termed holocellulose. Lignin, the cementing agent for the cellulose fibers, is a complex polymer of phenylpropane. Cellulose is a polymer formed from d (+)-glucose while the hemicellulose polymer is based on other hexose and pentose sugars. In woods, the cell wall fraction generally consists of lignin/cellulose in the ratio 43/57. Residues of the total wood, such as bark and sawdust, have differing compositions.

Table 3-12 presents some analyses of woods on a dry basis while Table 3-13 presents data for typical wood barks.

Table 3-12. CHEMICAL ANALYSES OF REPRESENTATIVE WOODS[a]
(wt %)

Sample	Ash	Extractables	Lignin	Holocellulose
Softwoods[b]				
Western white pine	0.20	13.65	26.44	59.71
Western yellow pine	0.46	15.48	26.65	57.41
Yellow cedar	0.43	14.39	31.32	53.86
Incense cedar	0.34	20.37	37.68	41.60
Redwood	0.21	17.13	34.21	48.45
Hardwoods[c]				
Tanbark oak	0.83	16.29	24.85	58.03
Mesquite	0.54	23.51	30.47	45.48
Hickory	0.69	19.65	23.44	56.22

[a]Encyclopedia of Chem. Tech. 1963, p. 358.
[b]Softwood refers to conifer woods.
[c]Hardwood refers to deciduous woods.

Table 3-13. CHEMICAL ANALYSES OF REPRESENTATIVE WOOD BARKS[a]
(% Dry Basis)

Species	Lignin	Extractables	Ash	Holocellulose[b]
Black spruce	45.84	24.78	2.1	27.28
Fir	39.16	30.37	3.1	27.37
White birch	37.8	21.6	1.5	39.1
Yellow birch	36.5	19.9	2.9	40.7
Beech	37.0	18.3	8.3	36.4

[a]From Wise 1946.
[b]By difference.

In comparing the ultimate analysis data for barks and whole woods in Table 3-4, there is no indication that the chemical makeup of the feedstocks is different. However, from the extractable and cell wall analyses it is evident that the lignin and extractable contents of barks are much greater than those of whole woods. It should be expected that these materials would exhibit different overall reactivities due to their chemical differences.

3.2.1 Cellulose

The carbohydrate fraction of plant tissues is composed of cellulose and hemicelluloses, which are moderate to high molecular weight polymers based on simple sugars. Cellulose itself is derived from d-glucose while the hemicelluloses are principally polymers of d-xylose and d-mannose. The hemicellulose composed of pectin generally is present in only very small quantities in woody material but can be a substantially abundant constituent of the inner bark of trees.

The cellulose polymer is shown in Fig. 3-2.

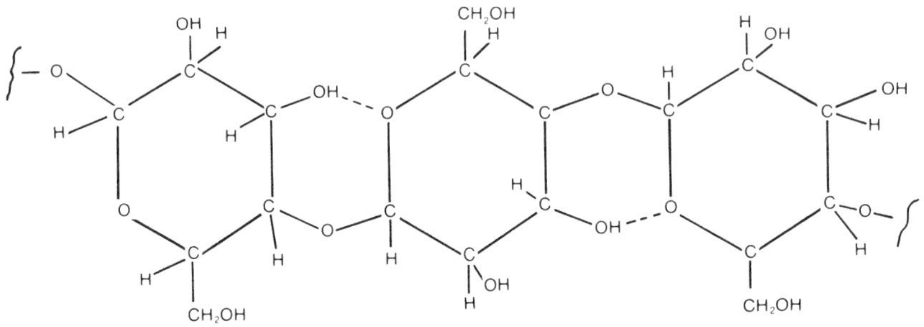

Figure 3-2. The Cellulose Molecule

Cellulose is composed of d-glucose units ($C_6H_{10}O_5$) bound together by ether-type linkages called glycosidic bonds. Glucose is a hexose, or six carbon sugar. In wood the polymers form thread-like chains of molecular weight greater than 100,000. In cotton, 3000 or more units with a combined molecular weight of 500,000 may be present in chains, yielding an extended length of 15,700 Å and cross section of 4 by 8 Å. These very long, thin molecules can be coiled and twisted but, because of the arrangement of the ether linkage, the chain is stiff and extended. An additional contribution to rigidity results from the hydrogen bonding between a hydroxyl hydrogen and the ring oxygen in the adjacent monomer. The threads are woven amongst each other in a random fashion, termed amorphous cellulose, and also fitted together in a crystalline arrangement. Strong van der Waals forces and hydrogen bonds between threads (termed secondary bonding) give rise to a lamellae structure. The weakest bond in the chain direction is the C-O glycosidic bond with an energy of 50 kcal. Cellulose fibers are thus very strong.

The dominant physical characteristic of cellulose is its extreme insolubility, which retards not only acid and enzymic hydrolysis but also the removal of lignins and hemicelluloses interspersed through the cellulose structures. The strong secondary bonding is responsible for the insolubility. Cellulose can be dissolved by strong acids such as hydrochloric, sulfuric, and phosphoric.

Pyrolysis of cell wall materials provides a mixture of volatile materials, tars, and char. The proportion of each fraction and its composition depends on the reaction conditions including temperature, pressure, heating rate, and atmospheric composition. Char results from the condensation of aromatic compounds formed from the primary decomposition products. Since aromatics are not present initially, the amount of char formed by condensation reactions is relatively small. Recent reviews of cellulose chemistry may be found in Shafizadeh and McGinnis (1971), Jones (1969), and references in Chapter 5.

3.2.2 Principal Hemicelluloses

Interlaced with cellulose in the cell walls are a number of other polymeric sugars termed hemicelluloses. These are generally differentiated from true cellulose by their solubility in weak alkaline solutions. Figure 3-3 shows a sequence employed by Timell (1967) for isolating softwood polysaccarides. Hemicelluloses are not precursors of cellulose; they are distinctly different compounds that contain acidic and neutral molecules of low and high molecular weight. In contrast to cellulose, which appears to be universal and invariant as the structural polysaccharide of higher land plants, the hemicellulose polysaccharides show a significant variation in composition and structure among species. Several reviews of hemicellulose chemistry have been presented by Polglase (1955), Aspinall (1959), and Whistler and Richards (1970).

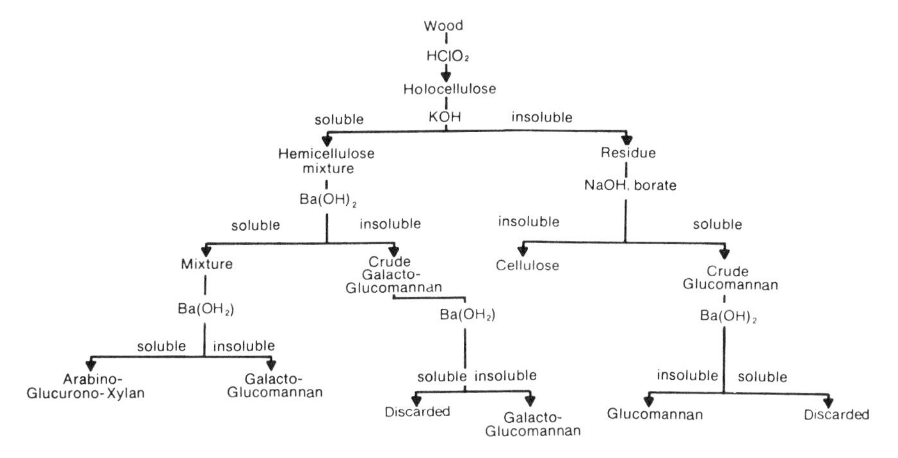

Figure 3-3. Extraction Sequence for Isolation of Softwood Polysaccharides

Most hemicelluloses contain two to four (and occasionally five to six) simpler sugar residues. D-xylose, d-glucose, d-mannose, d-galactose, l-arabinose, d-glucuronic acid, and 4-O-methyl-d-glucuronic acid residues constitute the majority of hemicellulose monomers as shown in Fig. 3-4. The structure is similar to that of cellulose except that the hemicellulose polymers generally contain 50 units to 200 units and exhibit a branched rather than a linear structure.

These structural characteristics, as well as the number and proportion of different sugar residues present (degree of heteropolymerization), largely determine the observed physical properties of hemicelluloses. The heteropolymerization decreases the ability to form regular, tight-fitting crystalline regions and thus makes hemicellulose more soluble than cellulose. Solubility is also increased due to the branching, which decreases the number of intermolecular hydrogen bonds, and the decreased degree of polymerization compared to cellulose.

3.2.2.1 Xylans

Xylans, the most abundant of the hemicelluloses, are polymers of d-xylose ($C_5H_{10}O$). Xylose is a pentose sugar. The xylan fraction of cellulose is often termed pentosan. They are most abundant in agricultural residues such as grain hulls and corn stalks. Hardwoods (deciduous) and softwoods contain appreciable amounts of xylans. Xylan chains are short, exhibiting molecular weights on the order of 30,000 or less. In addition, some xylans contain carboxylic acid and methyl-ether groups. Typical xylans are shown in Fig. 3-5. The acidic xylans contain d-glucuronic acid or the methylate acid as terminal branch units.

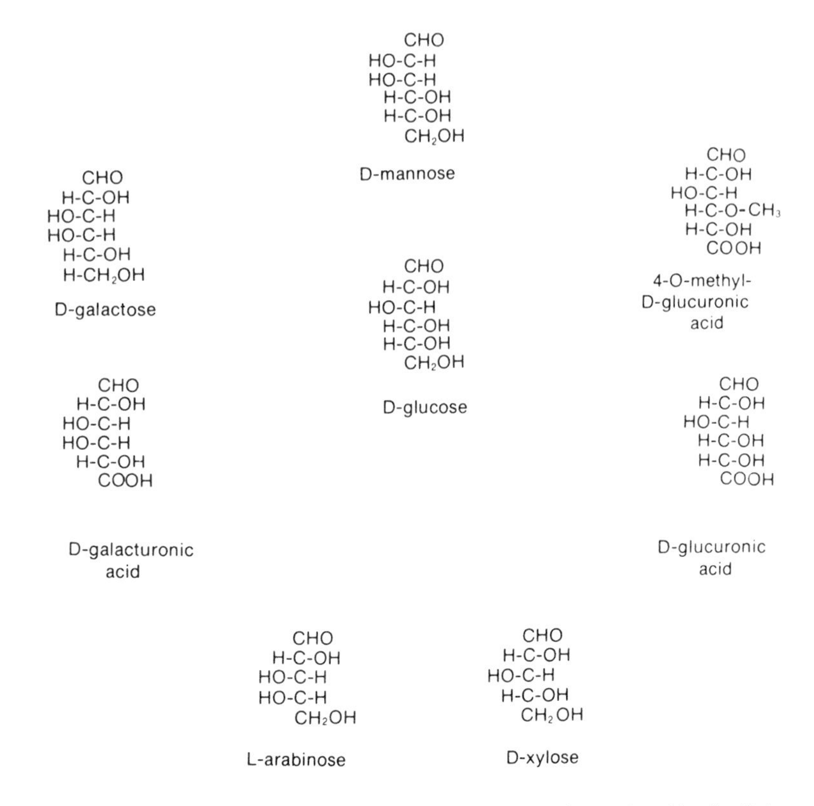

Figure 3-4. Structural Interrelationship of Commonly Occurring Hemicellulose Component Sugars

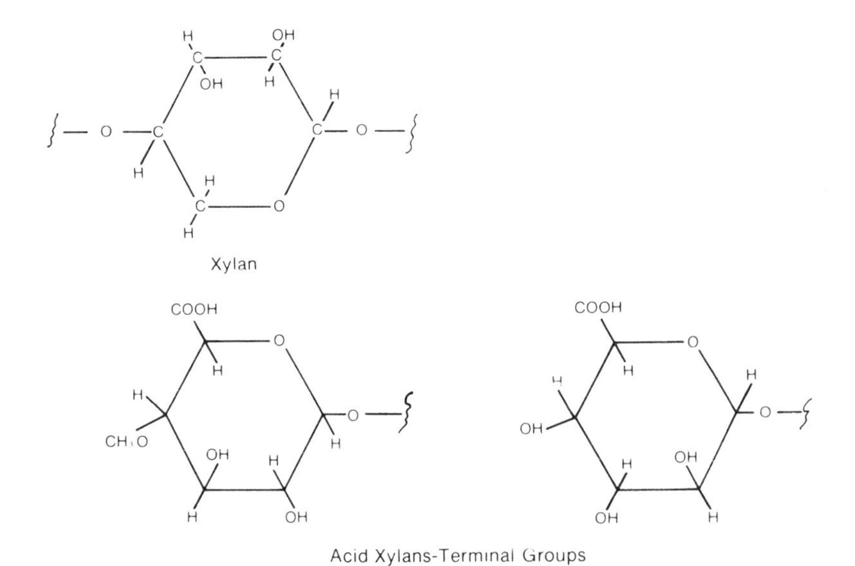

Figure 3-5. Xylan Hemicellulose Structures

Some of the acid xylans are of low molecular weight. They are known as hemicellulose-B and are differentiated from the normal xylans and other neutral hemicelluloses in that they are not precipitated from the alkaline extract by neutralization.

3.2.2.2 Mannans

Mannan-based hemicelluloses include glucomannans, which are built up from linked d-glucose and d-mannose residues in about a 30:70 ratio, and galactoglucomannans, made up of linked d-galactose, d-glucose, and d-mannose in 2:10:30 ratios. In softwoods, mannans are present in substantial amounts while in hardwoods there is generally very little mannan hemicellulose.

3.2.3 Cellulose Data for Woods

Table 3-14 presents some data on the cellulose content of woods. The holocellulose fraction of hardwoods is composed principally of cellulose and xylans. The total content of mannans and other hemicelluloses averages only 4.8% for the four samples. In softwoods, the cellulose fraction is about the same as in hardwoods. However, mannans are present to a much greater extent; the mannans equal or exceed the total xylans in the conifers. Other hemicelluloses are present at 5.4% on the average for the four samples.

Table 3-14. BREAKDOWN OF HOLOCELLULOSE FRACTION OF WOODS[a]

	Wt % in Holocellulose				
	Cellulose	Xylans	Acidxylans	Mannans	Others
Hardwoods					
Trembling aspen	71.5	20.0	4.1	2.9	1.5
Beech	64.5	23.8	6.5	2.9	2.3
Sugar maple	69.8	20.0	5.9	3.1	1.2
Southern red oak	59.8	28.3	6.6	2.9	2.4
Softwoods					
Eastern hemlock	69.0	6.1	5.0	17.1	2.8
Douglas fir	64.6	4.2	4.2	16.0	11.0
White spruce	65.2	9.5	5.0	16.3	3.9
Jack pine	65.1	10.1	5.6	15.1	4.0

[a]From Encyclopedia of Chem. Tech, 1963. p. 358.

3.2.4 Lignin

The noncarbohydrate component of the cell wall, termed lignin, is a three-dimensional polymer based primarily on the phenylpropane unit. Lignin is deposited in an amorphous state surrounding the cellulose fibers and is bound to the cellulose directly by ether bonds. Its exact structure is not known, although considerable information is available based on its chemical reactivity. In solubility analyses, lignin is defined as the cell wall portion not soluble in 72% sulfuric acid. Table 3-15 gives typical elemental analyses of wood lignins.

Table 3-15. ELEMENTAL ANALYSIS OF WOOD LIGNIN

Type	C (%)	H (%)	O (%)	OCH_3 (%)	Molecular Weight
Softwood	63.8	6.3	29.9	15.8	10,000
Hardwood	59.8	6.4	33.7	21.4	5,000

It is assumed, based on much evidence, that the lignins are composed of several monomer groups as shown in Fig. 3-6. These are combined to form the polymer by a variety of linkages involving the aromatic rings and functional groups. The polymer formed contains only single aromatic rings as shown in Fig. 3-7 (structural formula).

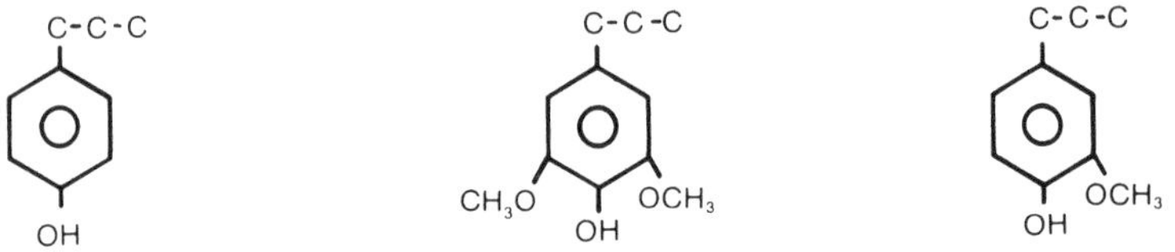

p-hydroxylphenylpropane syringylpropane guaiacylpropane

Figure 3-6. Several Monomer Units in Lignin

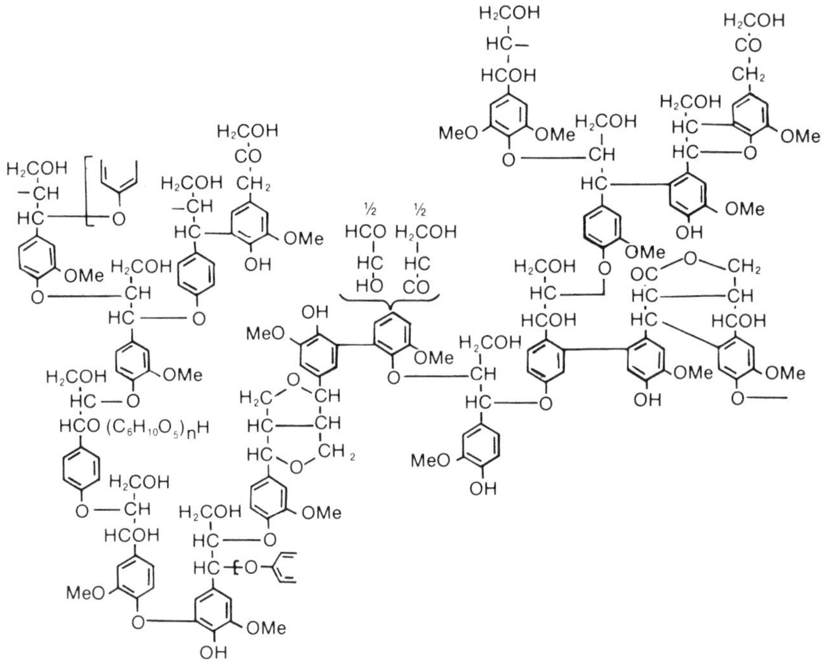

Figure 3-7. Representative Structure of Coniferous Lignin

The representative structure contains the phenylpropane substituted as sinapyl, coniferyl, and p-coumaryl alcohols as shown in Fig. 3-8. Lignification, as discussed by Freudenberg (1965), is thought to occur by dehydration-polymerization of these alcohol units. Thermal pyrolysis of lignin generally yields a considerable amount of char. It is likely that thermal pyrolysis and lignification follow the same route to yield a condensed polynuclear aromatic structure.

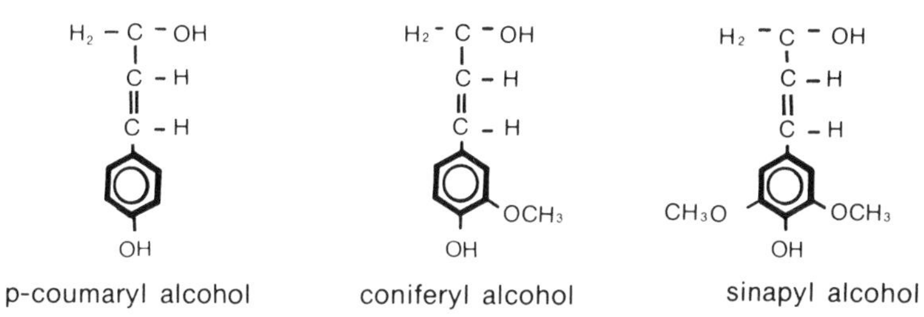

Figure 3-8. p-Hydroxycinnamyl Alcohols

The amount of lignin present varies among materials. Typical amounts for woods and barks are given in Tables 3-12 and 3-13. Table 3-16 gives data for a variety of other biomass materials.

Table 3-16. LIGNIN IN MISCELLANEOUS PLANT MATERIAL[a]

Material	Wt %, Dry Unextracted Material
Rice hulls	40.0
Bagasse	20.3
Peanut shells	28.0
Pine needles	23.9
Wheat straw	13.9
Corncobs	13.4

[a]From Encyclopedia of Chem. Tech. 1963, p. 361.

3.2.5 Extractables

The nature and quantity of extractables vary widely among woods. Table 3-17 lists the types of extractables found in a variety of woody materials. The resin and volatile oils are fragrant and found most abundantly in softwoods. Waxes, fatty acids, pigments, and carbohydrates are commonly found in all woods. Starches account for about 3% of the total wood. Since the quality and nature of extractables vary, the products after pyrolysis and gasification vary.

Table 3-17. EXTRACTABLE COMPONENTS OF WOOD

Volatile Oils (removed by steam or ether soluble)

- Terpenes ($C_{12}H_{16}$)
- Sesquiterpene ($C_{15}H_{24}$)
  - and their oxygenated derivatives

Resins and Fatty Acids (soluble in ether)

- Resin acids ($C_{20}H_{30}O_2$)
- Fatty acids (oleic, linoleic, palmitic)
- Glyceryl esters of fatty acids
- Waxes (esters of monohydroxy alcohols and fatty acids)
- Phytosterols (high molecular weight cyclic alcohols)

Pigments (soluble in alcohol)

- Flavonols, Pyrones, Anthranols } (multi-ring naphthenic and aromatic alcohols, chlorides, ketones acids)
- Tannins (amorphous polyhydroxylic phenols)

Carbohydrate Components (water soluble)

- Starch
- Simple sugars
- Organic acids

Table 3-18 presents some typical extraction data on woods. The bulk of the extractables may be removed by hot water and ether. The ether-soluble portion is usually much greater for the softwoods, showing the higher content of volatile oils and resins. The hot water extraction, which leaches some tannins as well as the carbohyrates, gives yields approximately the same for the soft- and hardwoods.

Table. 3-18. EXTRACTION DATA FOR WOODS[a]

Sample	Wt % of Solubles	
	Hot Water	Ether
Softwoods		
Western yellow pine	5.05	8.52
Yellow cedar	3.11	2.55
Incense cedar	5.38	4.31
Redwood	9.86	1.07
Western white pine	4.49	4.26
Longleaf pine	7.15	6.32
Douglas fir	6.50	1.02
Western larch	12.59	0.81
White spruce	2.14	1.36
Hardwoods		
Tanbark oak	5.60	0.80
Mesquite	15.09	2.30
Hickory	5.57	0.63
Basswood	4.07	1.96
Yellow birch	3.97	0.60
Sugar maple	4.36	0.25
Average—Softwoods	6.25	3.36
Average—Hardwoods	6.44	1.09

[a]From Encyclopedia of Chem. Tech. 1963, p. 358.

3.3 WOOD STRUCTURE

Wood is composed of cells of various sizes and shapes. Long pointed cells are known as fibers; hardwood fibers are about 1 mm in length, while softwood fibers vary in length from about 3 mm to 8 mm. The mechanical properties of wood depend largely on its density which, in turn, is largely determined by the thickness of the cell walls.

3.3.1 Physical Structure of Softwoods

Figure 3-9 shows a typical softwood structure taken from Siau (1971). In softwoods, the fluid conducting elements are the longitudinal tracheids and ray tracheids. Longitudinal and horizontal resin canals are also present in many species.

Longitudinal tracheids, shown in Fig. 3-10, make up the bulk of the structure of softwoods. These are long, hollow, narrow cells having no openings that are tapered along the radial surfaces for a considerable portion of the lengths where they are in contact with other tracheids. The surfaces of the tracheids are dotted with pits, minute depressions in the plant tissue wall which permit the movement of water and dissolved materials between tracheids. The pit is covered with a semipermeable membrane. Pits are oriented in softwoods as adjacent pairs (pit pairs); fluid flow occurs between tracheids in the direction normal to the principal direction of flow.

The tracheid diameter varies from 15 to 80 μm according to species, with a length ranging from 1200 to 7500 μm. Average values of diameter and length, respectively, are

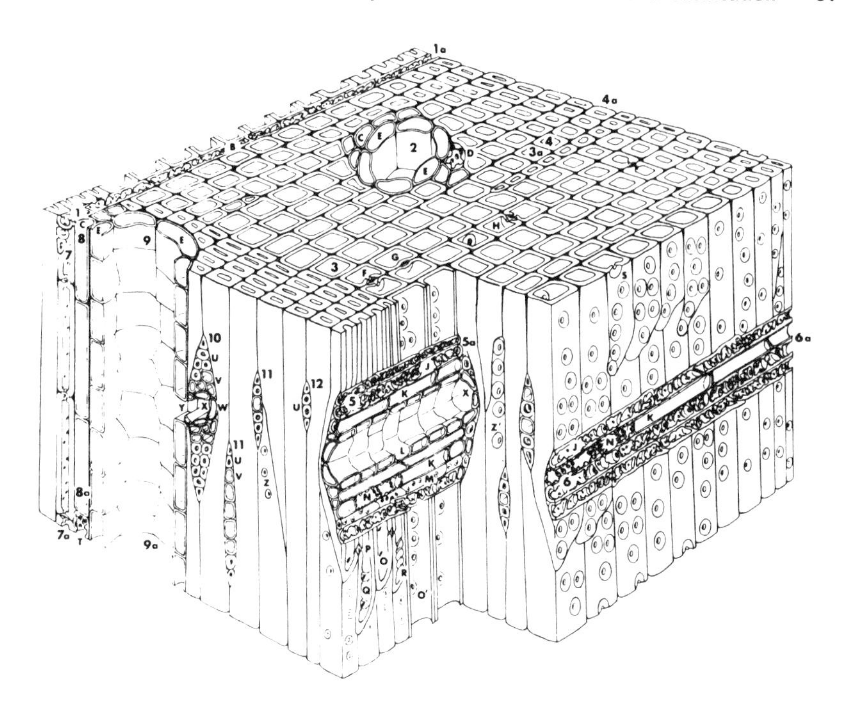

Figure 3-9. Gross Structure of a Typical Southern Pine Softwood

Transverse view. 1-1a, ray; B, dentate ray tracheid; 2, resin canal; C, thin-walled longitudinal parenchyma; D, thick-walled longitudinal parenchyma; E, epithelial cells; 3-3a, earlywood tracheids; F, radial bordered pit pair cut through torus and pit apertures; G, pit pair cut below pit apertures; H, tangential pit pair; 4-4a, latewood.

Radial view. 5-5a, sectioned fusiform ray; J, dentate ray tracheid; K, thin-walled parenchyma; L, epithelial cells; M, unsectioned ray tracheid; N, thick-walled parenchyma; O, latewood radial pit (inner aperture); O', earlywood radial pit (inner aperture); P, tangential bordered pit; Q, callitroid-like thickenings; R, spiral thickening; S, radial bordered pits (the compound middle lamella has been stripped away removing crassulae and tori); 6-6a, sectioned uniseriate heterogeneous ray.

Tangential view: 7-7a, strand tracheids; 8-8a, longitudinal parenchyma (thin-walled); T, thick-walled parenchyma; 9-9a, longitudinal resin canal; 10, fusiform ray; U, ray tracheids; V, ray parenchyma; W, horizontal epithelial cells; X, horizontal resin canal; Y, opening between horizontal and vertical resin canals; 11, uniseriate heterogeneous rays; 12, uniseriate homogeneous ray; Z, small tangential pits in latewood; Z', large tangential pits in earlywood.

33 μm and 3500 μm. The inner diameter which is available for flow is typically 20-30 μm. The effective radius of the pit openings is 0.01 to 4 μm due to the restriction created by the membrane. Typically, a tracheid contains 50 pits. In addition to pit pairs allowing longitudinal flow, there are also pit pairs leading from longitudinal tracheids to ray tracheids, permitting radial flow.

The volumetric composition of a typical softwood is as follows:

Longitudinal tracheids	93%
Longitudinal resin canals	1%
Ray tracheids	6%

Since the principal voidage is oriented longitudinally, the magnitude of the permeability in the longitudinal direction is much greater than the radial permeability. Figure 3-11 shows a schematic model for flow through a softwood.

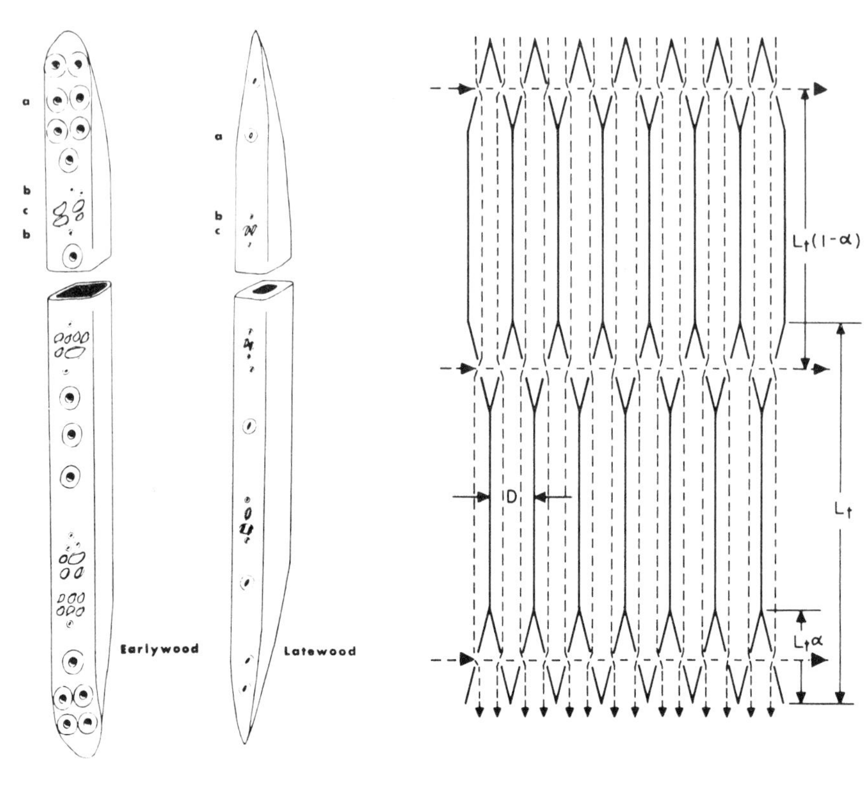

Figure 3-10. Radial Surfaces of Earlywood and Latewood Tracheids

(a,) intertracheid bordered pits: (b,) bordered pits to ray tracheids; (c,) 13 pinoid pits to ray parenchyma.

Figure 3-11. Softwood Flow Model

Tangential section showing pits on the radial surfaces of the tapered ends of the tracheids.

3.3.2 Physical Structure of Hardwoods

The structure of a typical hardwood is shown in Fig. 3-12. The dominant feature of the hardwood structure is the large open vessels or pores. Tracheids and pits are present but contribute significantly more resistance to flow. In a typical hardwood, the following structural composition is present:

Vessels	55%
Tracheids	26%
Woods rays	18%
Others	1%

Vessels are large, with diameters of 20 to 30 μm. The vessels are short, connected by "perforation plates" which offer very low flow resistance. Thus the vessels behave as long capillaries. Figure 3-13 shows the nature of flow through hardwoods.

3.3.3 Permeability

Permeability is important in pyrolysis. During heating, pyrolysis gases and liquids are generated within the particle and must pass through the porous stucture to the surroundings. Low permeability may significantly affect the product distribution by increasing

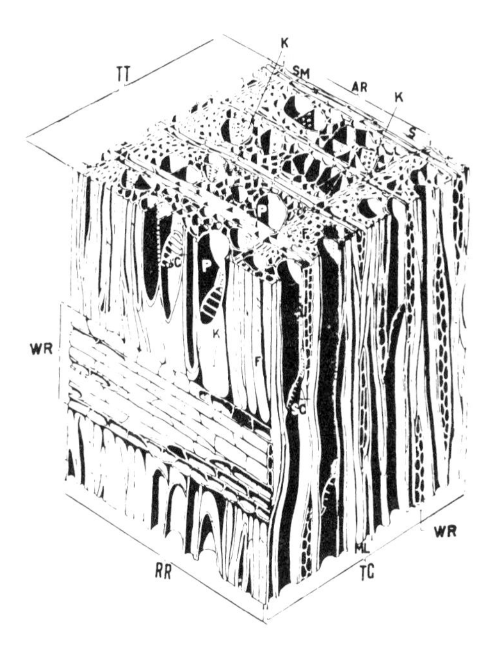

Figure 3-12. Gross Structure of a Typical Hardwood

Plane TT is the cross section. RR is the radial surface, and TG is the tangential surface. The vessels or pores are indicated by P, and the elements are separated by scalariform perforation plates, SC. The fibers, F, have small cavities and thick walls. Pits in the walls of the fibers and vessels, K, provide for the flow of liquid between the cells. The wood rays are indicated at WR. AR indicates one annual ring. The earlywood (springwood) is designated S, while the latewood (summerwood) is SM. The true middle lamella is located at ML.

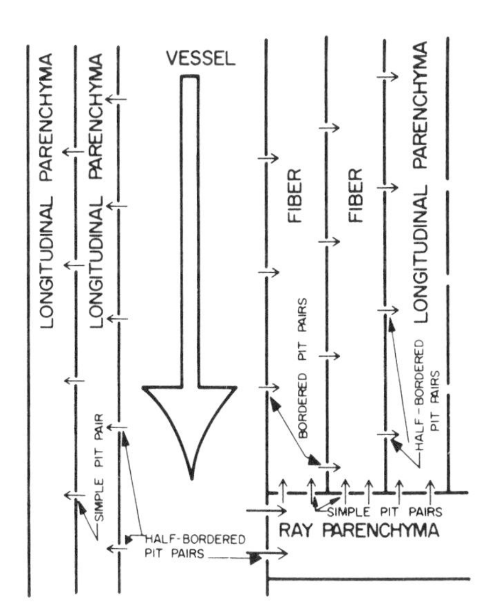

Figure 3-13. Generalized Flow Model for Hardwoods

The relative magnitude of the flow is indicated by the size of the arrow.

the residence time of the primary pyrolysis products in the hot zone, thereby increasing the probability that they will enter into secondary reactions. Pelletized, densified biomass will have a low permeability compared with natural woods. Table 3-19 shows the range of permeability for various natural woods.

In natural soft- and hardwood structures, it is evident that the porosity is directed principally in the vertical direction in the livewood. Physical properties such as thermal conductivity and diffusivity therefore depend on direction within fresh wood. Such a behavior is termed anistropic.

During densification, the voidage of the wood is greatly reduced and these physical properties become more uniform or isotropic. Other forms of compacted biomass, such as sawdust pellets or compacted municipal solid waste, can also be expected to be more or less isotropic.

3.4 PHYSICAL PROPERTIES

In addition to heating value, the other major physical data necessary for predicting the thermal response of biomass materials under pyrolysis, gasification, and combustion processes are thermal conductivity, heat capacity, true density, and diffusion coefficients.

Table 3-19. TYPICAL PERMEABILITY VALUES

Permeability $\left[\frac{cm^3 \text{ (air)}}{cm \text{ atm}}\right]$	Longitudinal Permeabilities	
10^4	Red Oak	$R \approx 150$ μm
10^3	Basswood	$R \approx 20$ μm
10^2	Maple, pine sapwood, Douglas fir sapwood (Pacific coast)	
10^1	Spruces (sapwood) Cedars (sapwood)	
10^0	Douglas fir heartwood (Pacific coast)	
10^{-1}	White oak heartwood Beech heartwood Cedar heartwood Douglas fir heartwood (intermountain)	
10^{-2} 10^{-3} 10^{-4}		Transverse permeabilities. (The species are in approximately the same order as those for longitudinal permeabilities.)

3.4.1 Thermal Conductivity

Thermal conductivity is defined in general terms as a proportionality factor which relates heat flow through a material to a temperature difference across a specified distance in that material. Mathematically, thermal conductivity is defined by Fourier's Law of Heat Conduction, given here for unidimensional heat flux in the x-direction in rectangular coordinates:

$$q_x = - k_x \frac{dT}{dx} \tag{3-21}$$

Most homogeneous materials are isotropic, and the thermal conductivity varies only with temperature

$$q = - k\nabla T \tag{3-22}$$

However, most naturally occurring biomass materials are anistropic. For wood, the thermal conductivity is a function of temperature and spatial direction. Modified biomass materials, such as densified wood, probably do not exhibit the same type of anisotropic behavior as the naturally occurring biomass materials. Thermal conductivity should be related to the various materials present in a substance. Thus in biomass thermal conductivity should be a function of the major constitutents, including moisture, cellulose, hemicellulose, and lignin.

Table 3-20 compiles available thermal conductivity data for biomass materials. No data are available for compacted biomass feedstocks. In general, no chemical analyses are presented with the data. The bulk of the data are probably effective thermal conductivities of powders rather than of the solids. The conductivities for solid woods, for example, are two to ten times greater than for many of the other biomass materials listed in Table 3-20 (e.g., sawdust and redwood shavings). Most data sources do not specify the state of the materials.

Table 3-20. THERMAL CONDUCTIVITY OF SELECTED BIOMASS MATERIALS

Material	Bulk Density (lb/ft^3)	Temperature (F)	Thermal Conductivity (Btu/ft-h-F)	Reference
Ashes, wood	—	32-212	0.040	Kern 1950
Cardboard	—	—	0.037	Kern 1950
Carbon, porous, with grain				
Grade 60 48% porosity	65.5[a]	Room Temp.	0.083	Perry and Chilton 1973
Grade 45 47% porosity	64.9[a]	Room Temp.	0.083	Perry and Chilton 1973
Grade 25 47% porosity	64.3[a]	Room Temp.	0.083	Perry and Chilton 1973
Carbon refractory brick 17% porosity	102.9[a]	Room Temp.	1.33	Perry and Chilton 1973
Celotex, sheet fiber from sugar cane	13.2	—	0.028	Handbook Chem. Phys. 1966
	14.8		0.028	Handbook Chem. Phys. 1966
	14.4	32	0.0253	McAdams 1954
	14.4	0	0.0242	McAdams 1954
	14.4	-100	0.0208	McAdams 1954
	14.4	-200	0.0175	McAdams 1954
	14.4	-300	0.0133	McAdams 1954
Charcoal - from maple, beech, and birch				
Coarse	13.2	—	0.030	Handbook Chem. Phys. 1966
6 mesh	15.2	—	0.031	Handbook Chem. Phys. 1966
20 mesh	19.2	—	0.032	Handbook Chem. Phys. 1966
Charcoal flakes	11.9	176	0.043	McAdams 1954
	15	176	0.051	McAdams 1954
		0 to 100	0.11	Perry and Chilton 1973
Coke powder				
Cork, regranulated				
Fine particles	9.4	—	0.025	Handbook Chem. Phys. 1966
3/16-in. particles	8.1	—	0.026	Handbook Chem. Phys. 1966
Corkboard	5.4	—	0.021	Handbook Chem. Phys. 1966
	7.0	—	0.022	Handbook Chem. Phys. 1966
	10.6	—	0.025	Handbook Chem. Phys. 1966
	14.0	—	0.028	Handbook Chem. Phys. 1966
	6.9	32	0.0205	McAdams 1954
	6.9	0	0.0200	McAdams 1954
	6.9	-100	0.0183	McAdams 1954
	6.9	-200	0.0142	McAdams 1954
	6.9	-300	0.0100	McAdams 1954
Cork, pulverized	10.0	32	0.035	McAdams 1954
	10.0	100	0.039	McAdams 1954
	10.0	200	0.032	McAdams 1954
Cotton	5.0	200	0.037	McAdams 1954
	5.0	100	0.035	McAdams 1954
	5.0	32	0.0325	McAdams 1954
	5.0	-100	0.0276	McAdams 1954

[a] Apparent density, defined in Section 3.4.3.

Table 3-20. THERMAL CONDUCTIVITY (concluded)

Material	Bulk Density (lb/ft^3)	Temperature (F)	Thermal Conductivity (Btu/ft-h-F)	Reference
Cotton	5.0	-200	0.0235	McAdams 1954
	5.0	-300	0.0198	McAdams 1954
Graphite				
2 3/4 in. diam., 3/4 in. thick				
30% porosity	98.6[a]	—	7.33	Handbook Chem. Phys. 1966
Porous, Grade 60 52% porosity	65.5[a]	—	4.17	Handbook Chem. Phys. 1966
Porous, Grade 45 53% porosity	64.9[a]	—	3.75	Handbook Chem. Phys. 1966
Porous, Grade 25 53% porosity	64.3[a]	—	3.33	Handbook Chem. Phys. 1966
Paper	—	—	0.075	McAdams 1954
Paper or pulp, macerated	2.5-3.5	—	0.021	Lewis 1968
Sawdust, various	12.0	—	0.034	Handbook Chem. Phys. 1966
Redwood	10.9	—	0.035	Handbook Chem. Phys. 1966
(and shavings)	8-15	—	0.0375	Lewis 1968
Sawdust (soft pine and oak)				
10-40 mesh	—	-295	0.016	Chow 1948
	—	-180	0.0195	Chow 1948
	—	-105	0.0235	Chow 1948
	—	-55	0.0265	Chow 1948
	—	-20	0.0295	Chow 1948
	—	+5	0.0325	Chow 1948
	—	+30	0.0335	Chow 1948
	—	+35	0.0385	Chow 1948
	—	+62	0.040	Chow 1948
Shredded redwood bark	4.0	32	0.0290	McAdams 1954
	4.0	-100	0.0235	McAdams 1954
	4.0	-200	0.0196	McAdams 1954
	4.0	-300	0.0155	McAdams 1954
	—	-50	0.0168	Rowley et al. 1945
	—	-25	0.0180	Rowley et al. 1945
	—	+25	0.0203	Rowley et al. 1945
	—	+75	0.0226	Rowley et al. 1945
Sheet Insulite, from wood pulp	16.2	—	0.028	Handbook Chem. Phys. 1966
	16.9	—	0.028	Handbook Chem. Phys. 1966
Wood fiber, mat	1.7	-50	0.016	Rowley et al. 1945
	1.7	0	0.018	Rowley et al. 1945
	1.7	+50	0.020	Rowley et al. 1945
	1.7	+100	0.023	Rowley et al. 1945
Blanket	3.5	-50	0.016	Rowley et al. 1945
	3.5	0	0.018	Rowley et al. 1945
	3.5	50	0.020	Rowley et al. 1945
	3.5	100	0.022	Rowley et al. 1945
Excelsier	1.64	-50	0.019	Rowley et al. 1945
	1.64	0	0.022	Rowley et al. 1945
	1.64	50	0.025	Rowley et al. 1945
	1.64	100	0.024	Rowley et al. 1945
Solid woods				
Balsa	8-12	-300	0.0151	Gray et al. 1960
	8-12	-285	0.0150	Gray et al. 1960
	8-12	-260	0.0167	Gray et al. 1960
	8-12	-207	0.0183	Gray et al. 1960
	8-12	-190	0.0192	Gray et al. 1960
	8-12	-160	0.0208	Gray et al. 1960
	8-12	-130	0.0233	Gray et al. 1960
	8-12	-95	0.0232	Gray et al. 1960
Balsa, across grain	7.3	—	0.028	Handbook Chem. Phys. 1966
	8.8	—	0.032	Handbook Chem. Phys. 1966
	20		0.048	Handbook Chem. Phys. 1966
Cypress, across grain	29	—	0.056	Handbook Chem. Phys. 1966
Mahogany, across grain	34	—	0.075	Handbook Chem. Phys. 1966
Maple, across grain	44.7	122	0.11	McAdams 1954
Fir	26	75	0.063	Chapman 1974
Oak	30-38	86	0.096	Chapman 1974
Yellow pine	40	75	0.085	Chapman 1974
White pine	27	86	0.065	Chapman 1974
Pine, white, across grain	34	59	0.087	McAdams 1954
	32	—	0.032	Handbook Chem. Phys. 1966
Pine, Virginia, across grain	34	—	0.082	Handbook Chem. Phys. 1966
Pine board, 1 1/4-in. thick	—	-50	0.0515	Gray et al. 1960
	—	0	0.054	Gray et al. 1960
	—	+50	0.0575	Gray et al. 1960
	—	+100	0.059	Gray et al. 1960
Pine, with grain	34.4	70	0.20	McAdams 1954
Oak, across grain	51.5	59	0.12	McAdams 1954

[a]Apparent density, defined in Section 3.4.3.

Steinhagen (1977) has summarized thermal conductivity data for several woods over the range -40 C to +100 C as a function of moisture content and has shown that moisture is an important parameter in wood conductivity. Since the moisture content is not known for the bulk of the entries in Table 3-20, the data presented are at best only semi-quantitative.

Completely lacking in the available data are thermal conductivities at higher temperatures. If thermal conductivity values are to be used in modeling pyrolysis or gasification processes, then new data over the actual range of processing conditions must be developed, including data for densified materials.

3.4.2 Heat Capacity

Heat capacity as normally reported is defined in terms of the enthalpy content of a material and represents the relative ability of a material to store energy. Enthalpy is a function of temperature and pressure.

$$H = H(T,P) \tag{3-24}$$

and

$$dH = \left(\frac{\partial H}{\partial T}\right)_P dT + \left(\frac{\partial H}{\partial p}\right)_T dp. \tag{3-25}$$

For solids and liquids $\partial H/\partial P$ is very small and

$$dH = \left(\frac{\partial H}{\partial T}\right)_P dT. \tag{3-26}$$

By definition the temperature dependency term $(\partial H/\partial T)_p$ called heat capacity at constant pressure, C_p, and is reported on a per unit weight basis. The resulting equation for the enthalpy change is as follows:

$$\Delta H = \int_{T_o}^{T} C_p \, dT. \tag{3-27}$$

This equation is normally used for materials of constant mass and no phase transitions. For example, if water is driven out of wood the apparent heat capacity may change very rapidly with temperature; the same is true for other phase transitions. Generally, if phase transitions are incorporated the enthalpy change will be:

$$\Delta H = \int_{T_o}^{T_p} C_p^{\alpha} dT + \Delta H_{\text{phase transition}} + \int_{T_p}^{T} C_p^{\beta} \, dT, \tag{3-28}$$

where

C_p^{α}, C_p^{β} = heat capacities of phases 1 and 2, and

T_p = temperature at which phase change occurs.

The heat capacity is a function of the composition and temperature but not the density of the material as long as compacting does not alter the chemical structure.

The data on heat capacity are limited. Some typical values are given in Table 3-21. No characterization data are reported for the samples.

Several C_p equations have been developed to predict the heat capacities of wood at temperatures to 100 C. As an example of specific heat equations for woods, Beall (1968) shows the equation in which moisture is an important parameter in estimating the heat capacity:

Table 3-21. HEAT CAPACITY

Material	F (F)	Btu/lb-F	Reference
Carbon	78-168	0.168	Perry 1973
	103-1640	0.314	Perry 1973
	132-2640	0.387	Perry 1973
Charcoal	50	0.16	Perry 1973
Cellulose	—	0.32	Perry 1973
Oak	—	0.57	Perry 1973
Fir	75	0.65	Chapman 1974
Yellow pine	75	0.67	Chapman 1974
Cork	68	0.45	Chapman 1974

$$C_p = 0.259 + (9.75 \times 10^{-4})M + 6.05 \times 10^{-4}\, T_1 + 1.3 \times 10^{-5} M\, T_1, \quad (3\text{-}29)$$

where

M = % moisture, up to 27%.

Other methods are available and generally are for the same temperature range.

As with thermal conductivity, no references were readily available for heat capacity of biomass materials for the temperature range of thermal processing conditions needed for pyrolysis or gasification; new data are needed for applicable temperature ranges.

3.4.3 Density

The density of the material is important in considering energy contents of fuels on a volumetric basis, such as for transporting, solids handling, and sizing reaction vessels. There are three ways of reporting solid material density: bulk density, apparent particle density, and skeletal density. These density values differ in the way in which the material volume is calculated. The bulk density volume basis includes the actual volume of the solid, the pore volume, and the void volume between solid particles. Apparent particle density includes solid volume and pore volume. Skeletal density, or true density, includes only solid volume. The three values are related as follows:

$$\rho_a = \rho_s (1-\epsilon_p) \quad (3\text{-}30)$$

$$\rho_b = \rho_a (1-\epsilon_b), \quad (3\text{-}31)$$

where

ρ_s = skeletal density, weight/volume
ρ_a = apparent density, weight/volume
ρ_b = bulk density, weight/volume

$$\epsilon_p = \text{particle porosity} = \frac{\text{volume of pores}}{\text{volume of pores and volume of solid}}$$

$$\epsilon_b = \text{bed porosity} = \frac{\text{volume of external voids}}{\text{volume of external voids and volume of particles}}$$

Densification of biomass is accomplished by reducing the particle porosity ϵ_p.

The density of biomass depends on the nature of the material, its moisture content, and degree of densification. Raw, oven-dry biomass (with 7% to 8% moisture) has an apparent density of about 40 lb/ft^3 (hardwoods) and 28 lb/ft^3 (softwoods). The density of woods with high moisture contents can be as high as 60 lb/ft^3. Densification produces particles with apparent bone dry densities of 55 lb/ft^3 to 75 lb/ft^3. The skeletal density of oven dry biomass has been reported to be 91 lb/ft^3 (Siau 1971).

3.4.3.1 Effect of Moisture Content on Density

The apparent density of wood and biomass depends on the moisture content. The dry and wet biomass apparent densities are related as follows from the moisture content obtained from the proximate analysis of the raw feedstock:

$$\rho_a(D) = (1 - M)\ \rho_a(R), \tag{3-32}$$

where

$\rho_a(D)$ = apparent density of dry biomass,
$\rho_a(R)$ = apparent density of raw biomass, and
M = proximate moisture.

For a typical raw biomass with 50% moisture and apparent dry density of 30 lb/ft^3, the raw biomass sample has a density of 60 lb/ft^3.

3.4.3.2 Densification

Densification by compaction reduces the internal voidage of the biomass material and reshapes the particles so that the bulk density is increased. The bulk density of green wood chips is typically 20 lb/ft^3, while the apparent density is on the order of 60 lb/ft^3. The typical external void fraction ϵ_b for chips is therefore about 0.67. The high voidage is due to the shape of the particles. Reshaping the particles to cylinders typically reduces the void fraction ϵ_b to about 0.5 and thus raises the bulk density to about 30 lb/ft^3. Thus the weight per unit volume is increased 50% by reshaping, and more material can be transported in the same carrier volume.

Densification of biomass by decreasing the particle porosity further improves handling. For raw, dry biomass of apparent density of 30 lb/ft^3, the particle porosity, ϵ_p, is typically 0.67 assuming 91 lb/ft^3 for the skeletal density. For densified samples, with reported apparent bone dry densities of 55 lb/ft^3 to 75 lb/ft^3, the particle porosity has decreased to 0.4 to 0.18. Thus in densification a large fraction of the internal voidage is removed.

Representative values of density are shown in Table 3-20 for uncompacted materials. As with thermal conductivity, the state of the material (and thus the type of density reported) is not specified for many solids.

3.4.4 Diffusion Coefficients in Biomass Materials

No data are readily available in the literature on gas diffusion coefficients in either natural or pelleted biomass materials or in their pyrolysis chars.

3.5 REFERENCES

Aspinall, G. O. 1959. Advances in Carbohydrate Chemistry. Vol 14: p. 429.

Beall, F. C. 1968. Specific Heat of Wood - Further Research Required to Obtain Meaningful Data. Madison, WI: Forest Products Laboratory; U.S. Forest Service Research Note FPL-0184.

Bituminous Coal Research, Inc. 1974. Gas Generator Research and Development, Phase II. Process and Equipment Development. OCR-20-F; PB-235530/3GI.

Boley, C. C.; Landers, W. S. 1969. Entrainment Drying and Carbonization of Wood Waste. Washington, D.C.: Bureau of Mines; Report of Investigations 7282.

Bowerman, F. R. 1969. Introductory chapter to Principles and Practices of Incineration. Corey, R. C., editor. New York: John Wiley and Sons.

Chapman, A. J. 1974. Heat Transfer. Third Edition. New York: McMillan Publishing Co.

Chow, C. S. 1948. "Thermal Conductivity of Some Insulating Materials at Low Temperatures." Proceedings Physics Society. Vol. 6: p. 206.

Encyclopedia of Chemical Technology. Kirk, R. E. and Othmer, D. F., editors. 1963. New York: Wiley Interscience.

Freudenberg. 1965. "Lignin: Its Constituents and Formation From p-Hydroxycinnamyl Alcohols." Science. Vol. 148: p. 30.

Gray, V. H.; Gelder, T. F.; Cochran, R. P; Goodykoontz, J. H. 1960. Bonded and Sealed External Insulations for Liquid Hydrogen Fueled Rocket Tanks During Atmospheric Flight. AD 244287; Div. 14; p. 51.

Handbook of Chemistry and Physics, 47th Ed. 1966. Cleveland, OH: Chemical Rubber Company.

Howlett, K.; Gamache, A. 1977. Forest and Mill Residues as Potential Sources of Biomass. Vol. VI. Final Report. McLean, VA: The MITRE Corporation/Metrek Division; ERDA Contract No. E (49-18) 2081; MTR 7347.

Institute of Gas Technology. 1978. Coal Conversion Systems Technical Data Book. DOE Contract EX-76-C-01-2286. Available from NTIS, Springfield, VA.

Jones, D. M. 1964. Advances in Carbohydrate Chemistry. Vol. 19: p. 219.

Kern, D. Q. 1950. Process Heat Transfer. New York: McGraw Hill Book Company.

Klass, D. L.; Ghosh, S. 1973. "Fuel Gas From Organic Wastes." Chemical Technology. p. 689.

Lewis, W. C. 1968. Thermal Insulation from Wood for Buildings: Effects of Moisture and Its Control. Madison, WI: Forest Products Laboratory; Forest Service; U.S. Dept. of Agriculture.

McAdams, W. H. 1954. Heat Transmission. New York: McGraw Hill Book Company.

Perry, R. H.; Chilton, C. H. (editors). 1973. Chemical Engineer's Handbook, 5th Ed., New York: McGraw Hill Book Company.

Pober, K. W.; Bauer, H. F. 1977. "The Nature of Pyrolytic Oil From Municipal Solid Waste." Fuels From Waste. Anderson, L. L. and Tillman, D. A., editors. New York: Academic Press. pp. 73-86.

Polglase, W. J. 1955. Advances in Carbohydrate Chemistry. Vol. 10: p. 283.

Rowley, F. B.; Jordan, R. C.; Lander, R. M. 1945. "Thermal Conductivity of Insulating Materials at Low Mean Temperatures." Refrigeration Engineering. Vol. 50: pp. 541-544.

Rowley, F. B.; Jordan, R. C.; Lander, R. M. 1947. "Low Mean Temperature Thermal Conductivity Studies," Refrigeration Engineering. Vol. 53: pp. 35-39.

Sanner, W. S.; Ortuglio, C.; Walters, J. G.; Wolfson, D. E. 1970. Conversion of Municipal and Industrial Refuse Into Useful Materials by Pyrolysis. U.S. Bureau of Mines; Aug.; RI 7428.

Shafizadeh, F.; McGinnis, G. D. 1971. "Chemical Composition and Thermal Analysis of Cottonwood." Carbohydrate Research. Vol. 16: p. 273.

Siau, J. F. 1971. Flow in Wood. Syracuse, NY: Syracuse University Press.

Spiers, H. M. 1962. Technical Data on Fuel. Sixth Edition. New York: Wiley; p. 291.

Steinhagen, H. P. 1977. Thermal Conductive Properties of Wood, Green or Dry, From -40 to + 100 C: Literature Review. Madison, WI: Forest Products Laboratory; U.S. Forest Service, Dept. of Agriculture.

Tillman, D. A. 1978. Wood as an Energy Resource. New York: Academic Press.

Timell, T. E. 1967. Wood Science Technology. Vol. 1: p. 45.

Wen, C. Y.; Bailie, R. C.; Lin, C. Y.; O'Brien, W. S. 1974. "Production of Low Btu Gas Involving Coal Pyrolysis and Gasification." Advances in Chemistry Series. Vol. 131. Washington, D.C.: American Chemical Society.

Whistler, R. L.; Richards, E. L. 1970. Chapter 37 in The Carbohydrates. Pigman, W. and Horton, D., editors. New York: Academic Press.

Wise, L. E. 1946. Wood Chemistry. American Chemistry Society Monograph Series No. 97.

Chapter 4

Beneficiation of Biomass for Gasification and Combustion

R. Bain
Colorado School of Mines

This section presents a relatively brief discussion of various methods of biomass beneficiation, with emphasis on methods that improve the properties of biomass materials with respect to suitability as gasification or combustion feedstocks. Beneficiation is very broad in scope, and this discussion is not meant as a comprehensive survey of the status of all beneficiation processing methods. (The use of a particular process as an example of a type of beneficiation process does not constitute an endorsement of that process.) Before discussing types of processes for biomass beneficiation, a brief discussion of basic definitions pertinent to beneficiation is required.

Beneficiation is defined as the treatment of some parent material, in this case biomass, so as to improve the physical and/or chemical properties of that material. Emphasis here is on improvement of gasification and combustion properties.

The major types of beneficiation processes to be discussed are drying, comminution, densification, physical separation, and chemical modification. In drying, physically bound water is driven off (the removal of chemically bound water is not included). By comminution, the particle size of a parent material is reduced to a desired range by shredding, cutting, grinding, or pulverization. In densification, the apparent particle density and the bulk density of a material are increased so as to lower transportation costs or processing equipment size by reducing the volume of material to be handled.

Physical separation involves the segregation of various components of a parent material into discrete subfractions. The purpose of this separation is varied; in some cases the separation may improve gasification or combustion properties, while in other cases the separation may be justified on economic grounds.

Chemical modification involves changing the chemical structure of the parent material to make the material more amenable to further processing. In many cases, the waste material or byproducts from a conversion process may also be considered to be a chemically modified biomass; for example, the furfural waste materials from a process producing furfural from corn silage could be considered an indirect beneficiation processing product (Lipinsky et al. 1977).

This section is divided into two subsections based upon two major types of biomass materials: wood and wood products (forestry biomass); and municipal solid wastes (MSW). In each section the advantages and disadvantages of various processes are discussed and available economic data are included. Agricultural biomass beneficiation has not been included due to lack of readily available data.

4.1 WOOD AND WOOD PRODUCTS

In this section various methods for beneficiation of wood and associated wood products are discussed. Emphasis is placed on processes such as comminution, drying, and densification of forestry biomass materials. Beneficiation processes for wood products aim to produce from the parent biomass a material that is a better quality feedstock for gasification or combustion, that has a higher volume energy density or higher specific surface area, and that has a higher gross heating value; if comminution adds heat and concurrently dries, all these goals have the same purpose—to make the use of wood and wood products economically viable.

4.1.1 Comminution

Size reduction processes are traditionally divided into four major classifications:

- compression—used for coarse reduction of solids;
- impaction—used for reduction to a broad range of particle sizes;
- cutting—used to produce solids of a definite size and shape, with few or no fines; and
- attrition—used to produce fine solids from nonabrasive materials.

The basic laws of comminution are given in many textbooks (McCabe and Smith 1967) and are used to estimate the energy requirements for crushing and the energy efficiency of size reduction. These laws are Kick's Law, which reflects the energy absorbed by a solid to the energy produced by crushing, and Rittinger's Law, which predicts that the work required for crushing is proportional to the change in surface area of the solid. These crushing laws were developed for the crushing or grinding of hard, friable solids such as coal, bauxite, and shale. The comminution of wood and wood products involves a process that Dornfield et al. (1978) call fiberization. There are basic differences between the grinding of wood and other biomass materials, and the grinding of hard materials that are caused by the fibrous, anisotropic, and compressive properties of wood. Also affecting comminution are the moisture content, the freshness (how long since harvesting has occurred), and type of wood (springwood, summerwood, etc.).

The comminution processes of interest for woods are mainly compression and cutting, although impaction and attrition are undoubtedly important in high-speed cutting operations. A representation of the two processes of interest was given by Dornfield (1978) and is shown as Fig. 4-1. It illustrates qualitatively the physical mechanisms taking place during wood size reduction.

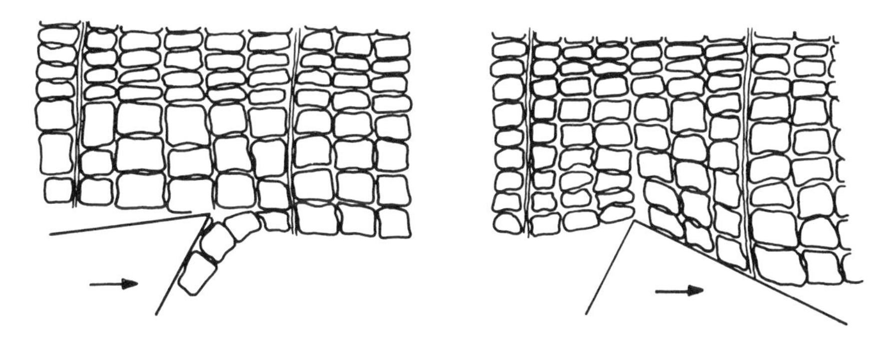

Fiber Cutting by Sharp Abrasive Grits

Compression-Relaxation of Fiber by Conditioned Grits

Figure 4-1. Comminution Mechanisms

Because little theoretical information has been published to predict energy requirements and power efficiencies in wood size reduction, the remaining discussion concerns specific types of equipment. The comminution equipment used is mainly cutting grinders. Systems used for reducing wood wastes are generally designed for field operation. The advantage of performing the size reduction in the field is in lowering transportation costs by increasing the wood bulk density and thereby the volume energy density. Cost data were compiled for ERDA by the MITRE Corporation (Bliss and Black 1977) for selected commercial comminution equipment; these data are shown in Table 4-1. The reported costs, fixed plus operating, ranged from $4.40 to $6.60 per dry ton equivalent. Qualitative discussions of various commercial systems are given below to indicate the types of equipment used in wood and wood waste size reduction.

The first system is the Morbark "Total Chipharvester" (Morbark Industries Product Bulletin), a portable, trailer-mounted harvesting machine designed to produce 5/8-in. to 1-in. chips from cut hardwood trees up to 22-in. diameter. The system also incorporates a separator to remove 90% of the dirt or sand and 50% of bark and foliage from the product chips. The chipped bark and foliage can also be recovered for fuel. The maximum throughput of the system is one ton per minute. The system uses knives mounted on a 75-in. diameter high speed disc for performing the actual chipping. No power consumption data per ton processed have been published.

The Mobile Harvestor, manufactured by Nicholson Manufacturing Company (Nicholson Mfg. Co. Product Bulletin) both fells trees up to 12 in. diameter and chips trees up to 19 in. diameter. The chipper is a three-knife, 48-in. diameter by 48-in., 550 rpm disc, and the nominal system capacity is 25 green tons/h. Again, no power consumption data were available.

Williams Patent Crusher and Pulverizer Company (Williams Product Bulletin) manufactures the "Hot Dog" shredder system that combines drying and shredding operations into one unit. The system can process 30 tons/h of wood or wood products. The shredding mill uses rotating hammers at high temperature. No information was given concerning power consumption.

Montgomery Hogs (Montgomery 1974) use a punch-and-die cutting action with fixed teeth rotating through fixed anvil slots. The units are designed to give minus 3/4-in. particles. Unit capacities vary from 7.5 tons/h to 100 tons/h. Reported horsepower requirements range from 100 to 500 hp. The units can be mounted as fixed or portable installations.

4.1.2 Drying

The general advantages of drying wood are well known. Removal of water reduces the weight of material that must be transported or handled in a processing plant, thereby lowering operating costs. In addition, the removal of water generally produces a feedstock of better quality for combustion and gasification processes. Table 4-2 shows the combustion efficiency for burning wood as a function of moisture content. Since most fresh woods contain considerable water (40 to 60 wt %) appreciable energy can be saved in later processing. This savings can be significant in processing energy requirements if waste heat from another processing step, such as the combustion step, is used to supply the thermal energy for drying. (An exception may be steam gasification in which water is one of the process feed materials.) The disadvantages of drying are also well known: in any processing step, equipment and operating costs must be considered in evaluating the usefulness of the process.

In general, drying of a biomass material means removing water from the solid to reduce the moisture content to an acceptably low value. In wood and wood products this moisture content reduction is usually accomplished by thermal drying, as opposed to mechanical drying done in centrifuges or presses. The major types of drying equipment used for particulate solids drying are screen conveyor dryers, screw-conveyor dryers, rotary dryers, and flash dryers, all of which are standard processing equipment. Detailed discussions of these dryers can be found in Perry's (1963) and McCabe and Smith (1967). Although the theoretical analysis of drying processes can become complicated, it can be divided into two parts to simplify the analysis: a steady-state drying process and a transient drying process. An example of rate drying curves is given in Fig. 4-2. The constant rate line is the steady-state portion of the drying process.

Table 4-1. EQUIPMENT ADAPTABLE TO COLLECTING AND/OR REDUCING FOREST RESIDUES[a]

Equipment	Slope Limitation[b] (%)	Size Diameter (in.)	Limitation Length (ft)	Cost per DTE[c] ($)	Support Equipment Needed[d]	Manufacturer
Morbark Chipharvester	limit of skidder	22	none	6.60	skidders and chain saws	Morbark Ind., Inc. Winn, Mich.
Precision Tree Harvester	limit of skidder	22	none	6.60	skidders and chain saws	Precision Chipper Corporation Birmingham, Ala.
Nicholson Ecolo Chipper	limit of skidder	24	none		skidders and chain saws	Nicholson Mfg. Co. Seattle, Wash.
Tree Eater	20	10	none	5.50 (11 DTE/h)	none	Tree Eater Corp. Gurdon, Ark.
Wagner-Bartlett Stump Splitter-Remover	limit of loader	96	none	3.50/stump	mounted on loader	Wagner Mfg. Co. Portland, Ore.
National Hydro-Ax	30	6	none	4.40 (11 DTE/h)	none	National Hydro-Ax Incorporated Owatoma, Minn.
Kershaw Klear Way	25	6	none	4.40 (11 DTE/h)	none	Kershaw Mfg. Co., Incorporated Montgomery, Ala.

[a]From Bliss and Black 1977.
[b]Based on working performance on firm soils.
[c]Includes all known costs; fixed and maintenance, move in and out, and necessary personnel.
[d]Does not include equipment needed for accumulation of reduced residues.

Table 4-2. THE EFFECT OF MOISTURE CONTENT ON HEAT RECOVERY AND COMBUSTION EFFICIENCY[a]

Moisture Content (%)	Recoverable Heat[b] (Btu/lb)	Combustion Efficiency (%)
0.00	7,097	82.5
4.76	7,036	81.8
9.09	6,975	81.1
13.04	6,912	80.4
16.67	6,853	79.7
20.00	6,791	78.9
23.08	6,730	78.3
28.57	6,604	76.8
33.33	6,482	75.4
42.86	6,178	71.8
50.00	5,868	68.2
60.00	5,252	61.1
66.67	4,639	53.9
71.43	4,019	46.7

[a]From Bliss and Black 1977.
[b]Theoretical values based on a maximum heating value of 8,600 Btu/lb, an initial wood temperature of 62 F, a flue gas temperature of 450 F, an initial air temperature of 62 F and 50% excess air.

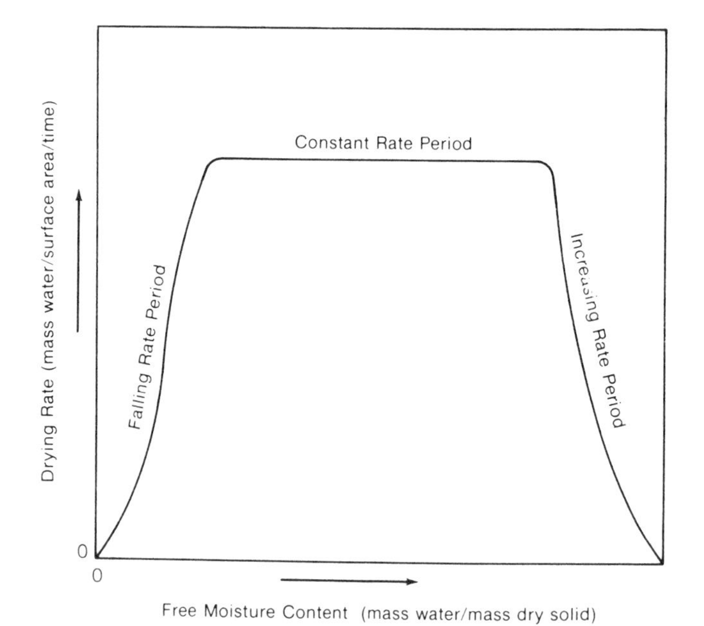

Figure 4-2. Typical Drying Rate Curve

Both steady-state and transient drying operations can be modeled by the appropriate heat and mass transfer equations. Detailed derivations of drying models can be found in references such as Perry's (1963), McCabe and Smith (1967), and Treybal (1968).

Reed and Bryant (1978) reported that although it theoretically requires about 1000 Btu to evaporate 1 lb of water, in drying wood it actually requires 1500 Btu to 2500 Btu to evaporate 1 lb of water, the precise value being dependent upon dryer efficiency.

Bliss and Black (1977) have presented information concerning the residual fuel value of hogged fuel as a function of moisture content (see Fig. 4-3). Miller (1977) has presented figures for energy requirements for conventional kiln drying in which he reports that it takes 96.2 MBtu to dry 25.4 thousand board feet of 2-in. southern pine from 50% moisture to 10% moisture. This reduces to 2.67 MBtu/ton of dry wood processed, or approximately 1500 Btu/lb of water evaporated.

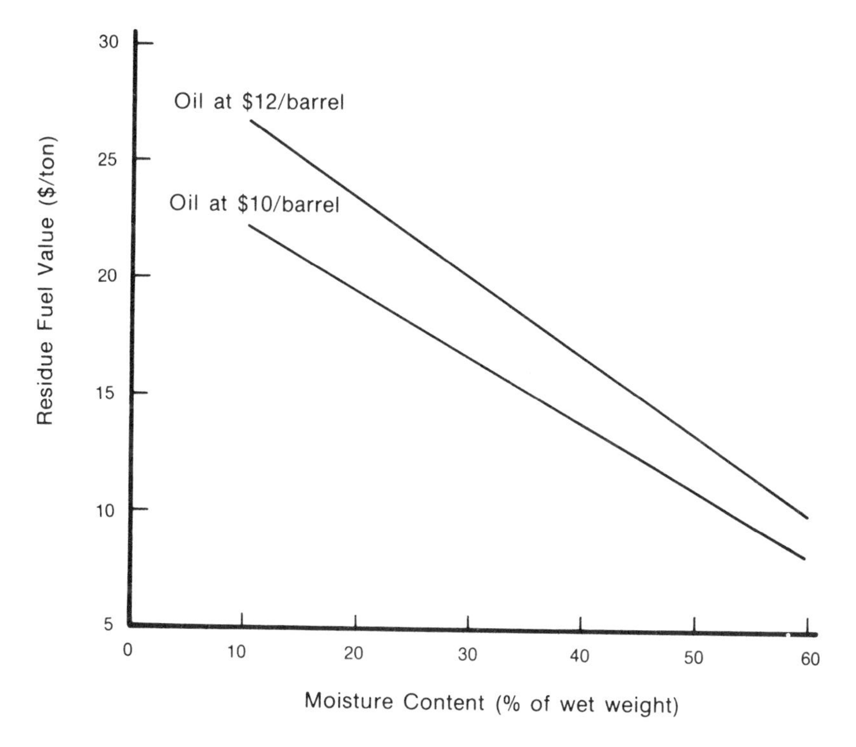

Figure 4-3. Relationship of Hog Fuel Value to Moisture Content Under Two Oil Price Assumptions

Gulf (1978) reports that a barrel of distillate fuel oil contains 5.82 MBtu of energy. This gives an estimated drying cost of $5.50/ton of wood if oil is sold at $12/barrel. The residual fuel value would increase by $12/ton of dried wood. These costs indicate possible economic feasibility for the drying process, although transportation, labor, and capital cost also would be needed to determine realistic feasibility estimates.

4.1.3 Densification

Reed and Bryant (1978) recently presented a comprehensive state-of-the-art evaluation of processes that produce densified biomass fuels (DBF). A review of their findings is presented here.

Five methods of densification for biomass materials are in commercial operation at the present time, with other processes in the development stage. The five processes are:

- pelleting—a die perforated with 1/4-in. to 1/2-in. holes rotates against pressure rollers, forcing feedstock through the holes at high pressure and densifying the feedstock;

- cubing—a modified form of pelleting producing a large size product (1-in. to 2-in.);
- briquetting—feed is compacted between rollers containing cavities; product looks like charcoal briquettes;
- extrusion—a screw forces a feedstock under high pressure into a die, forming 1-in. to 4-in. diameter cylinders; and
- rolling-compressing—employs a rotating shaft to wrap fibrous material and produce high density rolls of 5-in. to 7-in. diameter.

The densification process takes advantage of the physical properties of two of the major components of biomass materials, cellulose and lignin. Cellulose is stable to 250 C, while lignin begins to soften at temperatures as low as 100 C. Densification is carried out at temperatures that ensure that the cellulosic material remains stable but that soften the lignin fraction, making it act as a "self-bonding" agent that gives the final DBF its mechanical strength. Water content must be controlled in the range from 10% to 25% to minimize pressure requirements for densification.

Densification proceeds by heating a biomass material (of the proper moisture content) to 50 C to 100 C to soften the lignin, followed by mechanical densification that increases the biomass density to a maximum of 1.5 g/cm^3 and heats the material another 20 C to 50 C. The additional temperature increase liquefies waxes that act as additional binders when the product is cooled.

A list of manufacturers of densification equipment is given in Table 4-3. A detailed discussion of existing biomass densification plants was given by Reed and Bryant (1978) and is reproduced here, except for the ECO-FUEL II process, which is discussed in Section 4.2.4. Table 4-4 presents a list of DBF process developers.

A typical biomass compaction plant is shown in Fig. 4-4. The first step in the process is separation—stones and sand must be removed from forest or agricultural wastes and inorganics from municipal waste. The remaining biomass portion is then pulverized with hammer mills or ball mills to a size somewhat smaller than the minimum dimension of the pellets to be formed. This fraction is then dried in a rotary kiln or convection dryer. Finally, dried biomass is fed into the compactor which delivers pellets for storage or use.

Table 4-3. MANUFACTURERS OF DENSIFICATION EQUIPMENT FOR FEED AND FUEL[a]

Company	Type of Equipment
Agnew Environmental Products, Grants Pass, Ore.	Extruder
Agropack, Medina, Wash.	Roller-Compressor
Bonnet Co., Kent, Ohio	Wood and Wax Extruder
Briquettor Systems, Inc., Reedsport, Ore.	Extruder
California Pellet Mill Co., San Francisco, Calif.	Extruder and Pellet Mills, Cuber
Gear Cube Co., Moses Lake, Wash.	Cuber
Hawker Siddeley Canada Ltd., Vancouver, B.C.	Extruder
John Deere, Moline, Ill.	Cuber
Papakube Corp., San Diego, Calif.	Extruder Cuber
Reydco Machinery Co., Redding, Calif.	Extruder
Sprout Waldron, Muney, Pa.	Pellet Mills
Taiga Industries, Inc., San Diego, Calif.	Extruder

[a]From Currier 1977; Cohen and Parrish 1976. There may be other manufacturers unknown to the authors; this list in no way constitutes an endorsement by SERI or the authors.

Table 4-4. DBF PRODUCERS AND DEVELOPERS: PROCESS STATUS[a]

Company	Process Status	
	Commercial	Under Development
Bio-Solar Corp., Eugene, Ore. (Woodex)	X	
Combustion Engineering Corp.	X	
Guaranty Performance, Independence, Kans.	X	
Lehigh Forming Co., Easton, Pa.	X	
National Center for Resource Recovery (NCRR), Washington, D.C.		X
Papakube Corp., San Diego, Calif.	X	
SRI International		X
Taiga Industries, San Diego, Calif.	X	
Teledyne National, Cockeysville, Md.	X	
University of California Richmond Field Station		X
Vista Chemical and Fiber, Los Gatos, Calif.		X

[a]This list does not constitute an endorsement of particular processes by SERI or the authors. Furthermore, it is not exhaustive listing of processes.

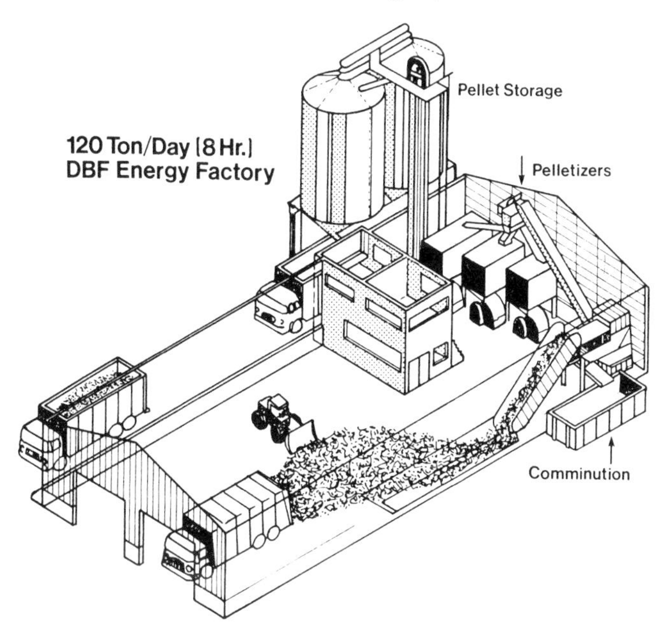

Figure 4-4. Typical Biomass Compaction Plant (PapaKube Corp.)

One of the more completely developed processes to date is R. Gunnerman's Woodex process (Gunnerman 1977), employing a hammer mill, dryer, and pellet mill. A 120-ton/day plant has been operating since 1976 in Brownsville, Ore. Gunnerman's company, Bio-Solar, recently installed a second, 300-ton/day plant in Brownsville; its dryer operates completely on pellets. Bio-Solar sells its products to customers in Oregon and Washington, where a major purchaser is the Western State Hospital near Tacoma. Two other Woodex plants are operating at the Sierra Power Corporation in Fresno, Calif., and the E. Hines Company in Burns, Ore. Three plants are under construction, and several business groups have acquired Woodex licenses.

A continuous flow extrusion technique is used by Taiga Industries* (Bremer 1975). Pulverized biomass with a moisture content of 10% is compressed by a screw, then fed into a prepressure chamber, where it is forced against a rotating spiral die-head with a cutting edge as shown in Figure 4-5. The frictional heat of the die face converts the biomass into a semifluid; the die-heat shears off a spiral slice of compressed biomass, forcing it into the die chamber. The densified product is expelled and cut to a specified length by a rotating flail. Taiga produces either a 10-cm by 30-cm log or 2.5-cm briquettes with a specific gravity of 1.2 to 1.45. The process expends 50 hph to 90 hph to produce 1 ton/h of DBF.

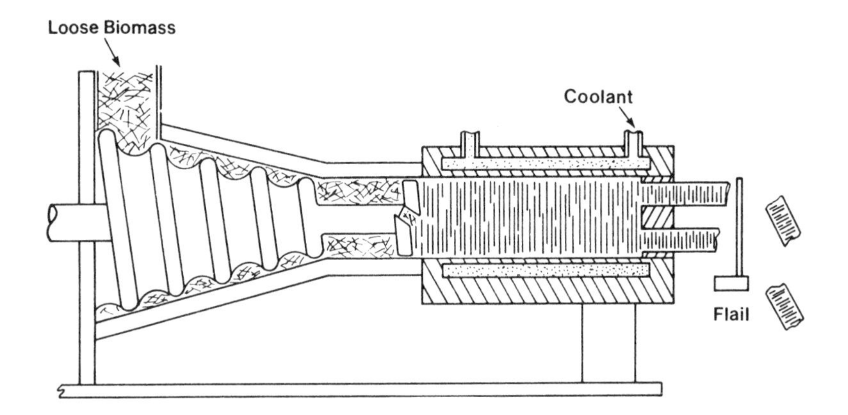

Figure 4-5. Taiga Extrusion Process

Another process, originally developed by Edward Koppelman to upgrade lignite, has been modified for biomass feedstocks and is now pending patent issuance (Koppelman 1977). SRI International, in cooperation with Koppelman, has constructed a pilot plant and tested various feedstocks. Details of the process are considered proprietary information but general features are: a water slurry feed system; a pyrolysis reactor; a water recovery system; and an output stream of a carbonaceous solid, a combustible gas, and a small amount of aromatic liquid. Product yields and composition depend on the feedstock and process variables (temperature, pressure, water content, and reaction time). SRI claims a process energy efficiency of 85% to 90%. The projected cost for an 1,800-ton/day plant is $10 to $15 million.

Solid waste densification is an attractive option because it helps solve two urban problems simultaneously: energy supply and waste disposal. Baltimore County and the Maryland Environmental Service, with Teledyne National as prime contractor, are operating a plant that separates combustibles from the solid waste stream, shreds that fraction, and then compacts it with a pellet mill (Herrman 1978). Ten tons per day are sold to a paper mill in Spring Grove, Pa., where the pellets are mixed with bark, ground in a

---

*Taiga publishes a Mod-Log sales brochure that describes the modified Bremer process, plant operations, cost, etc.

hog mill and blown into a boiler. A full-scale burn test program is now underway. Contracts with a utility and cement company are pending, following confirmation of performance.

Typical energy consumption values for pelleting of sawdust, fir bark, aspen, and municipal solid waste are given in Table 4-5, which shows 1% to 3% consumption of energy based upon the energy content of the product. Overall process efficiency for a 300-ton/day bark pelleting operation incorporating pulverization, drying, and pelleting steps has been estimated at 92.8%. The reported cost of this plant has been estimated to add $0.80/MBtu to the feedstock cost.

Table 4-5. ENERGY REQUIRED FOR PELLETING (300-HP PELLET MILL)[a]

Feedstock	Fraction Electrical Production Rate metric tonnes/h (tons/h)	of Product Energy Used kWh/metric tonne (kWh/ton)	Energy Consumed (%)
Sawdust	6.1	36.8	
	(6.7)	(33.5)	2.3
Aspen wood	8.2	27.2	
	(9.0)	(24.8)	1.7
Douglas fir bark	4.5	49.2	
	(5.0)	(44.7)	3.1
Municipal solid waste	9.1	16.4	
(MSW)	(10.0)	(14.9)	1.0

[a]From Reed and Bryant 1978.

NOTES:

(1) 11.6 kJ (11,000 Btu) thermal/kWh.

(2) The pelleting of MSW is volume limited in a 300-hp mill due to low density of feedstock—actual horsepower usage is 200 hp.

(3) The figures in this table are only representative; values are highly dependent on feed size, moisture content, etc.

4.2 MUNICIPAL SOLID WASTES

The major purpose of beneficiation of municipal solid wastes (MSW) has been to solve the disposal problems created by the extremely large volume of wastes generated by large metropolitan populations. This is done by creating a system that recycles the valuable materials and energy contained in the waste. Municipal solid waste processing operations can be divided into two major areas: the separation of an organic feedstock suitable for further processing and the actual conversion of this organic feedstock. This discussion focuses on the preparation operations, not the ultimate end use of the organic product.

Figure 4-6 presents a general flowsheet for various processing operations. In general, all MSW preparation plants use at least some of the steps in the following general outline.

- Preliminary (primary) shredding—the incoming raw refuse is reduced in size to 1 or 2-in. particles to allow further, more efficient processing.

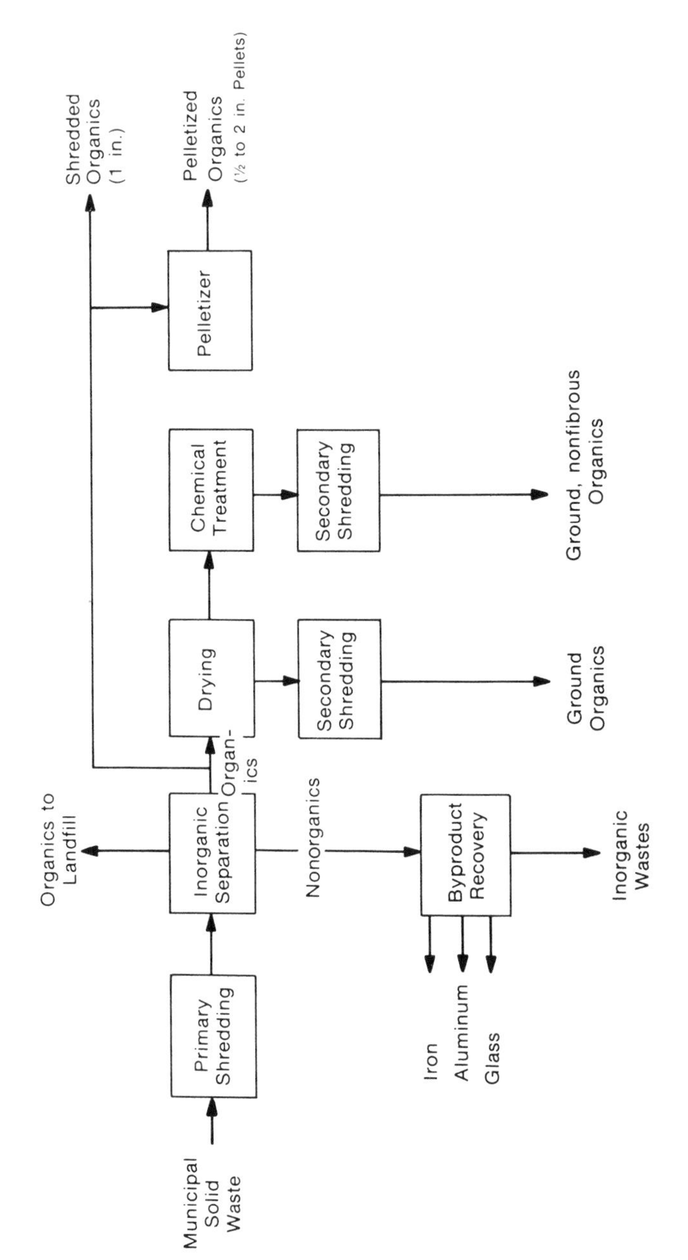

Figure 4-6. Processing of Municipal Solid Waste

- Separation of inorganics from organics—inorganic materials such as iron, aluminum, and glass are separated from the organic materials such as paper, cardboard, wood, and leaves. This operation produces an organic fraction which can be processed more efficiently and an inorganic fraction that is more amenable to byproduct recovery.
- Drying of organic fraction—this step also produces a product more amenable to processing and that has a larger gross heating value. In an integrated resource recovery plant, much of the energy for this operation is waste heat from a pyrolysis or combustion process.
- Secondary shredding—the particle size of the organic fraction is further reduced. This step is necessary for downstream processing in systems incorporating entrained flow pyrolysis or gasification operations.
- Densification—in some processes the organic fraction is densified by pelletization. This step makes a fuel with a higher volume energy density to both reduce the volume of material to be handled and to make the fuel compatible with existing materials handling facilities in power plants using coal as fuel.
- Chemical treatment—this process normally involves a chemical treatment to break down fibers chemically and therefore change shredding costs and the nature of the final product.

4.2.1 Primary Shredding

The general criteria for primary refuse grinders or shredders are that (1) the shredding should involve little or no addition of water to the feed, to minimize energy required for later drying operations; (2) material of the desired size should be removed as quickly as possible from the shredder to minimize production of fine inorganic material that would increase the difficulty of metals and glass recovery and would increase the ash content of the organic fraction by increasing the difficulty of ash-organic separation; and (3) the shredding process should be performed with little or no pretreatment of the feed refuse.

There are many types of size reduction equipment. Table 4-6 lists types of equipment (McCabe and Smith 1967) and their possible application in MSW processing.

Three general types of shredding equipment meet the criteria for raw refuse: hammer mills, vertical ring grinders, and flail mills. A hammer mill uses a high speed rotor on which are carried hammers of many different configurations (e.g., stirrups, bars, or fixed rings). The rotor runs in a housing containing grinding plates and the particle size of the product solid is governed by the clearance between the hammers and grinding plates. If a hammer mill is used for primary shredding, two or three stages may be required to obtain the necessary particle size range.

A vertical ring grinder consists of a large vertical rotor with peripheral grinding rings, usually gear-like, enclosed in a heavy casing. The feed material is ground mainly by attrition between the grinding rings and the protrusions on the casing. Each ring is mounted independently from the other rings, thus allowing shocks caused by particularly hard objects to be distributed over the entire machine. A flail mill operates on the same principle as a vertical ring grinder but instead of rings uses articulated arms which self relieve. Because of the strain-relieving properties of the latter two mills, the maintenance costs are normally lower than those of hammer mills.

Many companies have presented power consumption curves for primary grinding. Garrett Research and Development Company, Inc. (Garrett and Finney 1973) performed tests, the results of which are compared to data from Combustion Equipment Associates (Benningson and Rogers 1975) in Fig. 4-7. The differences in power requirements of the two systems result from the use of a chemical treatment step in the CEA process to produce "ECO-FUEL II." Therefore, the Garrett data are more applicable when only the energy consumption of shredding is to be estimated. To compare realistically the different power consumption curves, the overall cost (both economics and energy) of the two process schemes would need to be known. There is undoubtedly some tradeoff between shredding power consumption and chemical costs. A more detailed discussion of the chemical treatment step is given in Section 4.2.5.

Table 4-6. CURRENT SIZE-REDUCTION EQUIPMENT AND POTENTIAL APPLICATIONS TO MUNICIPAL SOLID WASTE[a]

Basic Types	Variations	Potential Applications to Municipal Solid Waste
Crushers	Impact	Direct application as a form of hammer mill
	Jaw, roll, and gyrating	As a primary or parallel operation on brittle or friable material
Cage disintegrators	Multicage or single cage	As a parallel operation on brittle or friable material
Shears	Multiblade or single blade	As a primary operation on wood or ductile materials
Shredders, cutters, and chippers	Pierce-and-tear type	Direct as hammer mill with meshing and shredding members, or parallel operation on paper and boxboard
	Cutting type	Parallel on yard waste, paper, boxboard, wood, or board plastics
Rasp mills and drum pulverizers		Direct on moistened municipal solid wastes, also as bulky item sorter for parallel line operations
Disk mills	Single or multiple disk	Parallel operation on certain municipal solid waste fractions for special recovery treatment
Wet pulpers	Single or multiple disk	Second operation on pulpable material
Hammer mills		Direct application or in tandem with other types

[a]From Weinstein and Toro 1976.

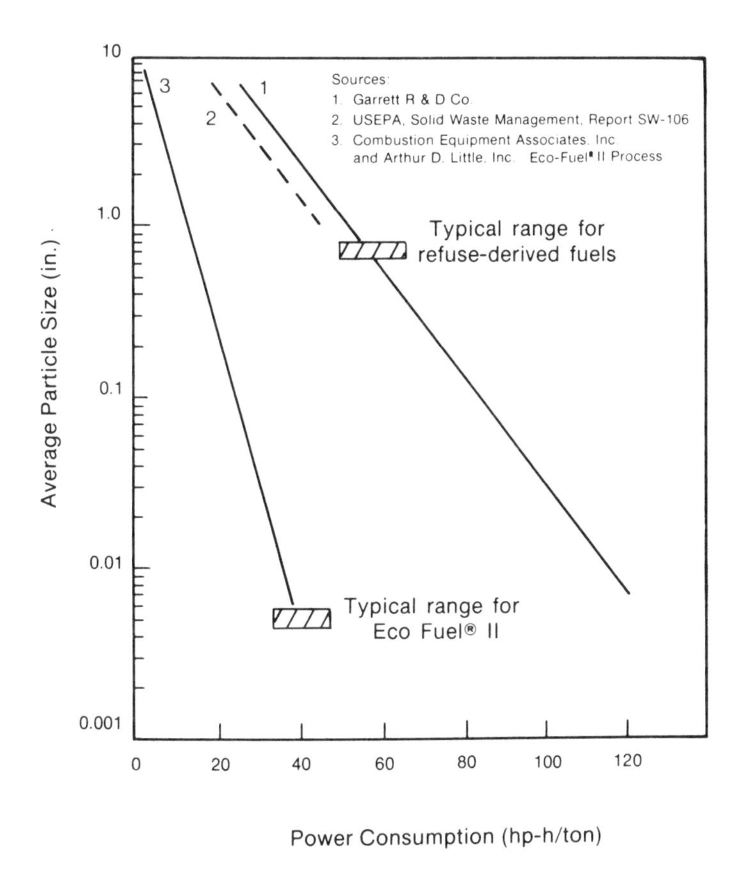

Figure 4-7. Average Particle Size Vs. Power Consumption for Size Reduction of MSW

4.2.2 Separating Inorganic Material From Organics

Depending upon the moisture content of the feed, a drying step may be required before the separation of inorganic from organic material. Among the more common methods of separation is the use of an air classifier. A large amount of research (Garrett and Finney 1973) has been performed in this area, and the concept has been incorporated into various resource recovery systems (Weinstein and Toro 1976, Section 4.0). Air classifiers may be zig-zag, straight-vertical, straight-horizontal, etc. All operate on the principle that the organics are low-density materials with large surface areas and that the inorganic materials are high-density with small surface areas. Classifiers are operated with the air flow rate maintained such that the superficial air velocity is larger than the terminal settling velocities of the organic particles but lower than the terminal settling velocities of the inorganic particles. Since there is overlap in settling velocities, the separation in an air classifier is not complete, and more than one classifier may be used, or the classification step may be followed by a screening step to give the desired degree of separation.

After the organic-inorganic separation has been accomplished the inorganic fraction may be further classified to recover iron, aluminum, and glass. Since this fraction does not contain a high proportion of the total biomass materials, no discussion is presented here. Detailed information on this topic can be found in many references (e.g., Garrett and Finney 1973; Weinstein and Toro 1976). A consideration that justifies the separation of these important byproducts is the fact that they can be sold to at least partially offset the cost of the beneficiation and conversion processes. Cheremisinoff and Morresi (1976) reported that these byproducts had a potential selling price of $3.70/ton of MSW in 1971. Garrett Research and Development (Chemical Week, 11 Dec. 1974) reported in 1971 that their resource recovery system cost $5.40/ton. By 1974 the processing cost had risen to $12.90/ton, but the revenues from byproducts and pyrolytic oil had risen to $10.36/ton. Therefore, the separation and recovery of byproducts is economically justified in a MSW plant.

4.2.3 Drying of Organic Fraction

The various types of drying processes were discussed in Section 4.1.2. The organic fraction of MSW normally has been ground and separated from the inorganic material, and its resulting density is such that the drying step is conducted in a rotary drier or an entrained flow-flash drier with direct solid-gas contact. The advantages of drying discussed for wood and wood products also apply to MSW. An example of downstream boiler efficiency (Kohlkepp 1974) with the organic fraction as a combustion boiler fuel (see Fig. 4-8) further reinforces the need for drying. No information is available showing the drying step costs versus downstream processing efficiency in resource recovery systems. These costs are normally lumped into total beneficiation (prep plant) costs.

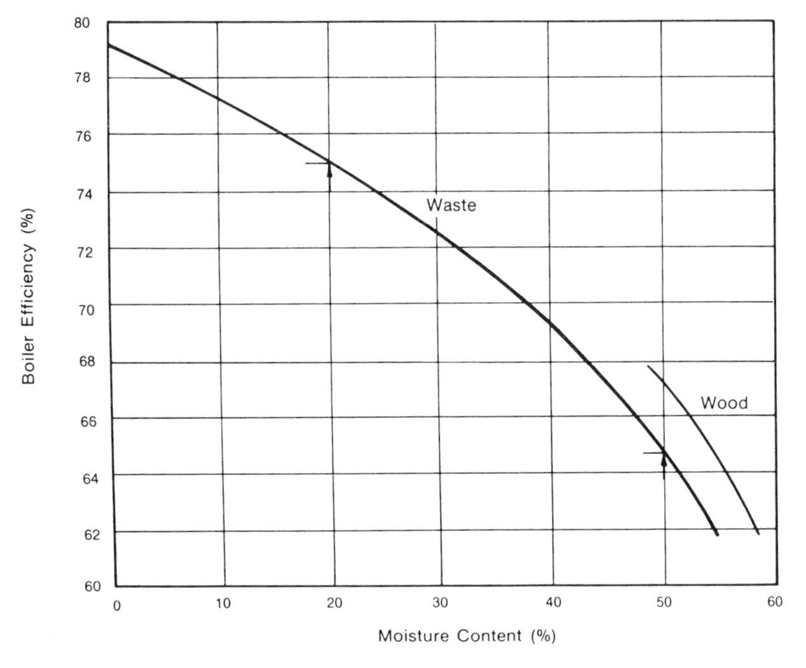

Figure 4-8. Boiler Efficiency Vs. Moisture Content

4.2.4 Densification

Alter and Arnold (1978) reported on a pilot plant operation to produce a densified refuse-derived fuel. The organic material leaving the secondary shredding process was fed to a small pellet mill manufactured by California Pellet Mill Company. In the pellet mill a die rotated past stationary rollers which formed a nip, forcing the feed material into the die. The product from the pelletizer in the pilot plant run had the following average properties:

Diameter	- 0.5 in.
Length	- 0.71 in.
Pellet Density	- 73 lb/ft^3
Bulk Density	- 39 lb/ft^3
Moisture Content	- 19 wt %
Ash Content	- 26.5 wt %

Alter and Arnold (1978) also presented data for pelletizer power consumption as a function of pelletizing rate, for rates as high as 9 tons/h (see Fig. 4-9). Energy consumption ranged from 16 kWh/ton at rates of 2 tons/h to 4 kWh/ton at rates of 7 tons/h.

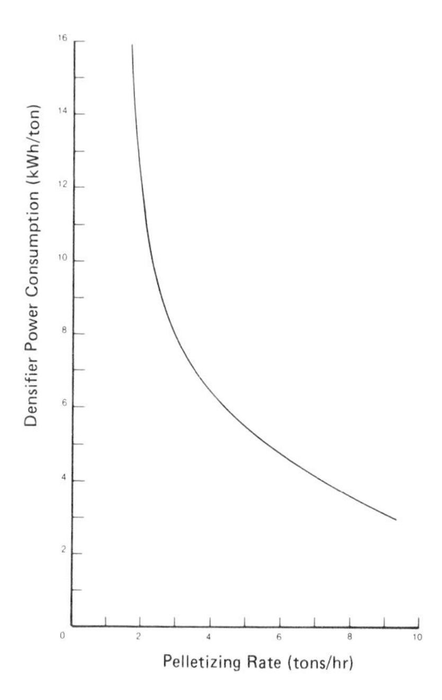

Figure 4-9. Densifier Power Consumption for NCPR Pelletizing Tests (Smoothed Data)

4.2.5 Chemical Modification

One chemical modification technique is discussed here, the "ECO-FUEL-II" process (Benningson and Rogers 1975) developed by Arthur D. Little, Inc. A production facility has been constructed to accomplish the primary shredding step, a ferrous metals separation step, a screening step, and a chemical treatment step.

In the chemical treatment step, a small amount of an inorganic acid, such as sulfuric acid (Combustion Equipment 1975) is added to the remaining refuse, mainly organics; the acid embrittles the cellulosic materials present, probably by rupturing the ether linkage structure of the cellulose molecules. The chemically treated material is then mixed with hot steel balls in a ball mill. This process both grinds and drys the organic materials at temperatures up to 400 F. The high temperature is said to enhance the embrittling action of the acid and therefore to lower the power requirements in the ball mill operation. The product of this process after residual inorganic separation is a nonfibrous dry solid with an average particle size of 0.006 in. The reported power requirements for grinding (see Figure 4-2) are much lower than for conventional shredding processes. The product has a high bulk density (30 lb/ft^3 to 35 lb/ft^3) in comparison to dried shredded fuels (3 lb/ft^3 to 5 lb/ft^3).

As a final comment on the discussion of beneficiation of municipal solid wastes, economic comparison among beneficiation processes are not presented here because capital and operating cost data for MSW plants generally have not been reported in the detail necessary to calculate the cost of the process steps. Schulz et al. (1976) compared all of the major resource recovery systems and compiled general costs for front-end plants (see Table 4-7). They show the cost of three of the beneficiation processing steps discussed here. Primary shredding costs $2.66/ton of MSW, air classification $1.73/ton of MSW, and secondary shredding $1.20/ton. Schulz et al. claim a $4.18 credit/ton of MSW for byproducts when sold. This means that the actual end usage of the organic fraction will determine process economics.

Table 4-7. MATERIALS RECOVERY: UNIT OPERATIONS COSTS[a]
(Basis: 1,000 ton/day Plant)

Unit Operations	Capital Cost ($/ton MSW)	Operating Cost ($/ton MSW)	Amortized Operating Cost ($/ton MSW)
Primary shredding (to -4 in.)	0.49	2.17	2.66
Air classification	0.31	1.42	1.73
Secondary shredding (to -1 in.)	0.16	1.04	1.20
Magnetic metals recovery	0.08	0.44	0.52
Rising current and heavy media separation	0.10	0.66	0.76
Roll crushing and electronic separation	180	0.06	0.53
Color sorting	425	0.14	0.56
Froth flotation	295	0.10	0.43
		Total	$8.39/ton MSW

[a]From Shulz et al. 1976.

Physical/chemical modification, while generally not required for gasification, is the goal of several new processes for disruption of the lignocellulosic complex structure. The processes provide an altered form of biomass in which the lignin, hemicellulose, and cellulose fractions are more readily separated by chemical and enzymatic means.

The Iotech process (Iotech 1979) uses high pressure (200-600 psi) steam to soften the biomass and then decompresses the biomass supersonically through a nozzle. The combination of shear and heat modifies the biomass matrix in a controllable, reproducible manner that can be designed for a specific process, i.e., ethanol fermentation, fast pyrolysis, etc. A similar process is used for explosive decompression (at approximately 250-psi steam) of municipal waste (Burke 1979). This process is said to require 60 Btu/lb trash.

4.3 REFERENCES

Alter, H. and Arnold, J. 1978. "Preparation of Densified Refuse-Derived Fuel on a Pilot Scale." Proceedings of the 6th Mineral Waste Utilization Symposium. Chicago, IL.

Benningson, R. M.; Rogers, K. J. 1975. Production of ECO-FUEL-II from Municipal Solid Waste. AIChE 80th National Meeting, Boston, MA.

Bliss, C.; Black, D. O. 1977. Silvicultural Biomass Farms, Vol. 5, Conversion Processes and Costs. McLean, VA: MITRE Corporation; ERDA Contract No. EX-76-C-01-2081.

Bremer, Allen R. 1975. U.S. Patent 3,904,340.

Burke, J. A. Jr. 1979. "Size Reduction Using Explosive Decomposition." Heniker, NH: Heniker Municipal Solid Waste Conference; July 22-27.

"Can Pyrolysis Put Spark into Refuse as a Fuel?" Chemical Week. pp. 53-54; 11 Dec. 1974.

Cheremisinoff, P. N.; Morresi, A. C. 1976. Energy from Solid Wastes. New York: Marcel Dekker, Inc.

Cohen and Parrish. 1976. Densified Refuse Derived Fuels. Washington, D.C.: National Center for Resource Recovery; Winter 1976; Bull. 6, No. 1.

Combustion Equipment Associates, Inc. 1975. Belgium Patent 845.249.

Currier, R. A. 1977. Manufacturing Densified Wood and Bark Fuels. Oregon State University Extension Service; July 1977; Special Report 490.

Dornfield, D. A., DeVries, W. R.; Wu, S. M. 1978. "An Orthomorphic Rheological Model for the Grinding of Wood." J. Engineering and Industry. Vol. 100: pp. 153-158.

Garrett, D. E.; Finney C. S. 1973. The Flash Pyrolysis of Solid Wastes. AIChE 66th Annual Meeting, Philadelphia, PA.

Gulf Oil Corp. 1978. Some Useful Facts on Energy.

Gunnerman, R. W. 1977. "Fuel Pellets and Methods for Making them from Organic Fibrous Materials." U.S. Patent 4,015,951.

Herrman, Robert H. 1978. Assistant Program Manager, Teledyne National. Personal Communication. June 1978.

Iotech Limited. 1979. Private communication from John Davis, Iotech Limited, 220 Laurier Ave. West, Ottawa, Ontario, Canada K1P 5Z9.

Kohlkepp, D. H. 1974. The Dynamics of Recycling. AIChE 78th National Meeting, Salt Lake City; UT.

Koppelman, Edward. 1977. "Process for Upgrading Lignite-Type Coal as a Fuel." U.S. Patent 4,052,168.

Lipinsky et al. 1977. Systems Study of Fuels from Sugarcane, Sweet Sorghum, Sugar Beets and Corn, Volume V: Comprehensive Evaluation of Corn. Columbus, OH: Battelle Columbus Laboratories; ERDA Contract W-7405-Eng-92.

McCabe, W. L.; Smith J. C. 1967. Unit Operations of Chemical Engineering. New York: McGraw Hill Book Company.

Miller, W. 1977. Energy Conservation in Timber Drying Kilns by Vapor Recompression." Forest Products. Vol. 27 (no. 9): pp. 9, 54-58.

Montgomery, K. C. 1974. "Model Eat-Rite Hog." Jacksonville, FL: Jackson Blow Pipe Company; Bulletin 86-10-73.

Morbark Industries, Inc. "Total Chipharvester." Product Bulletin.

Nicholson Manufacturing Co. Product Bulletin. "Harvesting Forest Biomass with the Nicholson Mobile Harvester and Chip Forewarder System." Seattle, WA.

Papa Kube Corp. Product Bulletin. San Diego, CA.

Perry, R. H., Chilton, C. H.; Kirkpatrick, S. D. Chemical Engineering Handbook, 44th Ed., New York: McGraw Hill Book Company; 1963.

Reed, T.; Bryant B. 1978. Densified Biomass: A New Form of Solid Fuel. Golden, CO: Solar Energy Research Institute. SERI-35.

Schulz, H. M. (Principal Investigators) et al. 1976. Resource Recovery Technology for Urban Decisionmakers. New York: Urban Technology Center, Columbia University.

Treybal, R. E. 1968. Mass-Transfer Operations. Second Edition. New York: McGraw Hill Book Co.

Weinstein, N. J.; Toro, R. F. 1976. Thermal Processing of Municipal Solid Waste for Resource and Energy Recovery. Ann Arbor, MI: Ann Arbor Science Publishers, Inc.

Williams Patent Crusher and Pulverizer Co. "Williams Hot Dog Shredders." Product Bulletin 871.

Chapter 5

Pyrolysis - The Thermal Behavior of Biomass Below 600° C

T. Milne
SERI

5.1 INTRODUCTION

This chapter focuses on pyrolysis as a precursor to gasification under both anaerobic conditions (steam, H_2, self-generated gas) and aerobic conditions (air, O_2). Pyrolysis of carbonaceous materials has been defined as incomplete thermal degradation, resulting in char, condensable liquids or tars and gaseous products, generally in the absence of air (Soltes and Elder 1979). Gasification generally refers to the combination of pyrolysis followed by higher temperature reactions of the char, tars, and primary gases to yield mainly low molecular weight gaseous products. In fast pyrolysis the distinction between pyrolysis and gasification becomes blurred.

Extensive literature exists pertaining to low temperature, slow pyrolysis where the emphasis is on char (carbonization), liquids (wood distillation), and both char and liquid (destructive distillation). For example, Soltes and Elder (1979) have reviewed pyrolysis with the emphasis on obtaining organic chemicals from biomass. Much information also exists on the mild thermal degradation of wood, papers, etc., in the context of structural integrity, aging, and other factors (Stamm 1956).

Two broad approaches to gasification can be distinguished: (1) gasifiers in which relatively large particles are subjected to inherently slow heating rates and long residence times, yielding gaseous products that approach equilibrium, and (2) gasifiers in which rather finely divided material is heated rapidly (fast pyrolysis), and the products are quenched after short residence times, to preserve high concentrations of nonequilibrium pyrolysis products. The following review of past work is divided into these two broad categories, treating biomass and its major components according to the kinds of study most often used.

5.2 SLOW PYROLYSIS

With macroscopic pieces of carbonaceous solids, the heating rate is controlled by heat transfer throughout the usually poorly conducting material. Heating rates of the order 0.01 C/s to 2 C/s are likely. This range of heating rates corresponds to the capability of commercially available thermal analytical instrumentation such as DSC, TGA, and DTA measuring devices. Much work has been done with very small samples under such slow heating rates, yielding data that may or may not be relevant to pyrolysis conditions in real gasifiers. Past work is summarized here for the three major components of biomass and for wood. For each type of material, the behavior is discussed under five headings: (1) thermogravimetric analysis (TGA), in which the sample weight loss is followed under both isothermal and dynamic heating; (2) kinetic analysis of pyrolytic data; (3) differential thermal analysis (DTA) and differential scanning calorimetry (DSC) in which latent and reaction heat effects are either inferred or measured directly; (4) gas and other product analyses, in which various techniques are used to determine primary and secondary decomposition products; (5) morphological and related studies, in which structural information is obtained as pyrolysis proceeds; and (6) molecular mechanisms, in which all of the above information is used to deduce the molecular course of the pyrolysis.

5.2.1 Thermogravimetric Analysis (TGA)

The thermal behavior of biomass is studied most often by measuring the rate of weight loss of the sample as a function of time and temperature. The rates observed are functions not only of time and temperature, but also of the size and the density of the sample. This complexity reflects the range of behavior in different kinds of gasifiers, but there is not necessarily an exact relationship between laboratory experiments and pyrolysis during gasification; nevertheless, TGA offers a semiquantitative understanding of the pyrolysis process under well-controlled laboratory conditions. A number of commercially available instruments of high sensitivity can measure weight loss versus time or temperature under such conditions.

Two types of results are found in the literature: isothermal TGA data showing the rate of pyrolysis at a fixed temperature (e.g., Fairbridge and Ross 1978) and dynamic TGA data showing weight loss at a fixed heating rate (e.g., Shafizadeh and McGinnis 1971, Fig. 5-1). Each type of result has a useful function (see kinetics discussion below).

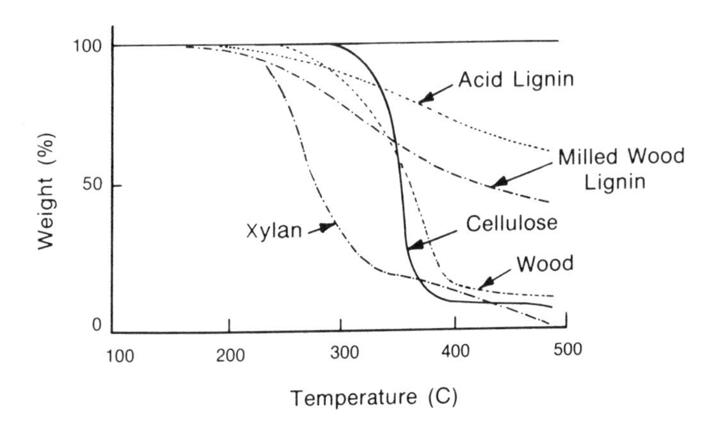

Figure 5-1. Examples of TGA Cottonwood and Its Components, Taken from Shafizadeh and McGinnis (1971)

In addition to yielding kinetic data, the dynamic TGA measurements can yield data equivalent to a proximate analysis; namely moisture content, volatile content, char, and ash, as shown for a sample of flax shives in Fig. 5-2.

5.2.1.1 Cellulose

In its many forms cellulose has received more extensive study than biomass or any of its other components. This stems from the fact that cellulose is the major component of most biomass, from its relevance in the context of fire research and municipal solid waste (MSW) utilization, and surely also because it is the least complicated, best-defined major component of biomass.

Extensive and detailed reviews of the thermal behavior of cellulose have been published, (Shafizadeh 1975 and 1968; Welker 1970; MacKay 1967; Broido and Kilzer 1963; Kilzer and Broido (1965); Antal et al. 1979) with the most recent and most extensive being that of Molton and Demmitt (1977). Both isothermal and dynamic TGA studies of cellulose have been made, often with small samples in commercial instruments. Vacuum, inert, steam, and air environments have been studied as well as the effect of impurities and added salts and the degree of crystallinity and polymerization. Only a few examples of the diverse studies can be discussed here. Aldrich (1974) studied the weight loss of rather large cylinders of α-cellulose under radiant heat fluxes of 0.4 cal/cm^2-s to 1.1

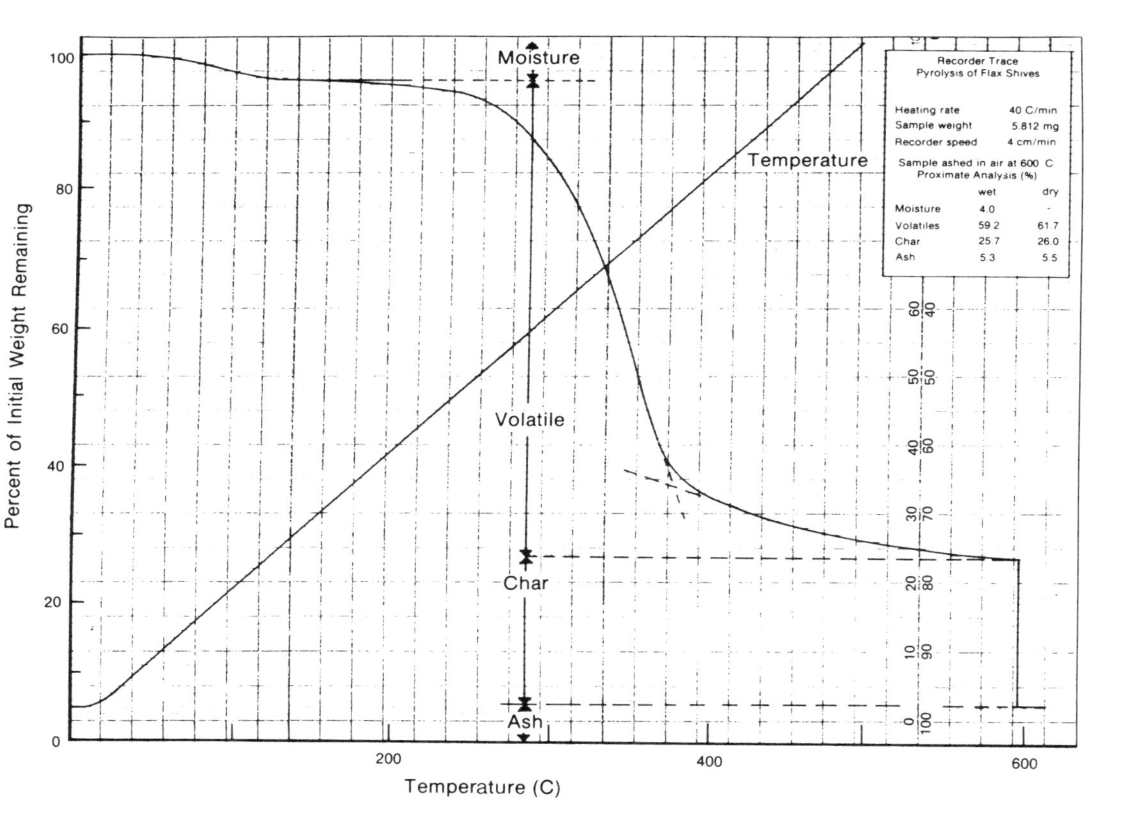

Figure 5-2. A Typical Dynamic TGA Result Obtained with Flax Shives and Showing Moisture, Volatile Matter, Char, and Ash Content

cal/cm^2-s. Fairbridge et al. (1978) studied fibrous cellulose powder in both isothermal and dynamic heating experiments in N_2 and air. Broido (1966) compared dynamic TGA curves for ash-free cellulose (0.01%), pure cellulose (0.15% ash), and cellulose with 1.5% $KHCO_3$ added. Lipska and Parker (1966) made isothermal TGA measurements on α-cellulose. Cardwell and Luner (1976) carried out isothermal TGA on two pulps. Basch and Lewin (1973) looked at the influence of fine structure on vacuum pyrolysis of cellulose. Antal et al. (1979) pyrolyzed cellulose from a number of sources at varying rates.

Weight loss experiments have also been carried out by: Van Krevelen et al. (1951); Stamm (1956); Corlateanu et al. (1974); Kosik et al. (1972); Akita and Kase (1967); Duvvuri et al. (1975); Barooah and Long (1976); Ramiah (1970); Madorsky et al. (1956, 1958); Shafizadeh and McGinnis (1971); Muhlenkamp and Welker (1977); Patel et al. (1970); McKay (1968); Parks (1971); Arseneau (1971); Mack and Donaldson (1967); Chatterjee and Conrad (1966, 1968); Tang and Neill (1964); Davidson and Losty (1965); Nunomura et al. (1975); Kato and Takahashi (1967); Shafizadeh and Bradbury (1979); Broido and Weinstein (1970, 1971); Chatterjee (1968); Cabradilla and Zeronian (1976); Murty and Blackshear (1966); McCarter (1972); Ainscough et al. (1972); and Ramiah and Goring (1967).

An example of a typical dynamic TGA curve for several kinds of cellulose, measured at SERI, is shown in Fig. 5-3. At the moderate heating rates shown here cellulose is stable

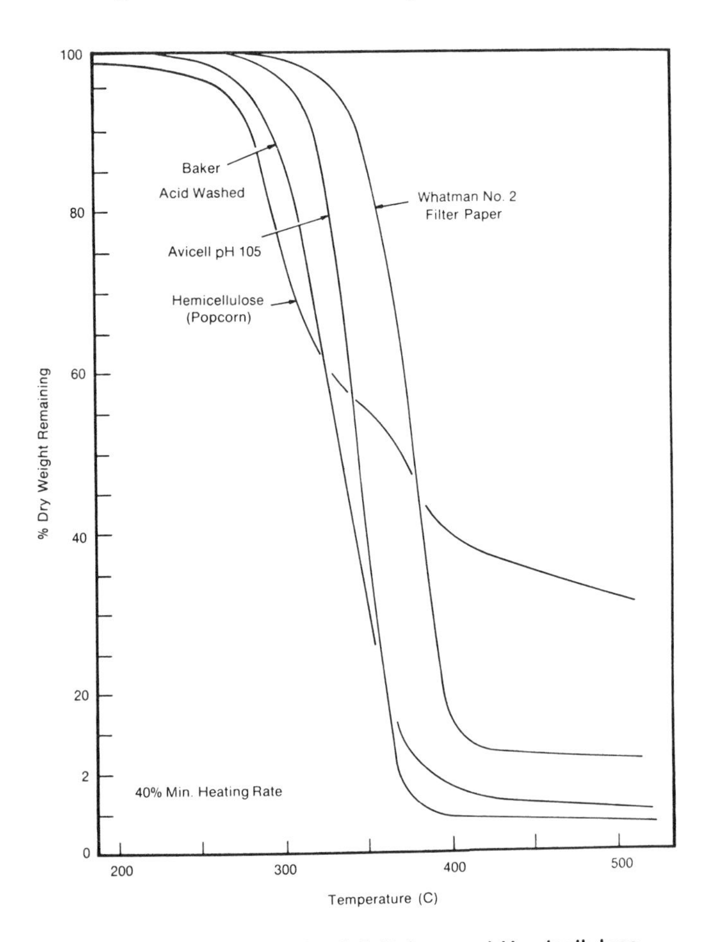

Figure 5-3. Pyrolysis of Cellulose and Hemicellulose

to temperatures over 300 C and then decomposes over a rather narrow range of about 50 C leaving a char residue of 5% to 15%, depending on the cellulose source, size, and heating rate. Isothermal weight loss curves have been reported by Stamm (1956) for lower temperatures.

5.2.1.2 Hemicellulose and Holocellulose

The hemicelluloses, partly because of their lesser abundance in wood and partly because of their variety of constituents, poorly defined degree of polymerization and crystallinity, and ambiguous extraction procedures, have received less study than cellulose. Work includes extracted hemicelluloses, pure components such as xylan, and holocellulose (lignin extracted and containing cellulose and hemicellulose).

TGA of holocellulose has been done by Duvvuri et al. (1975); Fang et al. (1975); and Domburg et al. (1969). Hemicellulose and xylan have been studied by Ramiah (1970); Browne (1958); Stamm (1956); Tang (1964); Domburg et al. (1969); Shafizadeh et al. (1972); and Shafizadeh and McGinnis (1971). Ramiah and Goring (1967) used dilatometry to follow pyrolysis. All studies indicate that the hemicelluloses are the least stable major component of wood, perhaps due to their lack of crystallinity. A typical TGA curve was shown in Fig. 5-1 (Shafizadeh and McGinnis 1971).

5.2.1.3 Lignin, Bark, and Black Liquor

Lignin is the most complicated, least understood, hardest to extract without change, and most refractory component of wood. Consequently, the interpretation of experiments with lignin is the most empirical and shows the most variable behavior of the wood constituents. Its behavior has been reviewed by Roberts (1970); Beall and Eickner (1970); Soltes and Elder (1979); Allan and Mattila (1971); and Tang (1964).

Weight loss experiments have been carried out on various lignin preparations by Van Krevelen et al. (1951); Duvvuri et al. (1975); Ramiah (1970); Shafizadeh and McGinnis (1971); Stamm (1956); Goos (1952); and Fang et al. (1975). Fairbridge and Ross (1978); Tran and Rai (1978) and Rensfelt et al. (1978) have done TGA on bark. A typical thermogram for two lignins is shown in Fig. 5-1. Minor decomposition appears to start at lower temperature than for cellulose, but most lignin pyrolysis occurs at higher temperatures. Large differences are seen in lignins prepared by different procedures. Acid lignin appears to be more stable than other derived lignins.

5.2.1.4 Wood and Other Biomass

It is reasonable to assume, at least qualitatively, that the pyrolysis of wood is closely related to the three major components of biomass, and several of the studies cited above reach this conclusion, though not with a quantitative demonstration (Antal et al. 1979). Reviews relevant to wood pyrolysis include Roberts (1970, 1971a,b); Beall and Eickner (1970); Tang (1964); Tran (1978); and Soltes and Elder (1979).

The pyrolysis of wood and related substances, measured through weight loss behavior, has been reported by Rensfelt et al. (1978); Babu (1979); Browne and Brenden (1964); Browne and Tang (1963); Corlateanu et al. (1974); Tang and Eickner (1968); Heinrich and Kaesche-Krischer (1962); Stamm (1956); Fairbridge and Ross (1978); Hileman et al. (1976); Shafizadeh and McGinnis (1971); Leu (1975); Muhlenkamp and Welker (1977); Duvvuri et al. (1975); Havens et al. (1971); Barooah and Long (1976); and Maa and Bailie (1978). The general features are what would be expected from the composition though quantitative comparisons are questionable. A typical TGA curve for hardwood is shown in Fig. 5-1.

TGA data on a few other forms of biomass such as manure, papers, and straw have been reported in many of the references listed above.

5.2.2 Kinetic Analysis of Pyrolysis

The sharp, well-defined TGA curves, especially for cellulose, suggest that a relatively simple reaction controls the decomposition kinetics prevalent at relatively slow heating rates, and a great deal of effort has gone into fitting classical kinetic theory to TGA data in general (Wendlandt 1974) and for biomass components in particular (Antal et al. 1979). Unfortunately, there is no generally accepted method for extracting kinetic data from dynamic TGA data, and the data can be fit quite well with a range of the adjustable constants. Whatever the theoretical merits of the resulting kinetic data, they serve to predict pyrolysis data over a range of conditions and thus should have engineering utility in designing and understanding gasifiers having slow heating rates. We will sketch here only enough of the kinetic background to make the results comprehensible.

The thermal decomposition curves can be fit using a general equation of the form:

$$dV/dt = k\, V^n\,, \tag{5-1}$$

where

$$k = A \exp(-E/RT)\,, \tag{5-2}$$

and V is the fraction of total volatiles remaining at temperature T. If the sample is heated at a constant rate, R = dT/dt, then Eq. 5-1 becomes

$$dV/dT = k\, V^n/R\,. \qquad (5\text{-}3)^*$$

Unfortunately, a wide variation of the activation energy E, the pre-exponential factor A, and the order n can give satisfactory fits to the data. Many investigators arbitrarily choose n = 1. Furthermore, it is not easy to measure sample temperature accurately in a free balance pan, particularly in vacuum. Table 5-1 (Antal et al. 1979) shows the variation of activation energy measured on cellulose and wood by a number of investigators. Antal et al. postulated that researchers who achieved the best temperature measurement found that E lay in the range 26-33. One of the most convincing aspects of these measurements is that they can predict the decomposition rate of cellulose quite well over a range of slow heating rates varying by a factor of 30 (Antal et al. 1979). However, present analyses do not predict the variation of char formation with pyrolysis conditions, and this would be especially useful for gasification.

The TGA curves for cellulose are relatively simple and can be fit using Eqs. 5-1 to 5-3. However, the TGA data for lignin, hemicellulose, and compound biomass are complex and will require a more complex theory for accurate description. Nevertheless, equations of this form can still be used for engineering prediction.

5.2.3 Differential Thermal Analysis (DTA) and Differential Scanning Calorimetry (DSC)

TGA records mass change during pyrolysis but not energy changes. In differential thermal analysis a thermocouple junction placed in the sample records the difference in temperature between the sample and another inert material. If an endothermic reaction occurs the sample temperature lags that of the reference, while an exothermic reaction causes sample temperature to lead the reference temperature. This gives a qualitative measure of the sign and degree of energy absorption or evolution during pyrolysis.

Recently this type of data has been made quantitative in the technique of differential scanning calorimetry. In this case, an electrical circuit adds heat to or subtracts heat from the sample to keep its temperature identical with that of the reference and records the amount of heat added or held back.

5.2.3.1 Cellulose

Many workers have used DTA to observe, semiquantitatively, the heat effects on pyrolysis of small samples (Patel et al. 1970; Parks 1971; Mack and Donaldson 1967; Tang and

---

*This equation cannot be solved analytically and various approximations and computer integrations are required to determine the constants.

Table 5-1. PYROLYSIS KINETICS DERIVED FROM EXPERIMENTS UTILIZING SMALL (~1 g) SAMPLES

Sample	Reference	Experiment	E(kcal/g-mole)
Cellulose	Akita and Kase (1967)	TGA, TC in Vacuum	53.5
Cotton	Madorsky, Hart and Straus (1956)	TGA, TC in Vacuum	50.
Cellulose	Ramiah (1970)	TGA, TC in Vacuum	36.-60.
Cellulose	Tang (1964)	TGA, TC in Vacuum	56.
Cellulose	Tang and Neil (1964)	TGA, TC in Vacuum	53.-56.
Cellulose	Arseneau (1971)	TGA, Flowing N_2	45.4
Wood	Browne and Tang (1963)	TGA, Flowing N_2	35.8
Cotton	Chatterjee and Conrad (1966)	TGA, Flowing N_2	33.
Cotton	Mack and Donaldson (1967)	TGA, Flowing N_2	48.8
Cellulose	Lipska and Parker (1966)	Fluidized Bed	50.
Cellulose	Chatterjee (data of Lipska and Parker) (1965)	Fluidized Bed	42.
Cellulose	Lipska and Woodley (1969)	Fluidized Bed	42.
Cellulose	McCarter (1972)	Evolved Gas	40.5
Cellulose	Murphy (1962)	Evolved Gas	39.4
Cellulose	Martin (1965)	Radiation	30.
Cellulose	Shivadev and Emmons (1974)	Radiation	26.
Cellulose	Lewellen, Peters and Howard (1976)	Electrically heated screen	33.4

Neill 1964; Herbert et al. 1969; Akita and Kase 1967; Broido 1966; Shafizadeh and McGinnis 1971; Arseneau 1961, 1963; Ramiah 1970; Berkowitz 1957; Tang and Eickner 1968; Breger and Whitehead 1951; Sandermann and Augustin 1963; Domansky and Rendos 1962; and Domberg et al. 1969). In a few cases the more quantitatively interpretable DSC has been used (Arseneau 1971; Mack and Donaldson 1967; Basch and Lewin 1973; Muhlenberg and Welker 1977). Finally, transient heat balances in pyrolyzing material have been used to estimate heat effects (Browne and Brenden 1964).

Typical curves obtained in DTA and DSC are shown in Figs. 5-4 (from Shafizadeh and McGinnis 1971) and 5-5. Qualitative agreement exists that the pyrolysis of cellulose, at least in the absence of very extensive secondary charring reactions in large samples, is entirely endothermic. The exact nature of the heat effects is quite sensitive to the extent of secondary reactions and to added impurities, as shown by the DSC curves for thin and thick specimens in Fig. 5-5.

5.2.3.2 Hemicellulose and Holocellulose

DTA data have been obtained by Ramiah (1970); Fang et al. (1975); Arseneau (1961); Sandermann and Augustin (1963); Domansky and Rendos (1962); Domburg et al. (1969);

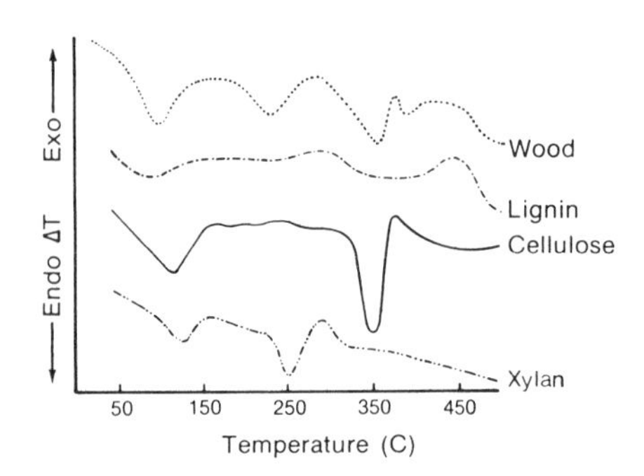

Figure 5-4. Differential Thermal Analysis of Cottonwood and Its Components

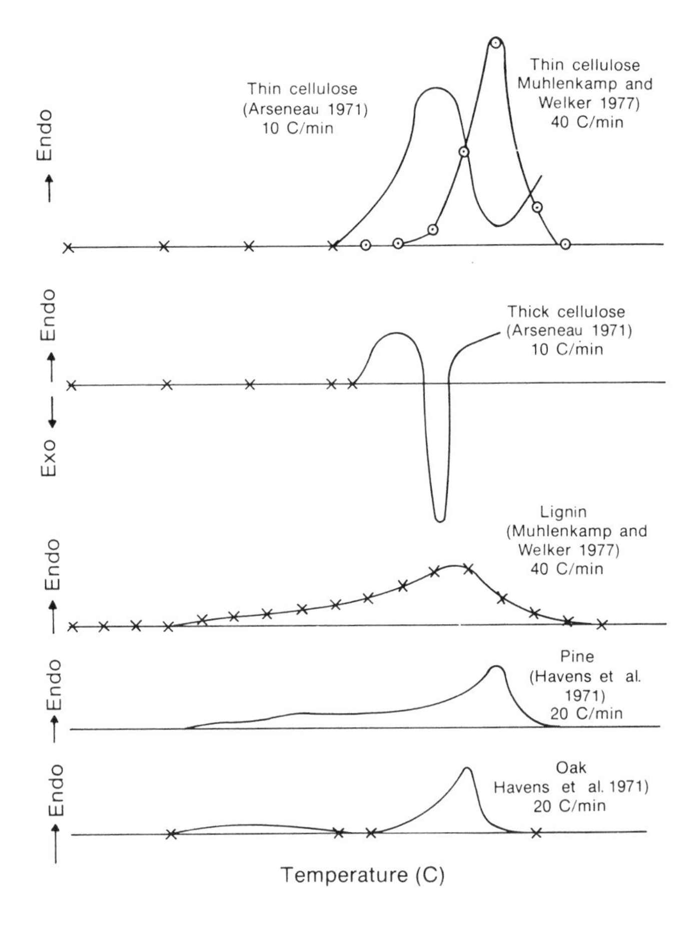

Figure 5-5. Selected Results of Differential Scanning Calorimetry of Wood and Its Components

and Shafizadeh and McGinnis (1971) for hemicelluloses and holocelluloses. No DSC measurements have come to our attention. As in cellulose, the decomposition of hemicelluloses appears to be endothermic, at least for the small samples usually employed. A representative DTA curve is shown in Fig. 5-4.

5.2.3.3 Lignin, Bark, and Black Liquor

DTA curves have been generated for lignins and bark by Fang et al. (1975); Arseneau (1961); Berkowitz (1957); Sandermann and Augustin (1963); Domansky and Rendos (1962); Domburg et al. (1969); Ramiah (1970); Shafizadeh and McGinnis (1971); Tang and Eickner (1968); and Breger and Whitehead (1951). The only DSC study discovered is that of Muhlenkamp and Welker (1977) on punky douglas fir. DTA and DSC curves are shown in Figs. 5-4 and 5-5.

5.2.3.4 Wood and Other Biomass

DTA analysis has been popular in spite of the ambiguities in its quantitative interpretation (Arseneau 1961; Shafizadeh and McGinnis 1971; Tang and Eickner 1968; Heinrich and Kaesche-Krischer 1962). In general, researchers have seen the features of component DTA curves in the whole wood thermogram (Fig. 5-4) (Breger and Whitehead 1951; Sandermann and Augustin 1963; Domansky and Rendos 1962; Domburg et al. 1969). The more interpretable DSC has been applied by Leu (1975); Muhlenkamp and Welker (1977); and Havens et al. (1971). Heats of wood pyrolysis have been deduced indirectly from measurements of temperature distributions in larger, pyrolyzing wood specimens (Roberts and Clough 1963; Bamford et al. 1946). The DSC results, and a reinterpretation of the data of Roberts and Clough by Kung and Kalelkar (1973), support the heat of pyrolysis of wood as endothermic, at least where secondary, char forming reactions are not extensive. DSC curves for pine and oak wood are shown in Fig. 5-5.

5.2.4 Gases and Other Products Evolved During Pyrolysis

5.2.4.1 Cellulose

A great deal of effort has been spent on analysis of the gaseous and condensable products of pyrolysis of cellulose, more often from the standpoint of deducing pathways of decomposition than from the relevance to subsequent gasification (McKay 1968; Byrne et al. 1966; McCarter 1972; Bolton et al. 1942; Min 1977; Madorsky et al. 1956, 1958; Davidson and Losty 1965; Robb et al., 1966; Halpern and Patai 1969; Tsuchiya and Sumi 1970; Glassner and Pierce 1965; Lipska and Wodley 1969; Smith and Howard 1937; Schwenker and Pacsu 1957; Greenwood et al. 1961; Venn 1924; Fairbridge et al. 1978; Schwenker and Beck 1963; Murphy 1962, Antal et al. 1979; and Goos 1952). Gas chromatography and mass spectrometry are widely employed. Occasionally, special techniques are used on the condensed phase, such as x-ray, density, measurements of degree of polymerization, ESR, IR, and vapor-phase thermal analysis.

Some examples of yields of char, tar, and gases are given in Table 5-2 for several pyrolysis experiments. The nature of the products depends on the rate of heating, the temperature, the degree to which primary pyrolysis products are confined in the char, and the presence of additives (catalysts), so that the results shown are not necessarily representative. For moderately fast pyrolysis the nature of the surrounding atmosphere (air, steam, H_2, inert) has little effect. Pressure is not a major variable either, except as it influences the escape of primary products.

5.2.4.2 Hemicellulose and Holocellulose

A few studies have been made on the gaseous and other products of holocellulose and hemicellulose pyrolysis (Fang and McGinnis 1975, 1976; Goos 1952; Shafizadeh et al. 1972; and Stamm 1956). Holocellulose and hemicellulose tend to yield more gases and less tar than cellulose. Table 5-3 gives examples of product compositions for pyrolysis of a holocellulose and a hemicellulose.

Table 5-2. PYROLYSIS PRODUCTS OF CELLULOSE REPORTED IN TWO DIFFERENT STUDIES

	Wt % of Sample		
	(Antal et al. 1979) 500 C	(Tsuchiya and Sumi 1970) 320 C	520 C
Total Accounted For	64%	89.5%	68.0%
Char	12	67.8	12.8
Tar[a]	35	10.3	28.4
Water	—	9.3	20.7
H_2 [b]	0.4		
CO [b]	18	0.5	2.6
CO_2 [b]	30	1.5	2.9
CH_4 [b]	0.5		
C_2H_4 [b]	0.5		
C_3H_6 [b]	0.5		
C_2H_6	0.5		
Other	1.3		
Hydrocarbons		—	0.3
Furan	0.03		0.04
2-Methylfuran		0.02	0.05
Furfural		0.06	0.08
5-Methylfural			
5-Hydroxymethyl Furfural		0.01	0.08
Levoglucosan[c]		3.8	18.2
1, 6-Anhydro-D-Glucofuranose		0.2	2.2
-D Glucose[c]		0.03	0.2
-D Glucose[c]		0.04	0.3
Dimers of Anhydroglucose		0.2	0.4
Unknown[c]		0.12	0.08
Unanalyzed Tar[c]		5.9	7.0

[a]Includes all tar fractions below.
[b]Upper limits - mass balance only 64%.
[c]Tar fraction.

Table 5-3. PYROLYSIS PRODUCTS FROM A HOLOCELLULOSE (FANG AND MCGINNIS [1976]) AND A HEMICELLULOSE (XYLAN) (SHAFIZADEH [1977])

	Wt % Product	
	Holocellulose (400 C)	Xylan (500 C)
Char	20.2	10
Tar	—	64
Water	37.3	7
CO_2	11.0	8
CO	5.3	—
Low Molecular Weight Hydrocarbons	0.4	
Methanol	1.1	1.3
Acetaldehyde	0.2	2.4
Acetic Acid	1.4	1.5
Furan	0.5	Trace
Acrylaldehyde	0.07	
Diacetyl	0.2	
1-Hydroxy-2-Propanone	0.06	0.4
2-Furaldehyde	0.5	4.5
Acetone-Propionaldehyde		0.3
2-3-Butanedione		Trace
3-Hydroxy-2-Butanone		0.6

5.2.4.3 Lignin, Bark, and Black Liquor

Much of the work yielding gas from lignin originates in gasification studies of black liquors (Liu et al. 1977; Brink 1976; Goheen et al. 1976; Prahacs et al. 1967a,b; 1971; Barclay et al. 1964; Rai and Tran 1975; and Schlesinger et al. 1973). In these studies pyrolysis and gasification were not always separated clearly so that probably only the lower temperature composition (500 C to 600 C) reflects lignin pyrolysis behavior. Vroom (1952); Fairbridge and Ross (1978); Schlesinger et al., (1973); and Rensfelt (1978) measured gas or product compositions from pyrolyzing bark. Goos (1952); Stamm (1956); Fletcher and Harris (1947, 1952); and Hileman et al. (1976) looked at products from lignin, Hileman et al. by subtracting the pyrolysis mass spectrum of cellulose from that of wood for comparison with the spectrum of extracted lignin.

Some of the products obtained in lignin pyrolysis are given in Table 5-4.

Table 5-4. EXAMPLES OF VOLATILE PRODUCTS FROM LIGNIN PYROLYSIS

	Douglas Fir Lignin[a] Pyrolyzed at 400-445 C for 7.5h	Kraft Black Liquor[b] Pyrolyzed at 490 C
Char	53-64.6%	Methyl mercaptan
Aqueous distilled	15-25%	Dimethyl sulfide
Tar	~9%	Benzene
Gases	—	Toluene
Organic acids	formic, acetic propionic plus traces of others	m and/or p-Xylene Anisol Phenol
Phenols	phenol, o-cresol p-cresol, guaiacol 2, 4-xylenol, 4-methyl and ethyl guaiacol, 4-n-propylguaiacol	o-cresol m and/or p-cresol 2, 5 and/or 2, 4 dimethyl 3, 5 dimethyl phenol
Catechols	catechol, 4-methyl and ethyl catechol, 4-n-propyl catechol	2, 3 dimethyl phenol 3, 4 dimethyl phenol

[a]Fletcher and Harris 1952.
[b]Brink et al. 1971.

5.2.4.4 Wood and Other Biomass

As expected, the products from wood pyrolysis are more complex than those from the wood components listed above. No study is known which demonstrates that the gaseous products of wood are the sum of its components under comparable pyrolysis conditions. Product analyses for wood, under conditions where pyrolysis may predominate, have been reported by Knight (1976); Babu (1979); Appell and Pantages (1976); Appell and Miller (1973); Stern et al. (1965); Schlesinger et al. (1973); Stamm (1956); Goos (1952); Rensfelt et al. (1978); Brink (1976); and Min (1977). Support studies for gasifier research tabulated in a later section also contain pyrolysis gas behavior. Catalysts have a large effect on the pyrolysis of cellulose and wood, but only a small effect on the pyrolysis of lignin. Table 5-5 gives some examples of product yields from the literature.

In general, pyrolysis of other forms of biomass gives similar products, except that manure contains a high proportion of volatile fatty acids and lignin products.

Table 5-5. SELECTED EXAMPLES OF WOOD PYROLYSIS PRODUCTS UNDER CONDITIONS RANGING FROM LONG RESIDENCE TIME TO FAST PYROLYSIS

	Wt % Products		
Species	White Fir (Fast Heating) (Brink & Massoudi 1978)	Pine (Slow Heating) (Knight 1976)	Douglas Fir (Fast Heating (Hileman et al. 1976)
Total Gases	69.0	25	
Char	5.7	32	
Tar	21.0	16	
Water	N/A	27	15.3
H_2	0.1		
CH_4	4.8[a]	2.0	2.0
CO	42[a]	7.3	21.2
CO_2	22[a]	14.2	5.7
C_2H_4			1.0
C_2H_6			0.3
C_2-C_4 HC		2.2	
C_3H_6			1.6
Methanol			0.4
Ethanol			1.4
2-Methyl Propene			0.3
Propenal			0.7
Furan			0.2
2-Oxopropanal			0.5
Hydroxyethanal			0.7
Ethanoic Acid			0.7
2, 3-Butanedione			0.3
2-Hydroxypropanol			0.5
Furfural			0.4
2, 3-Pentanedione			0.1
Furfurylalcohol			0.2
2-Methyl-2-Butenal			0.1
o-Methoxyphenol			0.2
2-Methoxy-4-Methylphenol			0.4
2-Methoxy-4-Methylanisole			0.1
4-Oxophentanoic Acid			0.4
4-Hydroxy Pentanoic Acid			0.2
p-Methoxyacetophenone			0.2
2-Methoxy-4-Propenylphenol			0.1
5-Hydroxymethyl-2-Furaldehyde			0.2

[a]Upper limits.

5.2.5 Morphology of Biomass During Pyrolysis

Surprisingly few investigations have involved time-dependent studies of the morphology of cellulose during pyrolysis (McCarter 1972). A recent exception is the study by Fairbridge et al. (1978) in which SEM and krypton absorption were used to characterize the developing chars.

No morphological studies have been found for the hemicelluloses and holocelluloses or lignin during pyrolysis.

Knudson and Williamson (1971) observed morphological changes in wood heated in air. Though much work has probably been done on the characteristics of chars produced in carbonization, we are not aware of systematic, time-resolved studies of the morphological changes in wood during the early stages of pyrolysis under conditions relevant to gasification.

5.2.6 Pyrolysis Mechanisms

5.2.6.1 Cellulose

All of the studies described here have been employed to infer the mechanisms for production of the large variety of products actually observed, including the study of model compounds. Several reviews have discussed the detailed organic reactions leading initially largely to levoglucosan, a primary product of cellulose pyrolysis, and subsequently to a wide variety of decomposition products (e.g., Shafizadeh 1968; Molton and Demmitt 1977). Mechanisms for formation of the lighter gaseous species are almost totally lacking; the time and space resolution of the gas sampling devices, if not the experimental pyrolysis arrangement itself, making intermediates hard to observe. A widely adopted scheme which provides a conceptual framework for many observations has been proposed by Kilzer and Broido (1965).

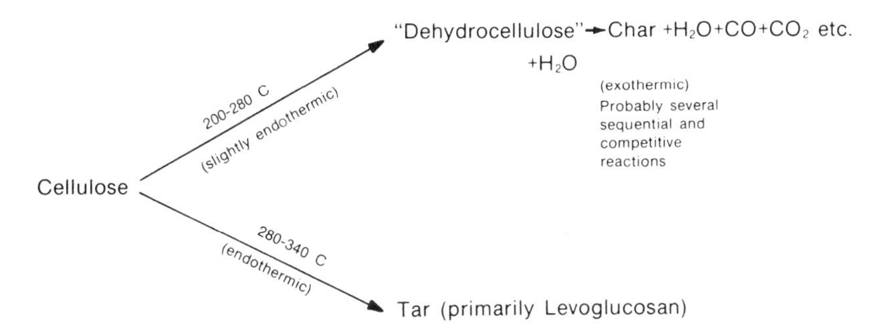

At low temperatures and slow heating rates, the upper path predominates. Under faster heating, the lower path becomes the more important. Shafizadeh (1968) elaborates on this scheme relative to biomass combustion as follows:

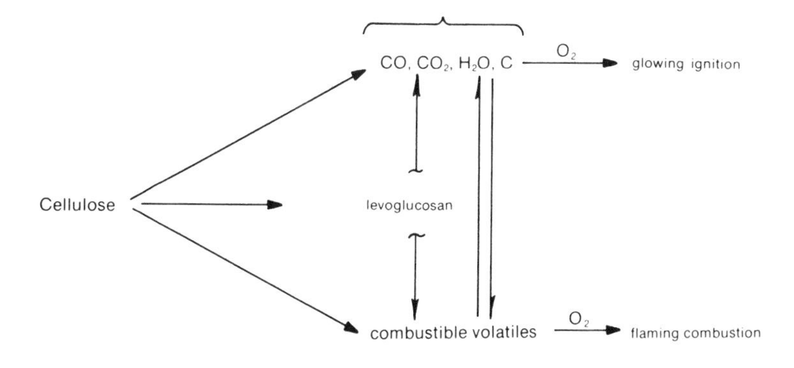

Quoting Shafizadeh:

> Thermal degradation of cellulosic materials proceeds through a complex series of concurrent and consecutive chemical reactions. The [above

scheme] provides an outline of the general sets of the degradation reactions of cellulose that could lead to the flaming combustion or glowing ignition of this material. The nature and extent of many individual reactions involved in this process are not known or insufficiently defined. However, it is known that these reactions are highly influenced by: the temperature and period of heating; the ambient atmosphere, oxygen, water, and other reacting or inert gases; and the composition and physical nature of the substrate, especially with respect to inorganic impurities and additives.

The general reactions can be divided into primary and secondary reactions, according to whether they directly affect the cellulosic substrate or one of the intermediate degradation products. Alternatively, two general pathways for degradation of cellulosic materials may be recognized. One involves fragmentation, and formation of combustible volatiles that could feed the flames, and the second mainly involves dehydration and the formation of carbonaceous char that could lead to localized, and relatively slower, glowing ignition. Since these two pathways compete for the same initial substrates, according to the prevailing conditions, one could predominate at the expense of the other.

Heating at the lower temperatures, as discussed later, favors the dehydration and charring reactions. Formation of levoglucosan, which is a principal intermediate compound, takes place at somewhat higher temperatures and leads to further decomposition reactions at the elevated temperatures.

This description, though in the context of combustion, is relevant to gasification as well.

5.2.6.2 Hemicellulose and Holocellulose

Speculations as to reaction pathways for the condensable organics from hemicellulose parallel those for cellulose. Soltes and Elder (1979) reported a postulated two-step decomposition. First, depolymerization to water-soluble fragments occurs, followed by decomposition to volatiles. The expected furan derivatives may be too reactive to survive the usual pyrolysis conditions. Browne (1958) discussed the older literature on hemicellulose pyrolysis behavior. Tang and Eickner (1968) postulated that early pyrolysis of hemicellulose to acetic acid and formaldehyde may affect pyrolysis of cellulose and lignin in wood. Goos (1952) indicated that the pentosans in hemicellulose give the most distinctive products, while little is known of hexosan behavior.

5.2.6.3 Lignin, Bark, and Waste Liquor

The rich structural variety in the hypothesized lignin macromolecule gives rise to many mechanistic pathways to observed condensable organic compounds. Furthermore, each lignin preparation gives a different substrate—with ambiguity as to the nature of the true "native lignin." Soltes and Elder (1979) note that lignin produces more aromatic compounds and char than cellulose. No product predominates as is the case with cellulose. Allan and Mattila (1971) assume that lignin pyrolysis is by homolytic cleavage with phenyl radicals important. Goos (1952) assumes that pyrolysis of lignin in H_2 may give a truer indication of primary fragments by minimizing secondary condensation reactions. The reader is referred to the references above for specific speculations.

5.2.6.4 Wood and Other Biomass

The general features of wood pyrolysis mechanisms usually have been discussed in terms of the behavior of wood's components, since few interactions or new products found only with wood have been observed. Roberts' (1970) review of the kinetics of wood pyrolysis is still timely. He accepts the prevailing practice of treating wood pyrolysis as a first order process following Arrhenius kinetics. The factors affecting pyrolysis are discussed in terms of composition, autocatalysis, physical structure, pressure, and wood type. His conclusions are:

- The use of a first-order reaction scheme to describe the complex process of wood pyrolysis is questionable theoretically but has empirical advantages.
- Hemicellulose, cellulose, and lignin have pyrolysis reactivities decreasing in the order in which the substances are listed. Most of the lignin will still be present after the bulk of the first two substances has decomposed.
- The cellulose component is extremely sensitive to catalytic and autocatalytic effects, with pure cellulose primary pyrolysis showing a high activation energy and impure or large samples exhibiting a much lower activation energy.
- Lignin pyrolysis shows much smaller effects due to additives or autocatalysis.
- Experiments with small samples may not be representative of large sample behavior to the extent that secondary reactions, autocatalysis, and physical structure play a role.
- Restraints on pyrolysis product movement due to the physical 1tructure of the wood are important at low temperature but largely disappear at temperatures of 300 C to 320 C.
- A fully developed pyrolysis wave in wood can be divided into four regions of increasing temperature:
  - wood structure is virtually intact with autocatalytic pyrolysis of most reactive components;
  - wood structure has failed, autocatalysis is reduced, pyrolysis of reactive components occurs;
  - pyrolysis of hemicellulose and cellulose complete and lignin pyrolysis is dominant;
  - all the wood is pyrolyzed to char; and
  - secondary reactions of primary volatiles occur with char residues.
- The choice of suitable kinetics for application to pyrolysis depends on the nature of the problem (e.g., ignition versus complete pyrolysis to char).

This type of sequence has been portrayed schematically for wood combustion by Kanury (1972) and is shown on the following page; in this scheme for gasification, the final step would be gasification and the heat flux might come from an external source as well as from partial combustion.

5.2.7 Discussion

The previous sections are little more than a guide to the kinds of studies that have been done on wood and its components. The reader is referred to the papers, and especially to the reviews, for details. Even had time and space permitted, it is not clear that a detailed comparison of these kinds of studies is warranted in the context of our gasification interests. Many of the studies just cited have involved small samples, slowly heated, with rapid escape of volatile products. In gasifier operation, two conditions will tend to prevail:

- In gasifiers intended to produce a gas tending toward equilibrium, large particles, with attendant slow heating rates, will be subjected to long residence times.

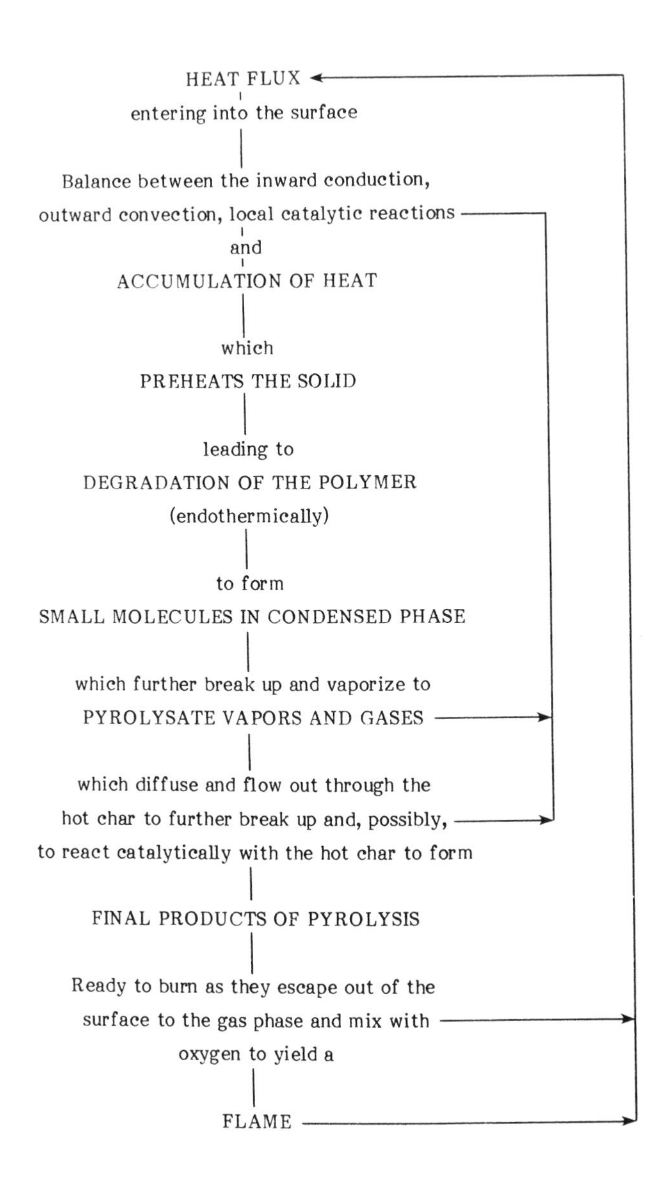
HEAT FLUX
entering into the surface
Balance between the inward conduction,
outward convection, local catalytic reactions
and
ACCUMULATION OF HEAT
which
PREHEATS THE SOLID
leading to
DEGRADATION OF THE POLYMER
(endothermically)
to form
SMALL MOLECULES IN CONDENSED PHASE
which further break up and vaporize to
PYROLYSATE VAPORS AND GASES
which diffuse and flow out through the
hot char to further break up and, possibly,
to react catalytically with the hot char to form
FINAL PRODUCTS OF PYROLYSIS
Ready to burn as they escape out of the
surface to the gas phase and mix with
oxygen to yield a
FLAME

- In gasifiers seeking to maximize production of nonequilibrium gas composition (e.g., olefins), small particles, with attendant fast heating rates, will be subjected to short residence times.

Thus, the conditions of relevance to gasification in laboratory studies will tend toward either large particles at slow heating rates or small particles at fast heating rates. It has been suggested that many of the studies cited above, involving small samples and intermediate heating rates, may have limited applicability to real fire (or gasification) situations (Kanury 1972).

The next section discusses studies relating to pyrolysis gasification of small particles with high heating rates and short residence time.

5.3 FAST PYROLYSIS

Less studied than slow pyrolysis, fast pyrolysis holds the possibility of direct production of products of high value such as olefins, especially ethylene and propylene. A number of studies, usually recent, partially characterize pyrolysis gasification under conditions of rapid heatup, high temperature, or short residence time. These studies can be grouped under four categories:

(1) slow-solid pyrolysis followed by short residence time for re-forming or secondary pyrolysis (Antal et al. 1979);

(2) fast-solid pyrolysis followed by almost instantaneous quenching of primary products in cold gas or vacuum (Lincoln 1965, 1974; Lincoln and Covington 1975; Martin 1965; Lewellen et al. 1976; Hileman et al. 1976; Broido and Martin 1961).

(3) fast-solid pyrolysis followed by relatively short residence times for re-forming or secondary pyrolysis (Brink and Massoudi 1978; Allan and Mattila 1971; Prahacs et al. 1971; Rensfelt et al. 1978; Berkowitz, Mattuck and Noguchi 1963; Diebold and Smith 1979; Kuester 1978; Brink et al. 1973; Mallon 1974).

(4) fast pyrolysis followed by relatively long residence times for re-forming or secondary pyrolysis (Brink 1976; Stern et al. 1965; Barber-Coleman 1975); or very high effective temperatures (Brown 1979; Krieger et al. 1979; Allan and Mattilla 1971).

These studies are summarized and results compared in the following sections, with the emphasis on production of olefins, since these high value products are observed only in fast pyrolysis.

5.3.1 Slow Pyrolysis, Short Residence Time

Antal et al. (1979) seem to be the only researchers to force separation of the slow, low temperature solid pyrolysis from the fast, high temperature gas re-forming/pyrolysis stages in gasification. They have reported results only for cellulose (Whatman filter paper) though studies on wood are in progress. Half gram samples of the cellulose were pyrolyzed at a heating rate of about 100 C/min to 500 C/min in a flow of steam or argon. The volatile pyrolysis products were then swept into a quartz reactor and allowed to react further for 1 to 10 s at temperatures to 750 C. Gaseous components CO, CO_2, H_2, CH_4, C_2H_6, C_2H_4 and C_3H_6 were reported as a function of temperature and residence time. Results in Ar and steam were essentially identical. The results were interpreted to give kinetics of formation of the products, though the composition of the intermediate gases and vapors was not measured completely. (Studies cited above could be used to estimate the likely pyrolysis products.) Table 5-6 shows representative gas compositions from this study and some processes discussed below.

5.3.2 Fast Pyrolysis, Very Short Residence Time

In these studies, pyrolysis was initiated by resistance or flash heating in vacuum or cold

Table 5-6. PRODUCT RESULTS IN FAST PYROLYSIS OF BIOMASS AND ITS CONSTITUENTS

Reference	Type and Form Biomass	Rate of Heating	Maximum Temp. of Pyrolysis/ Gasification	Residence Time at Temp.	Environment	Approximate Weight Percent of Organics: Char	Tar	Gases	H_2	CH_4	C_2H_4	CO	CO_2	C_2H_6	C_3H_6	H_2O	Other
Section 5.3.1 Antal (1978-79)	Whatman filter paper (cellulose) 0.125 g	100°-500°C/min	700°C	3.5 sec	Hot steam or argon	10	2	88	1.2	7.2	5.5	47.2	11.5	1.0	0.15	–	0.15
Section 5.3.2 Berkowitz-Mattuck & Noguchi (1963)	Cotton cellulose cloth	Carbon-arc radiant at 5 cal/cm²-sec	NA	1 sec irradiation	Cold helium	~20	–	–	–	–	0	~5	~0	–	–	–	*
		25 cal/cm²-sec				~8	–	–	–	–	0.86	~7.3	~2.1	–	–	–	**
Martin (1965)	α-cellulose + 2% carbon black	Carbon-arc radiant at cal/cm²-sec	NA	0.4-8 sec	Cold helium												
		4.4				20-35											
		11.6 early		~1 sec		4	80	–	~0.02	~0.02	~0.02	~3	~4	–	–	10	***
		11.6 late		~8 sec		4	55	–	~0.5	~0.9	~0.9	~13	~10	–	–	15	†
		10²-10³				0	–	–	–	–	–	–	–	–	–	–	–
Lincoln (1965)	α-cellulose + 2% carbon black	Carbon-arc and xenon lamp			Cold helium												Volatile organics
		1.5 cal/cm²-sec	~300°C	10 sec		33	19	48	–	–	–	3	9	–	–	32	3
		11.0 cal/cm²-sec	>600°C	4 sec		3	51	46	NA	NA	NA	13	11	–	–	16	6
		3,000 cal/cm²-sec	>600°C	½ ms		1	0	99	–	–	–	37	4	–	–	28	30
Lewellen et al (1976)	Cellulose filter paper 0.75 x 2.5 x 0.01 cm strip	Electrically heated mesh 400°C-10000°C/sec	250-1000°C	0.2-75,000 sec 400°C/sec	Vacuum to 1 atm cold He	No char	No gas analysis, weight loss vs time only										
Hileman et al (1976)	Douglas fir 1.5 mg samples	Pyroprobe at ~200°C/sec	550°C	3-4 sec	Cold argon or air	–	–	–	–	2.0	1.0	21	5.7	0.3	1.6	15.3	Propane 1.6 Many organics at 0.1-1 level
Section 5.3.31 Prahacs (1967)	Na, Ca, Mg-base spent liquors	Atomized spray into hot reactor	600°-900°C 5-45 psig	11-100 sec	Self-generated steam	Down to 3	–	–	qs 11	–	qs 7†††	–	–	–	–	–	–
Prahacs et al (1971)	Bark, slash wood, and spent liquors	NA. Various reactors qs pilot scale	600°-1000°C 0-25 psig	3-60 sec	N_2 or self-generated												
	Bark	Batch fed	810°C 0 psig	3.6 sec		–	–	89.5	–	–	6.2	–	–	–	0.6	–	C_2H_2 0.9 Benzene 2.1 Toluene 0.3
Rensfelt et al (1978)	Poplar wood	~1000°C/sec	400°-1000°C	~1 sec	Steam, H_2	–	–	–	–	~5	~5†††	–	–	–	–	–	–
Brink and Massoudi (1978)	White fir particles 20-40 mesh	~1000°C/sec	316°-871°C 843°C	3-5 sec 3.0 sec	N_2	2.5	7.1	92	~1	~10	~5	~62	~13	–	–	–	–
Diebold and Smith (1979)	Ecofuel II 200 μm	10^4-10^5°C/sec	500°-900°C	50-100 ms	Steam and CO_2	19	–	–	1	4	24§	36	16	–	–	–	–
Section 5.3.4 Stern et al (1965)	Sawdust 20-30 mesh	Fast	1000°C	Long (steel wool)	Self-generated	22.2	4.4	71.0	4.6	0.3	0	65.5	0.6	–	2.3	–	–
				(Alundum)		14.0	0.4	84.4	4.0	7.1	0.25	68.4	4.8	–	2.2	–	–
Brink (1976)	Wood, MSW and kraft	Probably fast	(475°-1125°)C 850°C	Uncertain	6.5 % moisture wood	–	–	–	1.8	11.2	7.0	73.0	13.2	–	45	–	–
	Black liquor				52.5% moisture wood	–	–	–	3.8	7.2	4.2	47.7	35.2	–	–	–	–

*Tar fraction, mainly levoglucosan.
**Variety of polar organics boiling below 187°C.
***Acetaldehyde 0.45; acrolein 0.15 acetone, furan, methanol ~0.7.
†Acetaldehyde 1.3, acrolein, acetone 0.25 methanol, furan 0.15.
††Illuminants.
†††Optimized.
§Unsaturates

transport gases so that the primary pyrolysis should have been rapidly quenched (milliseconds). Lewellen et al. (1976) heated thin strips of cellulose (~ 0.07% ash filter paper) by resistance heating in vacuum or helium. Residence times from 0.2 to 75,000 s, final temperatures from 250 C to 1000 C, and heating rates from 400 to 10,000 C/s were studied. Volatile products were not measured, but weight loss data could be fit over a surprisingly large range by a simple first-order equation. No char was formed. The rate data at the highest heating rates were interpreted to imply that the critical factor in pyrolysis is the residence time of volatiles in the cellulose matrix.

Lincoln (1965) used flash heating from both carbon arcs (1.5 and 11.0 cal/cm^2-s) and xenon flash lamps (up to 3,000 cal/cm^2-s) to pyrolyze α-cellulose (blackened with 2% carbon black) in both helium and vacuum. Gas chromatography and mass spectrometry were used for product identification. Comparison of slow versus fast pyrolysis showed a big change in primary products, with low energy flash heating producing tar or char. The importance of mineral impurities on pyrolysis also was stressed. Identified products were H_2O, CO_2, CO, two dozen volatile organic compounds, char, and tar (mainly levoglucosan). The higher energy flash heating (3000 cal/cm^2-s) produced virtually no tar or char. A mass balance on the reported products reveals that what Lincoln called "volatile organics" were in reality materials that had the overall empirical formula $CH_{1.24}$, which could have been 71% C_2H_2 and 29% C_2H_4 by volume. These values would represent 20.8 g C_2H_2 and 9.2 g C_2H_4 per 100 g cellulose. The mass spectrometer studies of directly emitted pyrolysis products gave evidence of short-lived intermediates not yet identified.

Berkowitz, Mattuck, and Noguchi (1963) used carbon arc radiation fluxes from 5 to 25 cal/cm^2-s to pyrolyze cotton cellulose in flowing helium. Products were classified into four ranges. Products boiling at -80 C, comprising 5% and 18% of the total at 5 and 20 cal/cm^2-s respectively, were CO, CO_2, CH_4, and C_2H_4. Products condensing between -80 C and room temperature comprised a dozen or so polar organics with boiling points between 14 C and 178 C, including H_2O, acetic acid, acetone, formic acid, formaldehyde, glyoxal, glycolic acid, lactic acid, and dilactic acid. The tar-like material condensing at room temperature was mainly levoglucosan. Chars, constituting about 20% and 10% at fluxes of 7 and 20 cal/cm^2-s respectively, were not characterized.

It was observed that the fast pyrolysis and slow pyrolysis produce similar kinds of products though yields may be very different.

Martin (1965) used a carbon arc to fast pyrolyze α-cellulose containing 2% carbon black (absorptivity 90%) and 0.15% ash. Pyrolysis products were swept away in helium directly to a gas chromatograph. Irradiation at 4.4 and 11.6 cal/cm^2-s for 0.4 to 8 s was used. Heating rates of the cellulose varied with depth of the rather thick specimens used. At the higher irradiation level only 4% char remained. Initially CO, CO_2, H_2O, and tar (mainly levoglucosan) are formed, with subsequent conversion of the tar to acetaldehyde, acrolein, acetone, furan, methanol, methane, ethylene, and H_2 as secondary products postulated to arise from the further cracking of the tar at the char layer. At 100 to 1,000 cal/cm^2-s no char is produced. Martin concluded that levoglucosan is the principal pyrolysis product.

Hileman et al. (1976) used rapid heating of 1.5-mg samples in a commercial pyrolyzer (Pyroprobe) coupled with the most sophisticated gas analysis equipment reported to date for such studies. Samples were heated in streams of Ar gas or air to 550 C in 3 to 4 s, with direct coupling to either a chemical ionization mass spectrometer or a gas chromatograph-mass spectrometer. Materials studied were Douglas fir, α-cellulose, and Bolker lignin, at estimated heating rates of 200 C/s. Tables of products are shown, with no single organic species dominating. At 400 C, fast pyrolysis gave the same product distributions as pyrolysis at 45 C/min. Also, fast pyrolysis at 400 C in air gave the same results as in argon. No levoglucosan is reported in the products, perhaps because it condensed before entering the mass spectrometer. Whether the Pyroprobe involves contact with metal and the possibility of catalytic effects is not known to us. The products of pyrolysis of lignin were deduced by subtracting the products of cellulose from those observed in wood. These derived products showed no resemblance to those produced by direct pyrolysis of an isolated lignin, raising questions about the effects of isolation and about the material interactions of components in pyrolyzing wood.

5.3.3 Fast Pyrolysis, Short Residence Time

In a number of studies both rapid heating and short residences times were employed in a single step pyrolysis/re-forming/gasification.

Allan and Mattila (1971) quote results of Goheen and Henderson on the extremely high-temperature pyrolysis of lignin. They blew powdered lignin, Douglas fir, and cellulose in He into an electric arc, achieving in each case about 14% conversion to C_2H_2. With a tungsten coil at 2000 C to 2500 C, 23% C_2H_2 was produced. In a small pilot plant as much as 40% C_2H_2 was formed but only about 12% could be quenched during extraction. It is possible that such extreme, costly heating conditions are reflecting very high temperature equilibrium compositions rather than pyrolysis kinetics. In fact, C_2H_2 is seldom reported in pyrolysis studies at lower temperatures. Recently plasma arc reduction of biomass has been reported (Brown 1979) as well as pyrolysis in microwave induced plasmas (Krieger et al. 1979).

Prahacs et al. (1971) pyrolyzed bark, slash, fir, pine, and various pulping liquors in several reactors. Conditions were: 0 psig to 25 psig; 600 C to 1000 C; residence times, 3 to 60 s. The exact conditions of heatup of particles and sprays are not given but probably exceed 100 C/s. Results of hydrocarbon gas production are shown as a function of operating variables. In general, although conversions were fairly low, olefins increased with increasing dilution in steam, with decreasing pressure, and with decreasing residence time. An optimal temperature existed for each set of other pyrolysis conditions. Bark gave a little more ethylene than α-cellulose and much more than black liquors. Investigation of continuous pyrolysis systems was recommended.

Prahacs (1967) reports pyrolysis results in an "atomized suspension technique" reactor in which pulping liquors are sprayed into a 1-ft diameter by 15-ft high reactor. Temperatures of 600 C to 900 C, pressures of 5 psig to 45 psig, water/organics ratios of 1.2 and 2.4, and residence times of 11 to 100 s were studied. The size distribution of the spray was not given, but it is assumed that heatup rates were quite fast. Ethylene and acetylene were maximized in Mg-based liquors while H_2 was maximized in Na-based liquors, presumably due to catalytic destruction of pyrolysis intermediates in the latter case.

Rensfelt et al. (1978) pyrolyzed powdered (500 μm) wood, peat, and municipal solid waste (MSW) in a vertical quartz tube in a furnace at 500 C to 1000 C. Heating rates of 1000 C/s were estimated with residence times of less than a second possible. Similar results were obtained in N_2, steam, and H_2. CH_4 and C_2H_4 production from MSW and wood are comparable. The results are interpreted as showing the importance of secondary reactions of the primary heavy hydrocarbons produced during rapid heating. For wood heated to 800 C, 70% conversion to gas occurs in 0.7 s.

Brink and Massoudi (1978) pyrolyzed fir-wood particles (400 μm to 840 μm) in a N_2 flow in an entrained flow furnace reactor. Suspension densities of 0.05 to 0.5 g/l were tested at 316 C to 871 C and residence times of 3 to 5 s. Calculations show that the largest particles reach 80% of the reactor temperature in 0.4 s. H_2, CO, CO_2, CH_4, C_2H_4, C_2H_6, char, and tar are shown as a function of system variables.

Kuester (1978) has obtained high olefin yields in a dual-fluidized bed pyrolysis reactor while Mallon (1974) reports the flash pyrolysis of a municipal solid waste rich in plastics.

Diebold and Smith (1979), using ECO FUEL-II in an entrained flow reactor, have obtained the most spectacular olefin production so far reported (see Table 5-6). The 250 μm powder (composition not reported but derived from MSW in a proprietary chemical comminution process) was entrained as a dilute phase in steam or CO_2, passing through a 1.9-cm diameter, 2- to 6-m long, externally flame heated tube in 50 ms to 150 ms. Heating rates of 10^4 C to 10^5 C/s are estimated. Gaseous products are shown as a function of temperature, residence time, and dilution. The latter parameter is especially important in olefin production. Under optimal conditions, from 700 C to 860 C, short residence time, and high degrees of dilution C_2+ hydrocarbons were 24% of the feedstock (by weight) and contained 53% of the original feedstock energy. Experiments are underway to test cellulose, lignin, and wood in this reactor to determine to what extent the olefin yields are an anomaly of the ECO FUEL-II, possibly due to the plastics content of the municipal solid waste from which ECO FUEL-II is derived (Diebold 1979).

Shock tube studies on biomass dust may provide information on pyrolysis (Lester 1979) as may the mostly older literature on dust explosions.

5.3.4 Fast Pyrolysis, Long Residence Time

Stern et al. (1965) pyrolyzed 20 to 30 mesh sawdust by dropping the material continuously on a packed bed of steel wool or Alundum maintained at 1000 C. The gases were forced through the bed and collected for analysis. Heating rates are probably fast, but residence times for secondary reactions are probably long. The steel wool experiments gave mainly H_2 and CO in 1:1 ratio at 1000 C, while the Alundum gave significant CH_4 yields but only a trace of C_2H_4.

Brink (1976) reports pyrolysis results for wet and dry wood, wet MSW, and Kraft black liquor. The exact reactor conditions are not given in this report nor are the particle size, heatup rate, or residence time. (Presumably, these are available in the primary references.) Gaseous compositions are given from 475 C to 1125 C, in some cases showing significant C_2H_4 yields.

Moderate olefin yields and large benzene yields were reported by Barber-Coleman (1975) in pyrolysis of simulated solid waste on a molten lead bath. An interesting study, the conditions of which are hard to classify, was carried out by Sanner et al. (1970), yielding moderate ethylene conversion from wet solid waste.

Table 5-6 presents an overview of typical product compositions for many of the above cited studies. It may be concluded that fast pyrolysis, coupled with dilute-phase, high-temperature, short residence time, secondary reactions can yield large quantities of olefins. Several of the cited studies continue to be active and new results can be expected to further shed light on the sequence of pyrolysis secondary reactions for a variety of biomass materials and components.

CONCLUSIONS

From all the evidence cited it appears that the products of primary pyrolysis are a sensitive function of physical size and state of the material, inorganic impurities, heating rate, and final temperature. Gaseous environment seems to be relatively unimportant. Secondary reactions are a function of contact with char, temperature, pressure, dilution, and residence time, with gaseous environment again perhaps secondary (air excluded). The "ideal" fast pyrolysis/gasification study should permit time resolved measurement of both the residue and gaseous species during the entire course of the reaction, with millisecond time resolution, in a realistic gaseous process environment at pressure and with particle sizes and loadings of practical interest. SERI's own approach to this ideal will involve coupling a high-pressure, free-jet, molecular beam, mass spectrometric sampling system with some form of entrained-flow laboratory reactor (Milne and Soltys 1979).

5.4 REFERENCES

Ainscough, A. N. ; Dollimore, D.; Holt, B.; Kirkham, W.; Martin, D. 1972. "The Thermal Degradation of Microcrystalline Cellulose." J. S. Anderson, ed. Reaction Solids, Proceedings of Seventh International Symposium. p. 543.

Akita, K.; Kase, M. 1967. "Determination of Kinetic Parameters for Pyrolysis of Cellulose and Cellulose Treated with Ammonium Phosphate by DTA and TGA." J. Polymer Science. Vol. 5 (Part A-1): p. 833.

Aldrich, D. C. 1974. "Kinetics of Cellulose Pyrolysis." Ph.D. Thesis. Cambridge, MA: Massachusetts Institute of Technology; February.

Allan, G. G.; Matilla, T. 1971. "High Energy Degradation." Chapter 14 in Lignins. Sarkanen, K. and Ludwig, C. H., ed. Wiley-Interscience.

Antal, M. J.; Edwards, W. E.; Friedman, H. C.; Rogers, F. E. 1979. "A Study of the Steam Gasification of Organic Wastes." Final Report; EPA University Grant No. R 804836010.

Appell, H. R.; Miller, R. P. 1973. "Fuel From Agricultural Wastes." Chapter 8 in Symposium: Processing Agricultural and Municipal Wastes. Inglett, G. E., ed. The AVI Publishing Co.

Appell, H. R.; Pantages, P. 1976. "Catalytic Conversion of Carbohydrates to Synthesis Gas." Thermal Uses and Properties of Carbohydrates and Lignins. Shafizadeh, Sarkanen, and Tillman, ed. Academic Press: p. 127.

Arseneau, D. F. 1961. "The Differential Thermal Analysis of Wood." Canadian J. Chemistry. Vol. 39: p. 1915.

Arseneau, D. F. 1963. "A DTA Study of Fire Retardants in Cellulose." Proceedings 1st Canadian Wood Chemistry Symposium. Toronto: pp. 155-162.

Arseneau, D. F. 1971. "Competitive Reactions in the Thermal Decomposition of Cellulose." Canadian J. Chemistry. Vol. 49: p. 632.

Babu, S. P. 1979. "Thermobalance Experiments." Private communication from IGT.

Bamford, C. H.; Crank, J.; Mahan, D. H. 1946. "The Combustion of Wood. Part I." Proceedings Cambridge Philosophical Society. Vol. 42: p. 166.

Barber-Coleman. 1975. Molten Lead-Bath Pyrolysis. Houston, TX: NASA LBJ Space Center; March; Final Report on Contract NAS 9-14305.

Barclay, H. G. ; Prahacs, S.; Gravel, J. J. O. 1964. "The AST Recovery Process. Pyrolysis of Concentrated NSSC Liquors." Pulp and Paper Magazine of Canada. pp. T553-564; Dec.

Barooah, J. N.; Long, V. D. 1976. "Rates of Thermal Decomposition of Some Carbonaceous Materials in a Fluidized Bed." Fuel. Vol. 55: p. 116.

Basch, A.; Lewin, M. 1973. "The Influence of Fine Structure on the Pyrolysis of Cellulose. I. Vacuum Pyrolysis." J. Applied Polymer Science. Vol. 11: p. 3071.

Beall, F. C.; Eickner, H. W. 1970. Thermal Degradation of Wood Components: A Review of the Literature. Madison, WI: USDA Forest Service; May; FPL 130.

Berkowitz, N. 1957. "On the DTA of Coal." Fuel. Vol. 36: p. 355.

Berkowitz-Mattuck, J. B.; Noguchi, T. 1963. "Pyrolysis of Untreated and APO-THPC Treated Cotton Cellulose During 1-sec Exposure to Radiant Flux Levels of 5-25 cal/cm^2-sec." J. Applied Polymer Science. Vol. 7: p. 709.

Bolton, K.; Cullingworth, J. E.; Ghosh, B. P.; Cobb, J. W. 1942. "The Primary Gaseous Products of Carbonization." J. Chemical Society (London). p. 252.

Breger, I. A.; Whitehead, W. L. 1951. "Thermographic Study of the Role of Lignin in Coal Genesis." Fuel. Vol. 30: p. 247.

Brink, D. L. 1976. "Pyrolysis-Gasification-Combustion: A Process for Utilization of Plant Material." Applied Polymer Symposium. No. 28: p. 1377.

Brink, D. L.; Massoudi, M. S. 1978. " A Flow Reactor Technique for the Study of Wood Pyrolysis. I. Experimental." J. Fire and Flammability. Vol. 9: p. 176.

Brink, D. L.; Massoudi, M. S.; Sawyer, R. F. 1973. "A Flow Reactor Technique for the Study of Wood Pyrolysis." Presented at Fall 1973 meeting of the Western States Section, Combustion Institute.

Broido, A. 1966. "Thermogravimetric and Differential Thermal Analysis of Potassium Bicarbonate Contaminated Cellulose." Pyrodynamics. Vol. 4: p. 243.

Broido, A.; Kilzer, F. J. 1963. "A Critique of the Present State of Knowledge of the Mechanism of Cellulose Pyrolysis." Fire Research Abstracts and Reviews. Vol. 5: p. 157.

Broido, A.; Martin, S. B. 1961. "Effect of Potassium Bicarbonate on the Ignition of Cellulose by Radiation." Fire Research Abstracts and Reviews. Vol. 3: p. 29.

Broido, A.; Weinstein, M. 1970. "Thermogravimetric Analysis of Ammonia-Swelled Cellulose." Combustion Science and Technology. Vol. 1: p. 279.

Broido, A.; Weinstein, M. 1971. "Low Temperature Isothermal Pyrolysis of Cellulose." Thermal Analysis. Proceedings Third ICTA DAVOS. Vol. 3: p. 285.

Brown, D. S. 1979. "Plasma Arc Reduction of Biomass for the Production of Synthetic Fuel Gas." Hawaii ACS meeting abstracts. April.

Browne, F. L. 1958. "Theories of the Combustion of Wood and Its Control." Madison, WI: Forest Products Laboratory. Report No. 2136. 69 pp.

Browne, F. L.; Brenden, J. J. 1964. "Heat of Combustion of the Volatile Pyrolysis Products of Fire-Retardant-Treated Ponderosa Pine." Madison WI: USFS Forest Products Laboratory. Research Paper FPL-19.

Browne, F. L.; Tang, W. K. 1963. "Effect of Various Chemicals on Thermogravimetric Analysis of Ponderosa Pine." Madison, WI: USFS Forest Product Laboratory. Research Paper FPL-6.

Byrne, G. A.; Gardiner, D.; Holmes, F. H. 1966. "The Pyrolysis of Cellulose and Action of Flame Retardants. II. Further Analysis and Identification of Products." J. Applied Chemistry. Vol. 16: p. 81.

Cabradilla, K. E.; Zeronian, S. H. 1976. "Influence of Crystallinity on the Thermal Properties of Cellulose." Thermal Uses and Properties of Carbohydrates and Lignins. Shafizadeh, F.; Sarkanen, K. V.; Tillman, D. A.; ed. Academic Press; p. 73.

Cardwell, R. D.; Luner, P. 1976. "Thermodynamic Analysis of Pulps. Part I: Kinetic Treatment of Isothermal Pyrolysis of Cellulose." Wood Science and Technology. Vol. 10: p. 131.

Chatterjee, P. K. 1968. "Chain Reaction Mechanism of Cellulose Pyrolysis." J. Applied Polymer Science. Vol. 12: p. 1859.

Chatterjee, P. K.; Conrad, C. M. 1966. "Kinetics of the Pyrolysis of Cotten Cellulose." Textile Research J. Vol. 36: p. 487.

Chatterjee, P. K.; Conrad, C. M. 1968. "Thermogravimetric Analysis of Cellulose." J. Polymer Science. Vol. 6: (Part A-1): p. 3217.

Corlateanu, E.; Mihai, E.; Simionescu, Cr. 1974. "Thermo-oxidative Destruction of Certain Fibrous, Cellulosic Materials and Their Components." J. Thermal Analysis. Vol. 6: p. 657.

Davidson, H. W.; Losty, H. H. W. 1965. "The Initial Pyrolysis of Celluloses." Second Conference on Industrial Carbon Graphite. London, 1965. pp. 20-28.

Diebold, J. P. 1979. Fast pyrolysis studies under contract to SERI.

Diebold, J. P; Smith, G. D. 1979. "Noncatalytic Conversion of Biomass to Gasoline." ASME Paper No. 79-Sol-29.

Domansky, R.; Rendos, F. 1962. "On the Pyrolysis of Wood and Its Components." Holz. Roh-Werkstoff. Vol. 20: p. 473.

Domburgs, G.; Sergeeva, V.; Kalminsh, A.; Koshik, M.; Kozmal, F. 1969. "New Aspects and Tasks of DTA in Wood Chemistry." Thermal Analysis Proceedings Second International Conference. Schwenker & Garr, ed. p. 623.

Durvuri, M. S.; Muhlenkamp, S. P.; Iqbal, K. Z.; Welker, J. R. 1975. "The Pyrolysis of Natural Fuels." J. Fire and Flammability. Vol. 6: p. 468.

Fairbridge, C.; Ross, R. A. 1978. "The Thermal Reactivity of Wood Waste Systems." Wood Science & Technology. Vol. 12: p.169.

Fairbridge, C.; Ross, R. A.; Sood, S. P. 1978. "A Kinetic and Surface Study of the Thermal Decomposition of Cellulose Powder in Inert and Oxidizing Atmospheres." J. Applied Polymer Science. Vol. 22: p. 497.

Fang, P.; McGinnis, G. D. 1975. "The Polyphenols from Loblolly Pine Bark." Applied Polymer Symposium No. 28. p. 363.

Fang, P.; McGinnis, G. D. 1976. "Flash Pyrolysis of Holocellulose from Loblolly Pine Bark." Thermal Uses and Properties of Carbohydrates and Lignins Symposium. Shafizadeh, Sarkanen and Tillman, ed. p. 37-47.

Fang, P.; McGinnis, G. D.; Parish, E. J. 1975. "Thermogravimetric Analysis of Loblolly Pine Bark Components." Wood & Fibre. Vol. 7: p. 136.

Fletcher, T. L.; Harris, E. E. 1947. "Destructive Distillation of Douglas Fir Lignin." JACS. Vol. 69: p. 3144.

Fletcher, T. L.; Harris, E. E. 1952. "Products From the Destructive Distillation of Douglas-Fir Lignin." TAPPI. Vol. 35: p. 536.

Glassner, S.; Pierce, A. R., III. 1965. "Gas Chromatographic Analysis of Products From Controlled Application of Heat to Paper and Levoglucosan." Analytical Chemistry. Vol. 37: p. 525.

Goheen, D. W.; Orle, J. V.; Wither, R. P. 1976. "Indirect Pyrolysis of Kraft Black Liquors." Thermal Uses and Properites of Carbohydrates and Lignins Symposium. Shafizadeh, Sarkanen, and Tillman, ed. Academic Press.

Goos, A. W. 1952. "The Thermal Decomposition of Wood." Chapter 20 in Wood Chemistry. 2nd Ed. Rheinhold.

Greenwood, C. T.; Knox, J. H.; Milne, E. 1961. "Analysis of the Thermal Decomposition Products of Carbohydrates by Gas-Chromatography." Chemistry and Industry (London). p. 1878.

Halpern, Y.; Patai, S. 1969. "Pyrolytic Reactions of Carbohydrates. Part V. Isothermal Decomposition of Cellulose in Vacuo." Israel J. Chemistry. Vol. 7: p. 673.

Havens, J. A.; Welker, J. R. ; Sliepcevich, C. M. 1971. "Pyrolysis of Wood: A Thermo-analytical Study." J. Fire and Flammability. Vol. 2: p. 321.

Heinrich H. J.; Kaesche-Krisher, B. 1962. "Contribution to the Explanation of the Spontaneous Combustion of Wood." Brennstoff Chemie. Vol. 43 (No. 5): p.142.

Herbert P. L.; Tryon, M.; Wilson, W. K. 1969. "DTA of Some Papers and Carbohydrate Materials." TAPPI. Vol. 52: p. 1183.

Hileman, F. D.; Wojcik, L. H.; Futrell, J. H.; Einhorn, I. N. 1976. "Comparison of the Thermal Degradation Products of α-Cellulose and Douglas Fir Under Inert and Oxidative Environments." Thermal Uses and Properties of Carbohydrates and Lignins Symposium. Shafizadeh, Sarkanen, and Tillman, ed. Academic Press: p. 49-71.

Kanury, A. M. 1972. "Thermal Decomposition Kinetics of Wood Pyrolysis." Combustion and Flame. Vol. 18: p. 75.

Kato, K.; Takahashi, N. 1967. "Pyrolysis of Cellulose. Part II. Thermogravimetric Analysis and Determination of Carbonyl and Carboxyl Groups in Pyrocellulose." Agricultural Biology and Chemistry. Vol 31: p. 519.

Kilzer, F. J.; Broido, A. 1965. "Speculations on the Nature of Cellulose Pyrolysis." Pyrodynamics. Vol. 2: p. 151.

Knight, J. A. 1976. "Pyrolysis of Pine Sawdust." Thermal Uses and Properties of Carbohydrates and Lignins. Shafizadeh, Sarkanen, and Tillman, ed. Academic Press; p. 158.

Knudson, R. M.; Williamson, R. B. 1971. "Influence of Temperature and Time upon Pyrolysis of Untreated and Fire Retardant Treated Wood." Wood Science and Technology. Vol. 5: p. 176.

Kosik, M.; Luzakova, V.; Reiser, V. 1972. "Study on the Thermal Destruction of Cellulose and Its Derivatives." Cellulose Chemistry and Technology. Vol. 6: p. 589.

Krieger, B. B.; Graef, M.; Allan, G. G. 1979. "Rapid Pyrolysis of Biomass/Lignin for Production of Acetylene." Hawaii ACS Meeting abstracts. April.

Kuester, J. L. 1978. "Urban Wastes as an Energy Source." Energy Systems: An Analysis for Engineers and Policy Makers. Marcel Dekker; Jan.

Kung, H. C.; Kalelkar, A. S. 1973. "On the Heat of Reaction in Wood Pyrolysis." Combustion and Flame. Vol. 20: p. 91.

Lester, T. 1979. Kansas State University. Private communication.

Leu, J. C. 1975. "Modeling of the Pyrolysis and Ignition of Wood." Dissertation Abstracts. Vol. 36 (Sect. B): p. 350.

Lewellen, P. C.; Peters, W. A.; Howard, J. B. 1976. "Cellulose Pyrolysis Kinetics and Char Formation Mechanism." Sixteenth Symposium (International) on Combustion. The Combustion Institute. p. 1471.

Lincoln, K. A. 1965. "Flash Vaporization of Solid Materials for Mass Spectrometry by Intense Thermal Radiation." Analytical Chemistry. Vol. 37: p. 541.

Lincoln, K. A. 1974. "A New Mass Spectrometer System for Investigating Laser-Induced Vaporization Phenomena." International J. Mass Spectrometry and Ion Physics. Vol. 13: p. 45.

Lincoln, K. A.; Covington, M. A. 1975. "Dynamic Sampling of Laser-Induced Vapor Plumes by Mass Spectrometry." International J. Mass Spectrometry and Ion Physics. Vol. 16: p. 191.

Lipska, A.; Parker, W. J. 1966. "Kinetics of the Pyrolysis of Cellulose in the Temperature Range 250-300 C." J. Applied Polymer Science. Vol. 10: p. 1439.

Lipska, A. E.; Wodley, F. A. 1969. "Isothermal Pyrolysis of Cellulose: Kinetics and GC-MS Analysis of the Degradation Products." J. Applied Polymer Science. Vol. 13: p. 851.

Liu, K. T.; Stambaugh, E. P.; Nack, H.; Oxley, H. 1977. "Pyrolytic Gasification of Kraft Black Liquors." Fuels from Wastes. Tillman, D. A. ed. p. 161.

Maa, P. S.; Bailie, R. C. 1978. "Experimental Pyrolysis of Cellulosic Material." Presented at 1978 AIChE 84th National Meeting. Atlanta, GA.

Mack, C. H.; Donaldson, D. J. 1967. "Effects of Bases on the Pyrolysis of Cotton Cellulose." Textile Research J. Vol. 37: p. 1063.

Madorsky, S. L.; Hart, V. E., Straus, S. 1956. "Pyrolysis of Cellulose in a Vacuum." J. Research NBS. Vol. 56 (No. 6): p. 343.

Madorsky, S. L.; Hart, V. E.; Straus, S. 1958. "Thermal Degradation of Cellulosic Materials." J. Research NBS. Vol. 60 (No. 4): p. 343.

Mallon, G. 1974. U.S. Patent 3,846,096. "Gasification of Carbonaceous Solid." Nov. 5.

Martin, S. 1965. "Diffusion-Controlled Ignition of Cellulosic Materials by Intense Radiant Energy." Tenth Symposium (International) on Combustion. The Combustion Institute. p. 877.

McCarter, R. J. 1972. "The Pyrolysis of Cellulose at Rates Approaching Those in Burning." Textiles Research J. Vol. 42: p. 709.

McKay, G. D. M. 1967. "Mechanism of Thermal Degradation of Cellulose - A Review of the Literature." Canada Dept. of Forestry and Rural Development, Forestry Branch Dept. Publ. No. 1201.

McKay, G. D. M. 1968. "Effect of Inorganic Salts on the Pyrolysis of Cellulose." Forest Products J. Vol. 18: p. 71.

Milne, T. A.; Soltys, M. 1979. Solar Energy Research Institute. In-house research task 3322.11.

Min, K. 1977. "Vapor-Phase Thermal Analysis of Pyrolysis Products From Cellulosic Materials. Combustion and Flame. Vol. 30: p. 285.

Molton, P. M.; Demmitt, T. F. 1977. "Reaction Mechanisms in Cellulose Pyrolysis: A Literature Review." Battelle Northwest Laboratories; Aug.; Report BNWL-2297.

Muhlenkamp, S. P.; Welker J. R. 1977. "The Pyrolysis Energy of Natural Fuels." J. Fire & Flammability. Vol. 8: p. 225.

Murphy, E. J. 1962. "Thermal Decomposition of Natural Cellulose in Vacuo." J. Polymer Science. Vol 58: p 649.

Murty, K. A.; Blackshear, P. L. 1966. "An X-Ray Photographic Study of the Reaction Kinetics of α-Cellulose Decomposition." Pyrodynamics. Vol. 4: p. 285.

Nunomura, A. ; Hidetake, I.; Akira, K.; Katsumi, K. 1975. "Thermal Degradation of Cellulose in Vacuo." Chemical Abstracts. Vol. 84: 61515.

Parks, E. J. 1971. "Thermal Analysis of Modified Cellulose." TAPPI. Vol. 54: p. 537.

Patel, K. S.; Patel, K. C.; Patel, R. D. 1970. "Study on the Pyrolysis of Cellulose and Its Derivatives." Die Makromolekular Chemie. Vol. 132: p. 23.

Prahacs, S. 1967. "Pyrolytic Gasification of Na-, Ca- and Mg- Base Spent Pulping Liquors in an AST Reactor." Advances in Chemistry Series. Vol. 69: p. 230.

Prahacs, S.; Barclay, H. G.; Bhaba, S. P. 1971. "A Study of the Possibilities of Producing Synthetic Tonnage Chemicals From Lignocellulosic Residues." Pulp and Paper Magazine of Canada. Vol. 72: p. 69.

Prahacs, S.; Gravel, J. J. O. 1967. "Gasification of Organic Content of Na - Base Spent Pulping Liquors in an Atomized Suspension Technique Reactor." Industrial and Engineering Chemistry, Process Design and Development. Vol. 6: p. 180.

Rai, C.; Tran, D. Q. 1975. "Recovery of Medium - to - High Btu Gas From Bark and Black Liquor Concentrates." AIChE Symp. Series No. 157. p. 100.

Ramiah, M. V. 1970. "Thermogravimetric and Differential Thermal Analysis of Cellulose, Hemicellulose and Lignin." J. Applied Polymer Science. Vol. 14: p. 1323.

Ramiah, M. V.; Goring, D. A. I. 1967. "Some Dilatometric Measurements of the Thermal Decomposition of Cellulose, Hemicellulose and Lignin." Cellulose Chemistry and Technology. Vol. 1: p. 277.

Rensfelt, E.; Blomkvist, G.; Ekstrom, C.; Engstrom, S.; Espenas, B-G.; Liinanki, L. 1978. "Basic Gasification Studies for Development of Biomass Medium - Btu Gasification Processes." IGT Biomass Symposium.

Robb, E. W.; Johnson, W. R.; Westbrook, J. J.; Seligman, R. B. 1966. "Model Pyrolysis - The Study of Cellulose." Beitrage zur Tabakforschung. Vol. 3: p. 597.

Roberts, A. F. 1970. "A Review of Kinetics Data for the Pyrolysis of Wood and Related Substances." Combustion and Flame. Vol. 14: p. 261.

Roberts, A. F. 1971a. "The Heat of Reaction During the Pyrolysis of Wood." Combustion and Flame. Vol. 17: p. 79.

Roberts, A. F. 1971b. "Problems Associated With the Theoretical Analysis of the Burning of Wood." 13th Symposium (International) on Combustion, The Combustion Institute.

Roberts, A. F.; Clough, G. 1963. "Thermal Decomposition of Wood in an Inert Atmosphere." Ninth Symposium (International) on Combustion. The Combustion Institute. p. 158.

Sandermann, W.; Augustin, H. 1963. "Chemical Investigations on the Thermal Decomposition of Wood - Part II. Investigations by Means of the DTA." Holz. Roh-Werkstoff. Vol. 21: p. 305.

Sanner, W. S; Ortuglio, C.; Walters, J. G; Wolfson, D. E. 1970. U. S. Bureau of Mines; August. RI7428.

Schlesinger, M. D.; Sanner, W. S.; Wolfsen, D. E. 1973. "Energy from the Pyrolysis of Agricultural Wastes." Chapter 9 in Symposium: Processing Agricultural and Municipal Wastes. Inglett, G. E., ed. The AVI Publishing Co.

Schwenker, R. F.; Beck, L. R. 1963. "Study of the Pyrolytic Decomposition of Cellulose by Gas Chromatography." J. Polymer Science. Part C (No. 2): p. 331.

Schwenker, R. F.; Pacsu, E. 1957. "Pyrolytic Degradation Products of Cellulose." Chemical Engineering Data Series. Vol. 2: p. 83.

Shafizadeh, F. 1968. "Pyrolysis and Combustion of Cellulosic Materials." Advances in Carbohydrate Chemistry. Vol. 23: p. 419.

Shafizadeh, F. 1975. "Industrial Pyrolysis of Cellulosic Materials." Applied Polymer Symposium No. 28. p. 153-174.

Shafizadeh, F.; Bradbury, A. G. W. 1979. "Thermal Degradation of Cellulose in Air and Nitrogen at Low Temperatures." J. Applied Polymer Science. Vol. 23: p. 1431.

Shafizadeh, F.; McGinnis, G. D. 1971. "Chemical Composition and Thermal Analysis of Cottonwood." Carbohydrate Research. Vol. 16: p. 273.

Shafizadeh, F.; McGinnis, G. D.; Philpot, C. W. 1972. "Thermal Degradation of Xylan and Related Model Compounds." Carbohydrate Research. Vol. 25: p. 23.

Shivadev, V. K.; Emmons, H. W. 1974. Combustion and Flame. Vol. 22: p. 223.

Smith, R. C.; Howard, H. C. 1937. "Aromatization of Cellulose by Heat." JACS. Vol. 59: p. 234.

Soltes, E. J.; Elder, T. J. 1979. "Pyrolysis." Preprint of Chapter in forthcoming book Organic Chemicals from Biomass. CRC Press.

Stamm, A. J. 1956. "Thermal Degradation of Wood and Cellulose." Industrial Engineering Chemistry. Vol. 48: p. 413.

Stern, E. W.; Logindice. A. S.; Heinemann, H. 1965. "Approach to Direct Gasification of Cellulosics." Industrial Engineering Chemistry Process Design and Development. Vol. 4: p. 171.

Tang, W. K-Y. 1964. "The Effect of Inorganic Salts on Pyrolosis, Ignition and Combustion of Wood, α-Cellulose and Lignin." Ph.D. Thesis. University of Wisconsin.

Tang, W. K.; Eickner, H. W. 1968. "Effect of Inorganic Salts on Pyrolysis of Wood, Cellulose, and Lignin Determined by Differential Thermal Analysis." Madison, WI: USFS Forest Products Laboratory. Research Paper FPC-82.

Tang, W. K.; Neill, W. K. 1964. "Effect of Flame Retardants on Pyrolysis and Combustion of α-Cellulose." J. Polymer Science. Part C (No. 6): pp 65-81.

Tran, D. Q. 1978. "Kinetic Modeling of Pyrolysis and Hydrogasification of Carbonaceous Materials." Ph.D. Thesis. University of Wyoming.

Tran, D. Q.; Rai, C. 1978. " A Kinetic Model for Pyrolysis of Douglas Fir Bark." Fuel. Vol. 57: p. 293.

Tsuckiyn, Y.; Sumi, K. 1970. "Thermal Decomposition Products of Cellulose." J. Applied Polymer Science. Vol. 14: p. 2003.

Van Krevelen, D. W.; Van Heerden, C.; Huntjens, E. J. 1951. "Physicochemical Aspects of the Pyrolysis of Coal and Related Organic Compounds." Fuel. Vol. 30: p. 253.

Venn, H. J. P. 1924. "The Yield of B-glusosane Obtained from Low-Pressure Distillation Cellulose." J. Textile Institute. Vol. 15: p. 414.

Vroom, A. H. 1952. "Bark Pyrolysis by Fluidization Techniques - a New Approach to Bark Utilization." Pulp and Paper Magazine of Canada. Vol. 53: p. 121.

Welker, J. R. 1970. "The Pyrolysis and Ignition of Cellulosic Materials: A Literature Review." J. Fire and Flammability. Vol. 1: p. 12.

Wendlandt, W. 1974. "Thermal Methods of Analysis." 2nd Edition.

Chapter 6

Thermodynamics of Gas-Char Reactions

R. Desrosiers
SERI

6.1 INTRODUCTION

The products of the pyrolytic reactions described in Chapter 5 do not conform to chemical equilibrium because gas phase reactions are very slow below 500 C. However, at temperatures above about 500 C, chemical equilibrium is approached fast enough so that thermodynamic calculations can predict important trends and in some cases the gas compositions to be expected.

The temperatures, residence times, and gas-solid contacting methods employed in gasification equipment strongly affect the degree of attainment of equilibrium. In downdraft fixed bed gasifiers, products of pyrolysis and combustion are drawn over a bed of charcoal at temperatures between 700 and 1000 C and approach equilibrium closely. In updraft fixed bed gasifiers, initial combustion gases also filter through a hot char bed, but then they are mixed with the products of low temperature pyrolysis and the exit gas analysis bears little relation to equilibrium. The uniformly high temperatures in a fluidized bed offer favorable conditions for equilibrium, but the degree attained depends on gas residence time.

Reliable predictions of product compositions for any gasifier may be obtained only from a detailed kinetic model for that reactor incorporating global reaction rates, which may be strong functions of gas velocities, particle sizes, etc. These kinetic restrictions limit the quantitative significance of an equilibrium calculation; nevertheless, such an exercise is of considerable predictive value in estimating the effects of changes in the major thermodynamic variables: temperature, pressure, and composition.

In this chapter, the results of equilibrium calculations are presented which illustrate the predicted effects of temperature, pressure, feed moisture content, and oxidant/fuel ratio on gasifier performance; the results have been plotted and are discussed in detail below. The purpose of the discussion is not to present an exhaustive parametric study but to extract as much information as possible from a set of salient examples.

6.2 MAJOR PROCESSES AND REACTIONS

The processes occurring in any gasifier are oxidation, reduction, pyrolysis, and drying. The unique feature of the updraft gasifier is the sequential occurrence of these processes: they are separated spatially and therefore temporally. For this reason, the operation of an updraft gasifier will be used in the following discussion. The reaction zones and a schematic temperature profile for an updraft gasifier are illustrated in Fig. 6-1. Several reactions of importance in char gasification are listed in Table 6-1.

In the lowest zone, oxidation of char with oxygen occurs; the heat released here drives subsequent processes. In this zone, the oxygen pressure is high enough to favor CO_2 formation. This reaction (e) is very fast, probably being mass-transfer limited, and the thickness of this zone may vary in magnitude from one to tens of centimetres.

The gas stream issuing from Zone One is hot and rich in CO_2 (and H_2O if the blast contains steam). The high temperatures favor, kinetically and thermodynamically, the

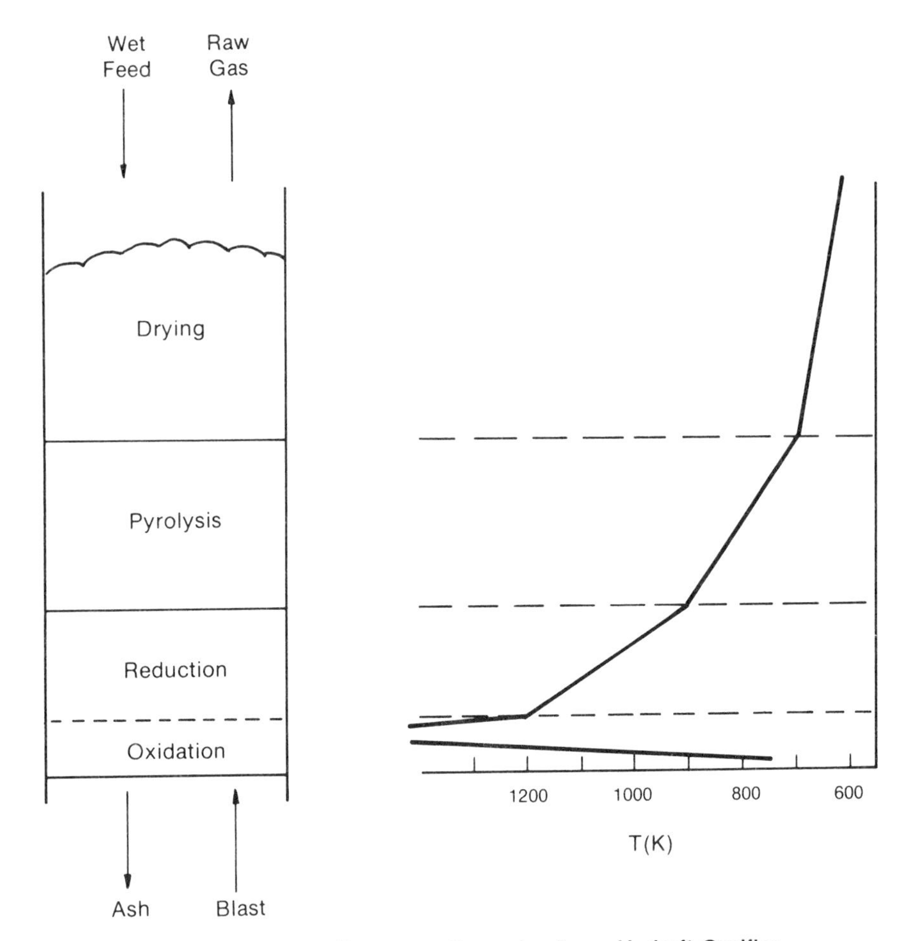

Figure 6-1. Major Processes Occurring in an Updraft Gasifier

Table 6-1. IMPORTANT REACTIONS IN GASIFICATION

	Reaction	H (kJ/mole) 298 K	1000 K
(a)	$CO + H_2O = CO_2 + H_2$	-41.2	-34.77
(b)	$C + 2H_2 = CH_4$	-74.93	-89.95
(c)	$C + H_2O = CO + H_2$	131.4	136.0
(d)	$C + CO_2 = 2CO$	172.6	170.7
(e)	$C + O_2 = CO_2$	-393.8	-394.9

Boudouard (d) and water-gas (c) reactions which are highly endothermic. These control the upper limit of temperature in the reduction zone. (A dramatic example of this effect has been observed in a downdraft gasifier. If pure oxygen is used in the blast, the temperature does not rise more than a hundred degrees or so above that for the air blast except in the immediate vicinity of the tuyeres.)

This temperature stabilization phenomenon may be explained with reference to Fig. 6-2 in which the log of the mass action expression (Q)* for the reactions (1) through (4) is plotted against the reciprocal of temperature.

In Fig. 6-2 (a), consider a point near the intersection of curves (b), (c), (d). If the system attempts to attain a temperature higher than this, then the endothermic reduction reactions (c), (d) are thermodynamically favored and begin to moderate the temperature. These reactions are too slow to be controlling, however, until the temperature nears 1400 K. Above this temperature, they constitute an effective energy sink and limit further rise in temperature. Conversely, the exothermic methanation reaction could provide a temperature floor for the reduction zone of a fixed bed gasifier. However, this reaction is probably too slow to be important in controlling bed temperature. Figure 6-2 (b) illustrates the effect of increased pressure, which is to raise the level of the "stable" temperature interval.

As the gases rise beyond the reduction zone, they come into contact with cooler, solid feed. The temperature falls below 900 K and the reduction and shift reactions are frozen. The gas composition at this point may be reasonably close to the equilibrium composition for some temperature within the reduction zone.

The partially dried feed above the char bed is pyrolyzed by the rising, hot gas stream; the immediate products are low molecular weight hydrocarbons, alcohols, acids, oils, and tars, as well as CO, H_2, CO_2, H_2O, and CH_4. The hydrocarbons undergo cracking and re-forming to H_2, CO, and CO_2. The temperature near the top of the bed is too low for this re-forming to be completed, and the raw gas stream exiting the reactor is laden with products which are not characteristic of the equilibrium established in the reduction zone or of the primary pyrolysis products.

The downdraft gasifier is operated so that the final gas-solid contact is one involving hot char rather than volatile-laden feed, and a near-equilibrium product distribution is achieved. The degree to which other types of gasifiers approach equilibrium is related to residence time. Thus, a fluidized bed with recycle can approach equilibrium very closely while equilibrium concepts may have no relation to a fast pyrolysis process with millisecond residence times.

6.3 THE EQUILIBRIUM CALCULATION

The equilibrium calculations were made using a computer program called "GASEQ". The algorithm is based on that developed by D. R. Cruise (1964) at the Naval Ordnance Test Station at China Lake. A large thermodynamic data file compiled from the JANAF (Stull and Prophet 1971) tables is required. The program will "burn" any feed for which a composition is specified. All gaseous products are assumed to behave ideally, and all condensed products are treated as pure phases. The user can specify the temperature or allow an iterative calculation of the adiabatic flame temperature; in the latter case, a heat of formation for the feed must be supplied. Usage of the program is described in detail in Desrosiers (1977).

The calculations were based on the typical analysis for dry, sulfur- and ash-free wood shown in Table 6-2.

---

* For reaction $\Sigma \nu_i \chi_i = 0$ among ideal gases χ_i, we have

$$K = P^{\nu} \Pi y_i^{\nu_i} = P^{\nu} Q$$

$$\text{or } \log Q = \log K - \nu \log P$$

where

$K = K(T)$ equilibrium constant

$Q = Q(P,T)$ mass action quotient

ν_i = stoichiometric coefficient

$\nu = \sum_{\text{gases}} \nu$

y_i = mole fraction

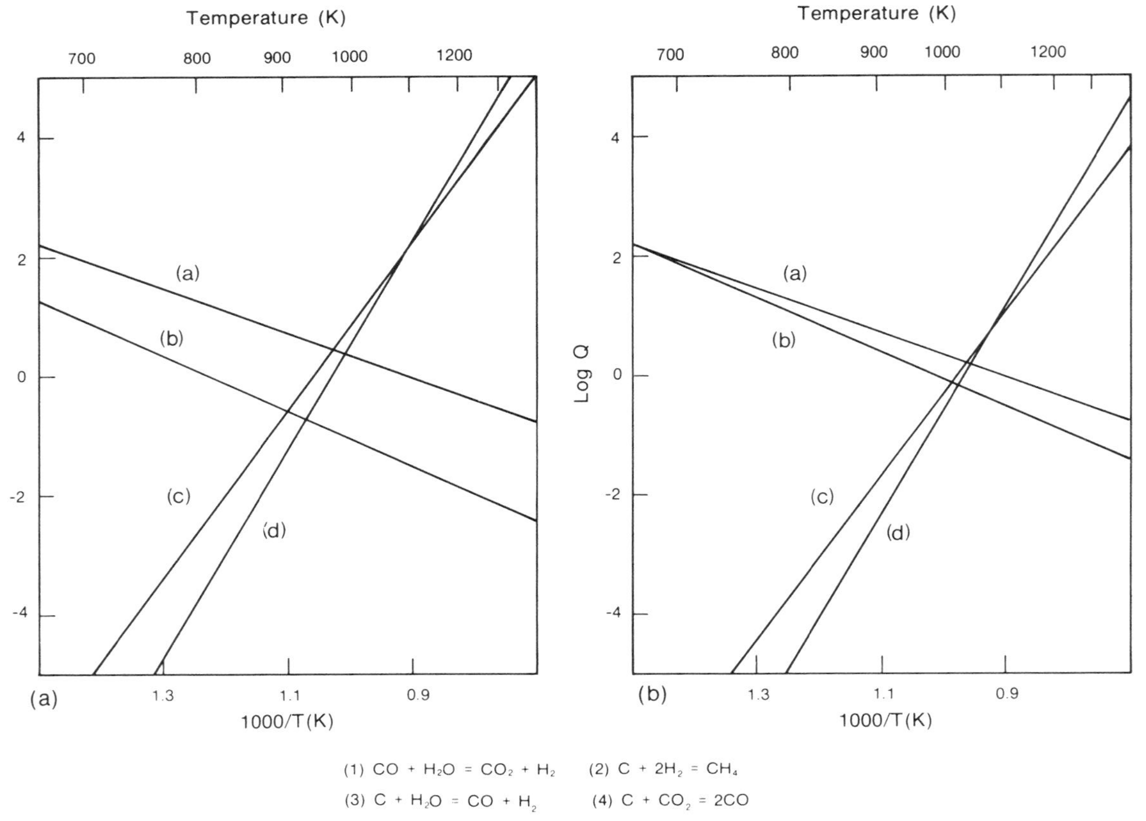

Figure 6-2. Mass Action Quotient Vs. Reciprocal Temperature for Gasification Reactions at (a) 1 and (b) 10 atm

Table 6-2. TYPICAL ANALYSIS FOR DRY SULFUR- AND ASH-FREE WOOD

Composition	C	52.50wt %		
	H	6.16		
	O	41.24		
	N	0.10		
High Heating Value (HHV)[a]		-22.21	kJ/g	(-9550 Btu/lb)
Low Heating Value (LHV)		-20.9	kJ/g	(-8987 Btu/lb)
Heat of Formation		- 3.74	kJ/g	(-1609 Btu/lb)
Formula				
C_6 basis	$C_6H_{8.39}O_{3.54}N_{0.1}$	(FW = 137.27)		
C_1 basis	$C H_{1.4} O_{0.59} N_{0.017}$	(FW = 22.86)		
Stoichiometric Oxidant Ratio				
Molar O_2/dry wood C_1 basis		1.055		
Weight O_2/dry wood		1.476		
Weight air/dry wood		6.364		

[a]HHV (LHV) = Heat of Combustion with product water in liquid (vapor) form.

LHV (kJ/g) = HHV (kJ/g) - 0.2122 X wt % H

In each calculation the following species were considered as possible products:

H, H_2
N, N_2, NH_3, NO, NO_2, CN, HCN
O, O_2, H_2O, OH
C, CO, CO_2, CH_4, C_2H_2, C_2H_4, CH_2O
C_s

No attempt was made to model "char" or "tar". Carbon (as graphite) was the only solid product considered. For all conditions investigated, the only products present in significant amounts ($>10^{-4}$ mole %) were C_s, H_2, H_2O, CO, CO_2, CH_4, and N_2. It is important to note that no hydrocarbon other than CH_4 is thermodynamically stable under gasification conditions. Acetylene, ethylene, and higher hydrocarbons (as well as oils and tars) are produced by most gasifiers: these are nonequilibrium products. A gasifier can be designed to inhibit or promote the production of these materials, and this behavior may be correlated roughly with residence time and temperature in the pyrolysis and reduction zones.

6.4 RESULTS

The calculations are organized into five related sets in which the effects of one or two variables are investigated. The results are presented visually in the form of plots; each plot is described separately. The entire set of plots for a series immediately follows the discussion in the text.

6.4.1 Series 1 — Pyrolysis, Gasification, and Combustion Partitioned by the Equivalence Ratio

A concept widely used in the study of hydrocarbon fuel combustion is the equivalence ratio (ER), which is defined as the oxidant to fuel weight ratio divided by the stoichiometric ratio. Thus ER must be greater than or equal to 1.0 for complete combustion of the fuel to carbon dioxide and water. The equivalence ratio is used here to describe wood pyrolysis, gasification, and combustion. Complete combustion of wood (as defined in Table 6-2) with oxygen requires 1.476 g O_2/g wood or 6.364 g air/g wood.

The first series of plots introduces the kind of information which is readily calculable from the equilibrium composition. The results of 20 calculations for an adiabatic system of dry wood with varying quantities of air are illustrated in Figs. 6-3 (a) through 6-3 (e). Along the abscissa in each case is plotted the equivalence ratio:

$$ER = \frac{\text{weight oxidant/weight dry wood}}{\text{stoichiometric oxidant/wood ratio}}$$

The curve in Fig. 6-3 (a) is the adiabatic flame temperature (AFT) as a function of equivalence ratio. The intersection with the axis at ER = 0 occurs at 913 K (640 C). This point corresponds to pyrolysis, the reaction of wood in the absence of oxygen. Notice that as air is added to an ER of 0.255 (1.62 g air/g wood) the AFT rises very slowly from 913 K to 1025 K. As air is added beyond this point, however, the AFT rises dramatically to combustion temperatures (2300 K). This break in the curve corresponds to the point at which carbon disappears. Carbon formation in g/g dry wood is plotted in Fig. 6-3 (b). (Note the expanded scale for ER). Since carbon is the only condensed product formed, the gas production in g/g dry wood is obtained easily from

$$\text{Gas production (g/g dry wood)} = 1 + \text{ER X (stoich. oxidant ratio)} - C_s \text{ formation}$$

The dry gas composition appears in Fig. 6-3 (c). The curves for each component display an extremum or an inflection point at ER = 0.255.

The reaction corresponding to the calculated product distribution at ER = 0 (pyrolysis) is:

$$CH_{1.4}O_{0.59} = 0.64C_s + 0.44H_2 + 0.15H_2O + 0.17CO + 0.13CO_2 + 0.005\ CH_4$$

For a point just beyond the carbon stability region at ER = 0.275 (gasification), the stoichiometry is:

$$CH_{1.4}O_{0.59} + 0.29O_2 + 1.1N_2 = 0.63H_2 + 0.07H_2O + 0.90CO + 0.10CO_2 + 1.1N_2$$

Further addition of air results in consumption of H_2 and CO until combustion conditions are approached at ER = 1.0:

$$CH_{1.4}O_{0.59} + 1.05O_2 + 3.99N_2 = 0.15H_2 + 0.67H_2O + 0.11CO + 0.89CO_2 + 3.99N_2$$

The low heating value (LHV) of the dry gas is plotted in Fig. 6-3 (d). The initial rapid decrease in LHV correlates with the disappearance of CH_4. Beyond ER = 0.255, the LHV approaches zero as CO and H_2 are consumed. [To convert from Btu/SCF (60 F, 1 atm) to MJ/Nm^3 (0 C, 1 atm), divide by 25.39].

The three curves in Fig. 6-3 (e) illustrate the variation of chemical, sensible, and total energy in the gas. The chemical energy stored in the gas is maximal at ER = 0.255, corresponding to complete carbon uptake. This is the point at which one should operate an air-blown gasifier. (Note that in most of the figures, calculated points are simply connected by straight lines. The gas composition data were interpolated in some cases to yield smooth curves).

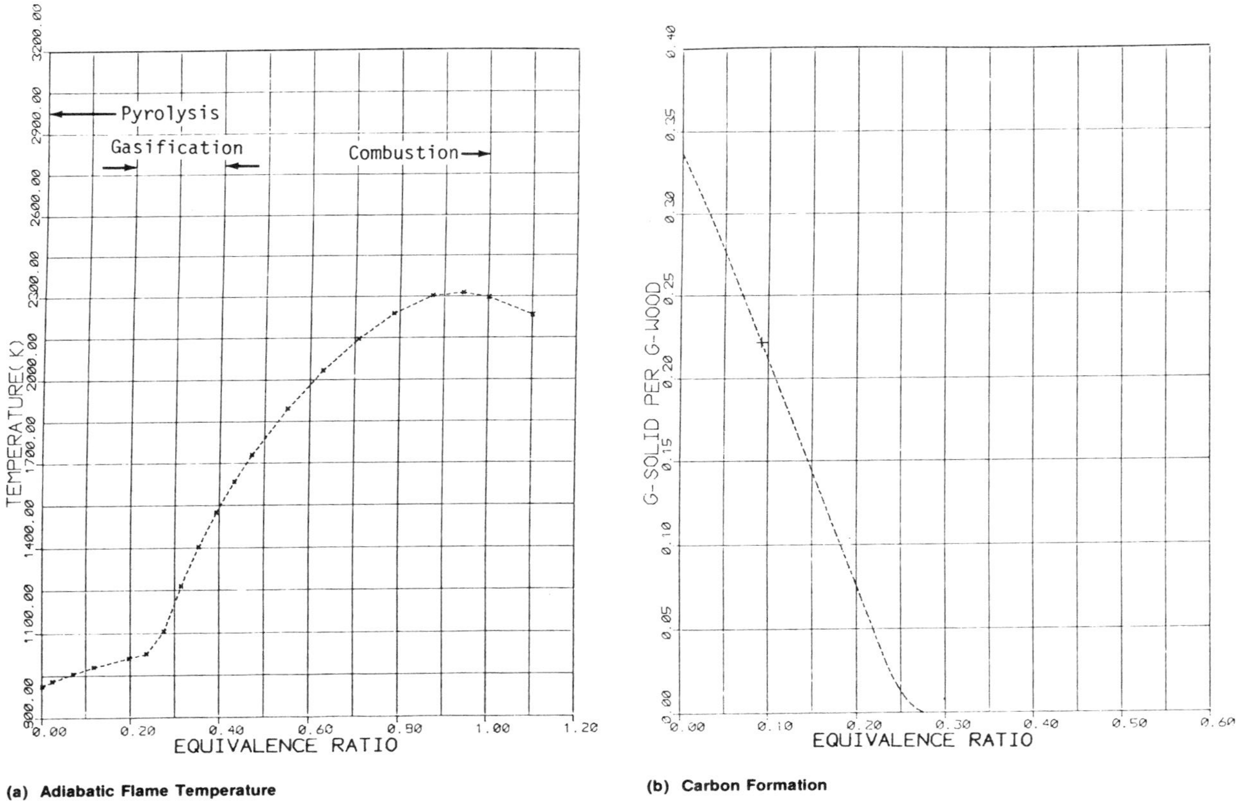

(a) Adiabatic Flame Temperature

(b) Carbon Formation

Figure 6-3. Adiabatic Air Gasification of Dry Wood at 1 atm

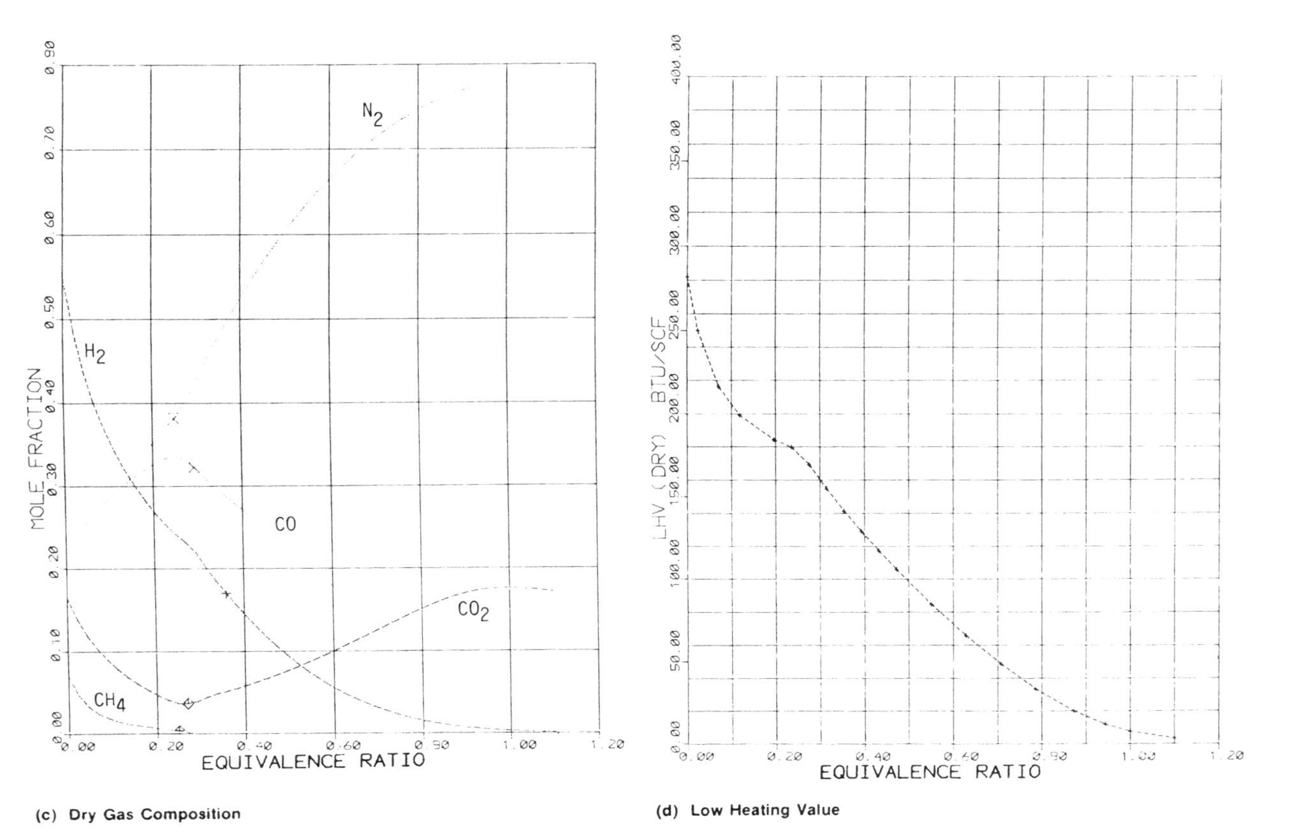

(c) Dry Gas Composition

(d) Low Heating Value

Figure 6-3. Adiabatic Air Gasification of Dry Wood at 1 atm

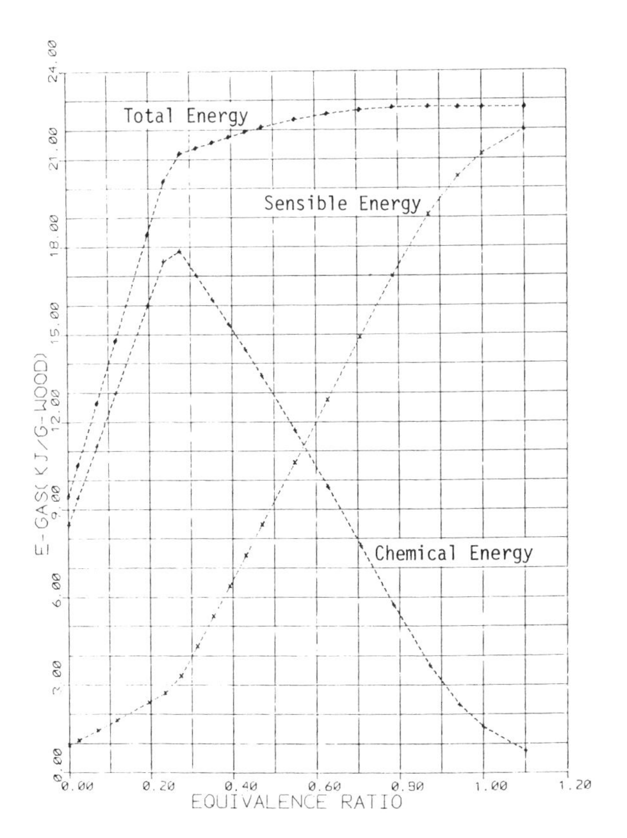

(e) Energy in Product Gas

Figure 6-3. Adiabatic Air Gasification of Dry Wood at 1 atm

The AFTs for a variety of systems are plotted in Fig. 6-4. (The lines appear more broken because fewer points were plotted.) Curve f is the AFT for air. The more dramatic temperature rise with oxygen (a) is readily apparent for ER = 0.26 and higher. The very small AFT difference (< 150 K) for equivalence ratios below 0.26 is a consequence of the temperature stabilizing reduction of CO_2 and H_2O in the presence of hot carbon. This effect is extended to higher ER values by the addition of excess char, as illustrated by curve e (Fig. 6-4).

Curves a, b, c (Fig. 6-4) correspond to oxygen gasification of dry wood at 0, 100, 300 psig (1, 6.8, 20.4 atm). In the combustion region (high ER) significant temperature differences are observed. In the region of interest, gasification (ER = 0.2 to 0.3), however, negligible changes in AFT are induced by a twentyfold change in pressure. In fact, the dry gas compositions for oxygen gasification at 1 atm and at 300 psig [Figs. 6-8 (b,d)] can almost be superimposed for ER < 0.50. Pressure has a negligible effect on the gasification of dry wood. Curves d and g (Fig. 6-4) are AFT versus ER profiles for the adiabatic gasification of wet wood. The lowering of both pyrolysis and combustion temperatures is significant as expected.

The carbon formation curves d, f (Fig. 6-5) for these two cases illustrate the extreme effect of water addition on carbon consumption.

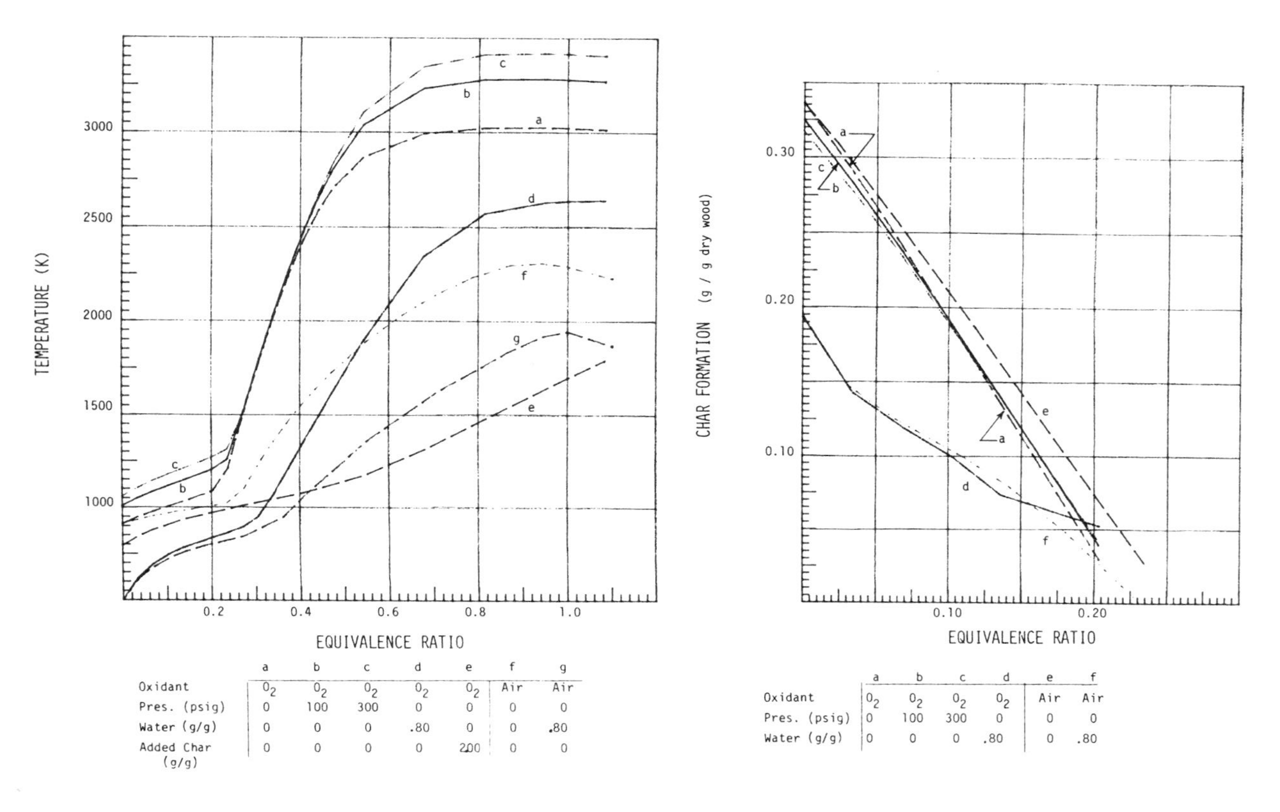

Figure 6-4. Adiabatic Flame Temperature Vs. Equivalence Ratio

Figure 6-5. Char Formation Vs. Equivalence Ratio

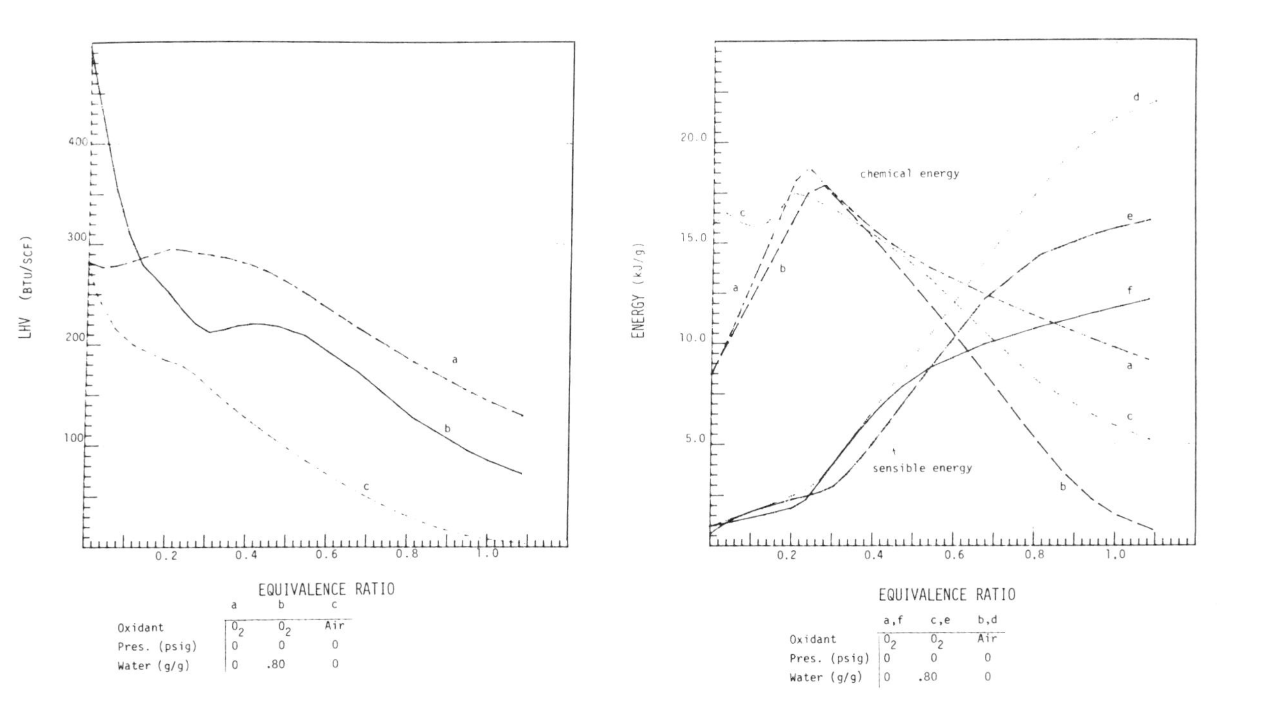

Figure 6-6. Low Heating Value Vs. Equivalence Ratio

Figure 6-7. Energy Distribution Vs. Equivalence Ratio

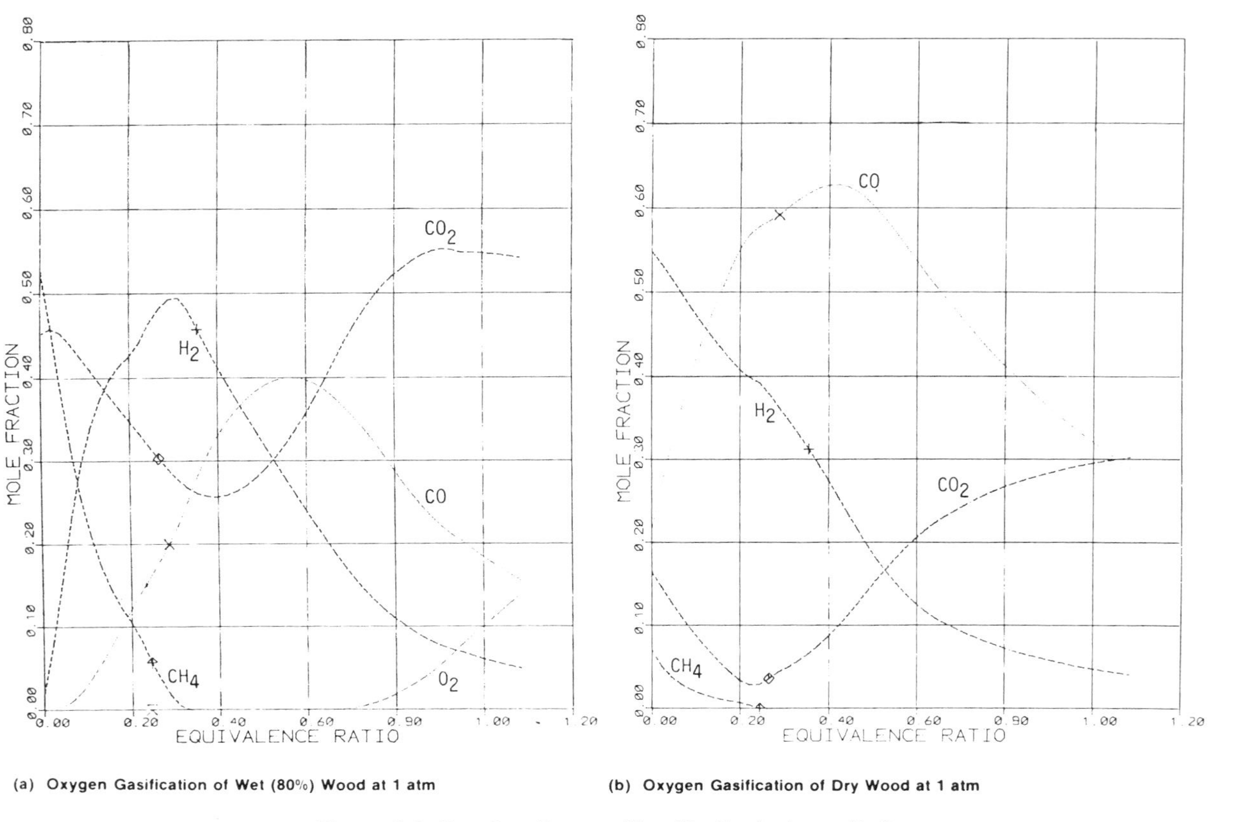

(a) Oxygen Gasification of Wet (80%) Wood at 1 atm

(b) Oxygen Gasification of Dry Wood at 1 atm

Figure 6-8. Dry Gas Composition Vs. Equivalence Ratio

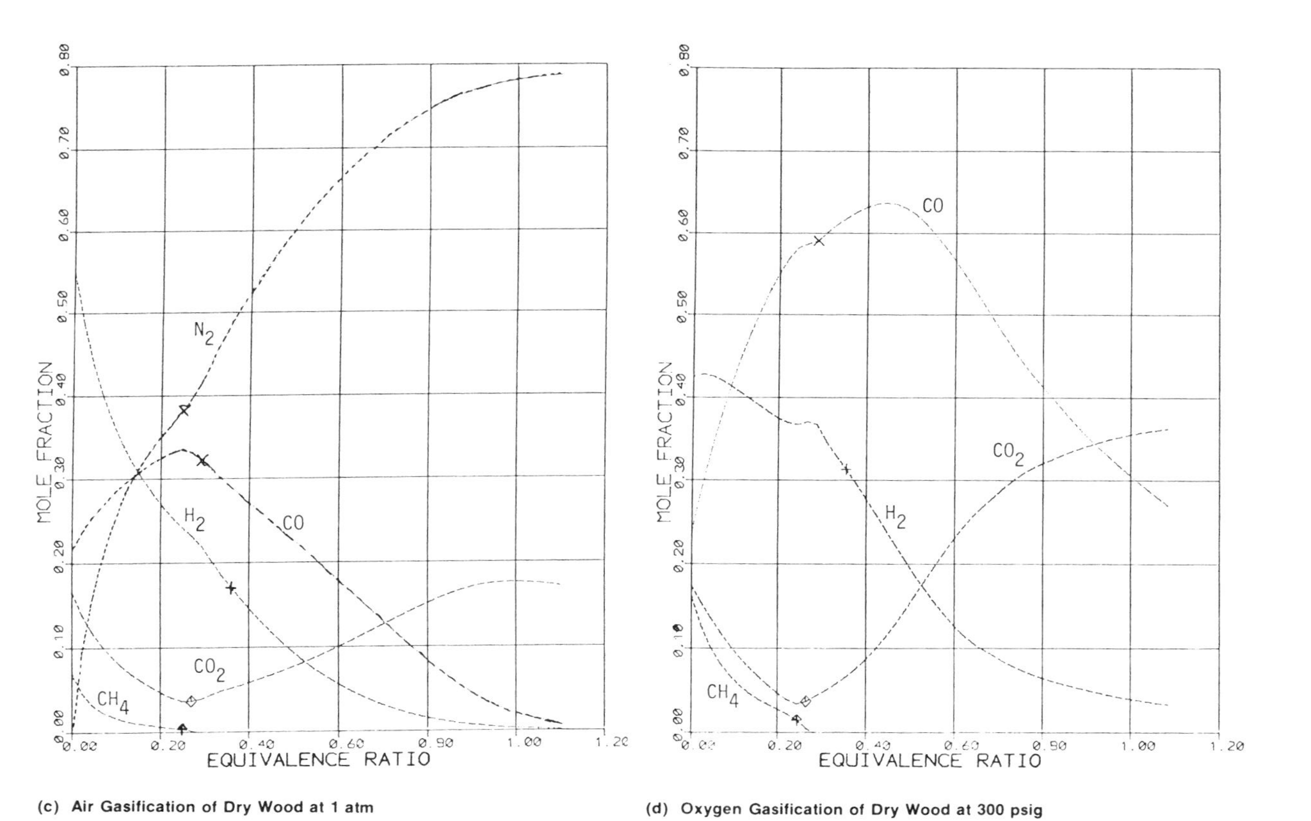

(c) Air Gasification of Dry Wood at 1 atm

(d) Oxygen Gasification of Dry Wood at 300 psig

Figure 6-8. Dry Gas Composition Vs. Equivalence Ratio

Table 6-3. ADIABATIC REACTIONS OF WOOD UNDER PYROLYSIS AND GASIFICATION CONDITIONS

$$CH_{1.4}O_{0.59} + yO_2 + zN_2 + wH_2O = x_1C_s + x_2H_2 + x_3H_2O + x_4CO + x_5CO_2 + x_6CH_4 + x_7N_2$$

Oxidant	Pressure (psig)	Water (g)	ER	T (K)	y	z	w	x_1	x_2	x_3	x_4	x_5	x_6	x_7
O_2	0	0	0	913	0	0	0	0.64	0.44	0.15	0.17	0.13	0.005	0
O_2	0	0	0.1016	1005	0.11	0	0	0.36	0.57	0.08	0.52	0.11	0.023	0
O_2	0	0	0.2709	1497	0.29	0	0	0	0.60	0.10	0.94	0.062	0	0
Air	0	0	0.2750	1105	0.29	1.1	0	0	0.63	0.07	0.90	0.10	0	1.1
O_2	0	80	0	502	0	0	1.02	0.37	0.014	1.03	0	0.29	0.34	0
O_2	0	80	0.2709	897	0.29	0	1.02	0	0.95	0.66	0.36	0.59	0.059	0
O_2	300	0	0	1060	0	0	0	0.61	0.30	0.19	0.17	0.12	0.11	0
O_2	300	0	0.2709	1502	0.29	0	0	0	0.60	0.10	0.94	0.06	0	0

The LHV and energy distribution curves for the cases of oxygen/air gasification of dry wood and oxygen with wet wood are plotted in Figs. 6-6 and 6-7. Curve b for wet wood initially lies above curve a for dry wood due to enhanced methane formation at the lower temperatures. Beyond ER = 0.15, however, the LHV for wet wood is lowered due to shifting of CO to CO_2 with added water. (The LHV is calculated for the dry gas composition.) The initial increased uptake of carbon with added water results in more chemical energy being stored in the gas (curve c, Fig. 6-7).

The dry gas compositions for four of the cases discussed above appear in Figs. 6-8 (a) through 6-8 (d). (Dry gas compositions are more easily compared with the gas analyses reported by investigators.) Stoichiometric reactions for some cases are listed in Table 6-3. The extent of water formation, which is not apparent in the gas composition plots, is readily inferred from the table.

6.4.2 Series 2 — Oxygen Gasification of Dry Wood at Fixed Temperature and Pressure

The calculations described here are for systems held at constant temperature and pressure. In Figs. 6-9 (a) and 6-9 (b), dry gas compositions are plotted versus equivalence ratio for several temperatures. The range of ER values extends from 0 to 0.30, the region of interest in gasification. In Fig. 6-9 (a), curves a and b are for H_2 at 900 and 1400 K, respectively. The weak temperature dependence is evident. Curves c, d, e, f are CO_2 concentrations at 900, 1000, 1100, and 1200 K. In Fig. 6-9 (b), the curves a, b, c, d are for CO at 900, 1000, 1100, 1400 K in the order listed; curves e, f, and g are for CH_4 at 900, 1000, and 1100 K. Methane is a minor component above 1100 K.

In a fluidized bed gasifier, temperature and equivalence ratio may be adjusted nearly independently: externally heated and recirculated inert bed material can influence the temperature level, and bleeding a variable amount of oxidant into a recycle stream will affect the ER. In a fixed-fuel-bed gasifier, however, the ER is not adjusted easily. Simply increasing the air rate, for example, will not necessarily have any effect on the ER. Introducing more air may simply expand the active portion of the bed, resulting in more throughput and leaving unchanged the ratio of air to wood consumed. One of the most sensitive tests of any kinetic model will be to predict the effect of air rate on bed temperature and equivalence ratio.

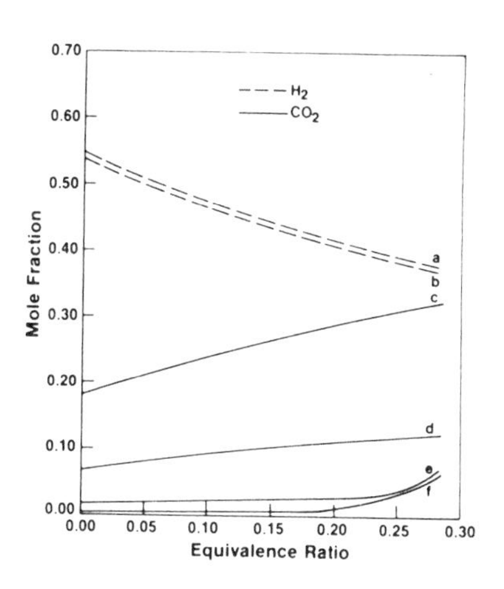

(a) H_2 and CO_2

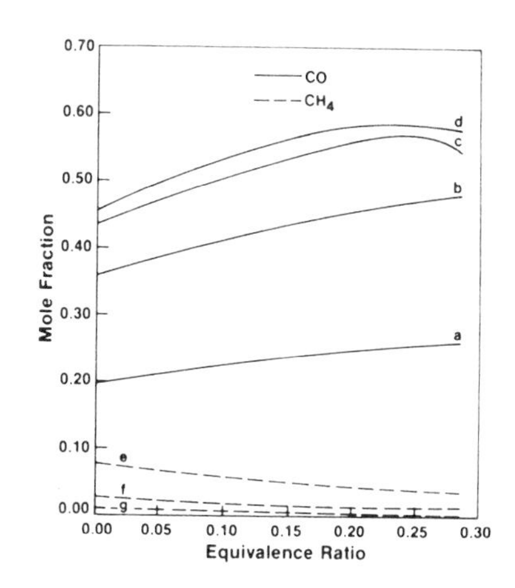

(b) CO and CH_4

Figure 6-9. Gas Composition Vs. Equivalence Ratio at Specified Temperatures and 1 atm

Given that the ER for a fixed bed gasifier may not be an adjustable parameter, it is interesting to observe the variation of composition with temperature for fixed ER. Figures 6-10 (a) and 6-10 (b) are plots of this type:

Figure 6-10 (a): H_2 Curves a, b, c, d: ER = 0.00, 0.068, 0.169, 0.284
CO_2 Curves h, g, f, e: ER = 0.00, 0.068, 0.169, 0.284

Figure 6-10 (b): CO Curves d, c, b, a: ER = 0.00, 0.068, 0.069, 0.284
CH_4 Curves e, f, g : ER = 0.00, 0.115, 0.284

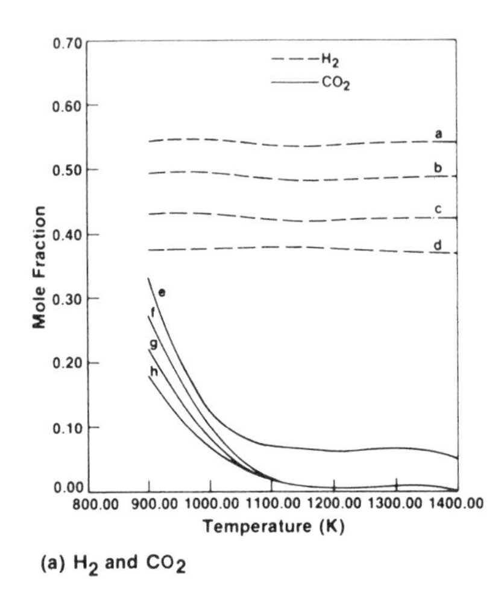

(a) H_2 and CO_2

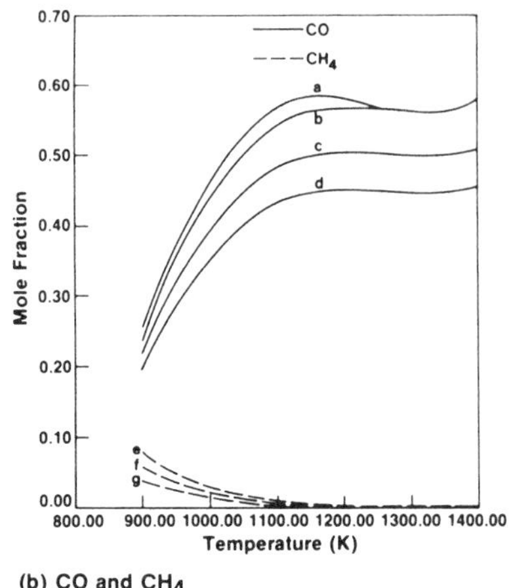

(b) CO and CH_4

Figure 6-10. Gas Composition Vs. Temperature at Specified Equivalence Ratio and 1 atm

Carbon formation in g/g dry wood and low heating value in Btu/SCF are plotted versus ER in Figs. 6-11 and 6-12:

Figure 6-11: C_s Curves a, b, c, d, e: T = 900, 1000, 1100, 1200, 1400 K

Figure 6-12: LHV Curves d, c, b, a : T = 900, 1000, 1100, 1400 K

Figures 6-13 through 6-16 are also fixed temperature runs but at 200 psig:

Figures 6-13 (a), 6-13 (b): Dry gas compositions versus ER are plotted for H_2, CO_2, CO, and CH_4. There are five curves for each species corresponding to T = 900, 1000, 1100, 1200, and 1400 K. The direction of increasing temperature for each set of curves is indicated by an arrow.

Figures 6-14 (a), 6-14 (b): Dry gas concentrations are plotted as functions of temperature. There are seven curves for each species corresponding to ER = 0.000, 0.034, 0.068, 0.115, 0.169, 0.224, 0.284, with an arrow indicating the direction of increasing oxygen input.

Figures 6-15, 6-16 Carbon formation and LHV are plotted versus ER for five temperatures: T = 900, 1000, 1100, 1200, 1400 K.

Stoichiometric reactions for several conditions are listed in Table 6-4.

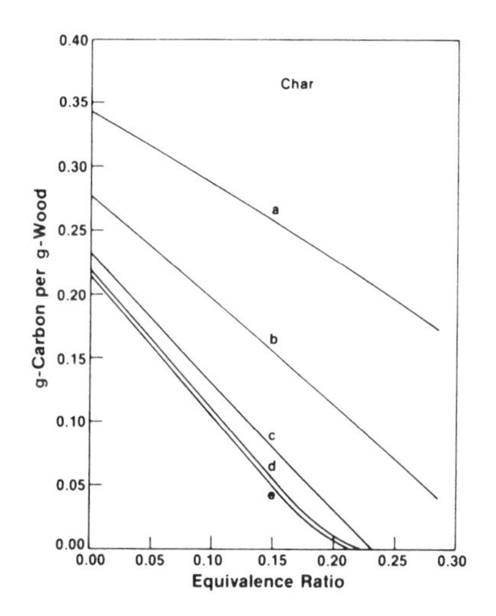

Figure 6-11. Carbon Formation Vs. Equivalence Ratio at Specified Temperatures, 1 atm

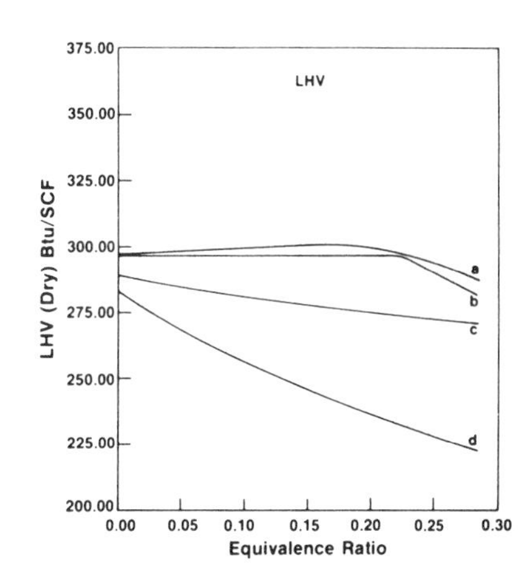

Figure 6-12. Low Heating Value Vs. Equivalence Ratio at Specified Temperatures, 1 atm

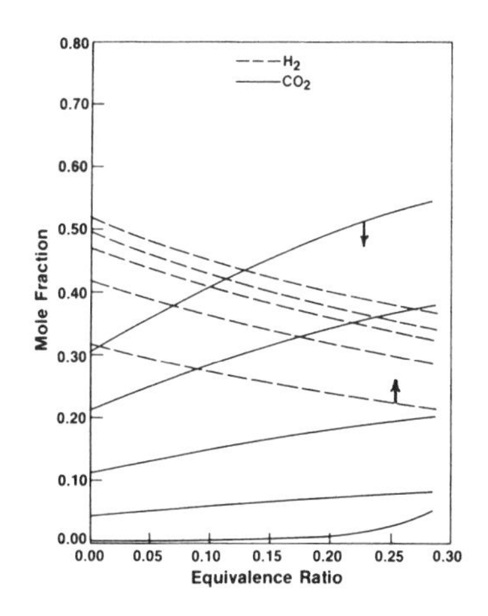

(a) H_2 and CO_2

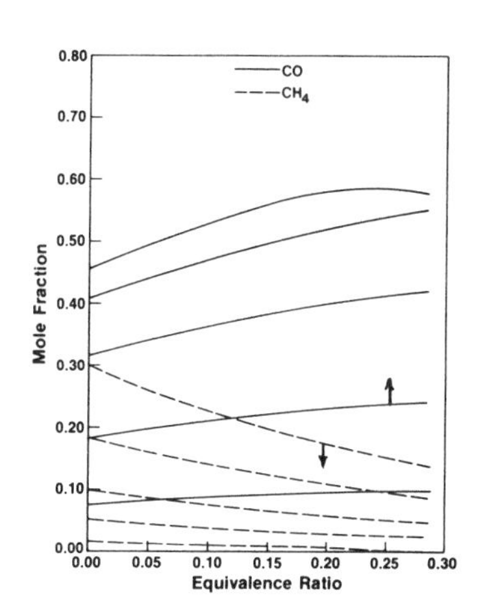

(b) CO and CH_4

Figure 6-13. Gas Composition Vs. Equivalence Ratio at Specified Temperatures and 200 psig

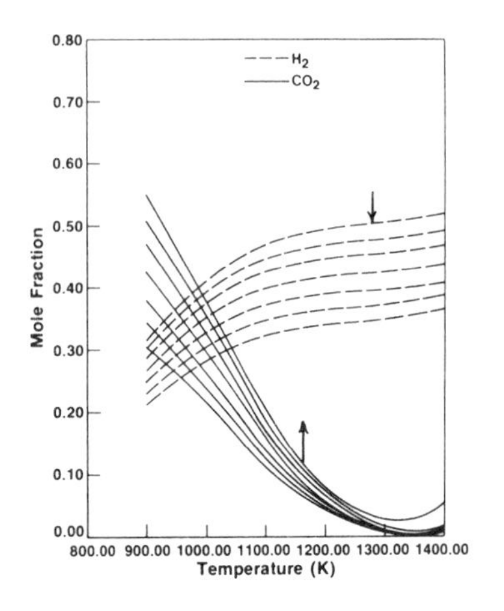

(a) H_2 and CO_2

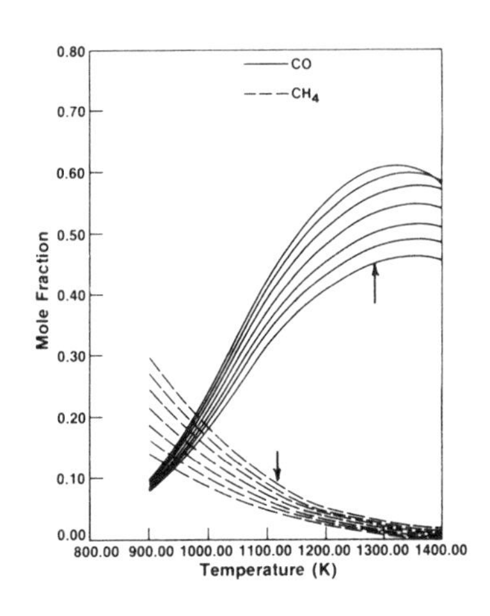

(b) CO and CH_4

Figure 6-14. Gas Composition Vs. Equivalence Ratio at Specified Temperatures and 200 psig

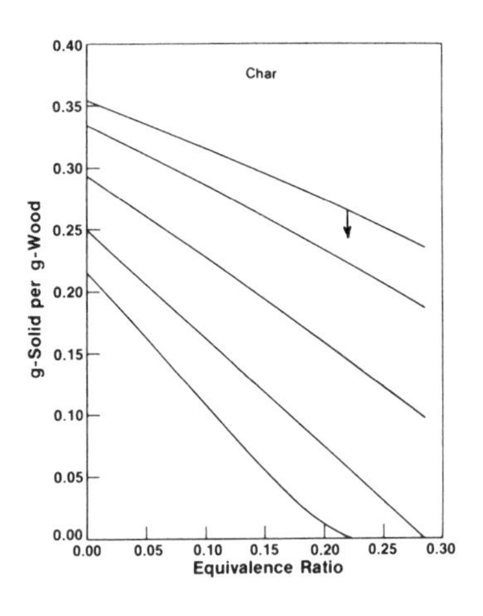

Figure 6-15. Carbon Formation Vs. Equivalence Ratio at Specified Temperatures, 200 psig

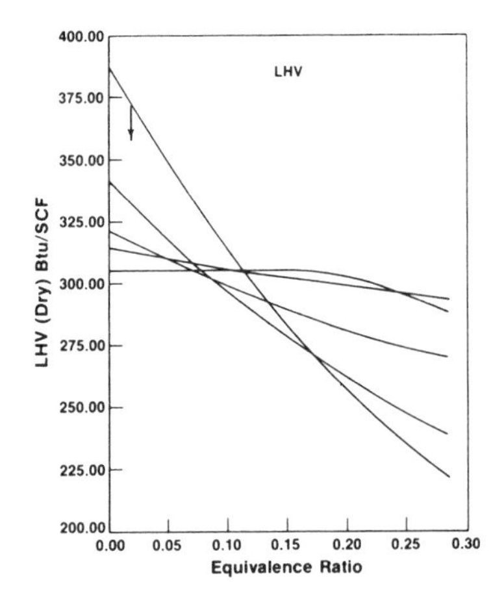

Figure 6-16. Low Heating Value Vs. Equivalence Ratio at Specified Temperatures, 200 psig

Table 6–4. OXYGEN GASIFICATION OF DRY WOOD AT FIXED TEMPERATURE AND PRESSURE

$$CH_{1.4}O_{0.59} + yO_2 = x_1C_s + H_2 + x_3H_2O + x_4CO + x_5CO_2 + x_6CH_4$$

Temperature (K)	Pressure (psig)	ER	y	x_1	x_2	x_3	x_4	x_5	x_6
	pyrolysis								
900		0		0.65	0.42	0.16	0.15	0.14	0.059
1000		0	0	0.53	0.57	0.08	0.37	0.07	0.028
1200	0	0		0.42	0.68	0.008	0.57	0.005	0.006
1400		0		0.41	0.69	0.001	0.59	0.0005	0.002
	gasification								
900		0.284	0.30	0.33	0.40	0.22	0.28	0.35	0.040
1000	0	0.284	0.30	0.08	0.56	0.10	0.72	0.19	0.019
1200		0.284	0.30	0	0.61	0.09	0.90	0.10	0.0003
1400		0.284	0.30	0	0.59	0.11	0.92	0.08	—
	pyrolysis								
900		0		0.67	0.15	0.26	0.036	0.15	0.14
1000	200	0	0	0.64	0.26	0.21	0.11	0.13	0.11
1200		0		0.48	0.51	0.075	0.42	0.045	0.055
1400		0		0.41	0.64	0.016	0.56	0.006	0.021
	gasification								
900		0.284	0.30	0.45	0.15	0.35	0.07	0.38	0.10
1000	200	0.284	0.30	0.35	0.26	0.28	0.22	0.34	0.08
1200		0.284	0.30	0	0.52	0.10	0.84	0.12	0.04
1400		0.284	0.30	0	0.58	0.11	0.92	0.08	0.002

6.4.3 Series 3 — Water Addition to Gasification

Nearly all gasifiers produce char. Some processes use all effluent char to raise steam, while others produce a surplus. The char is not a particularly attractive boiler fuel since it is low in volatiles. If the char is not recycled in the gasification plant, it cannot be considered a product of high value; in these cases, complete gasification of char should be promoted.

The controlling variables in char formation are moisture and temperature. Biomass feedstocks occur with varying amounts of moisture, depending on extent of pretreatment and method and duration of storage. For some gasification schemes, this inherent moisture may be an advantage. Those reactors with the capability of recycling a portion of the product gas may be particularly suited for handling wet feeds, because the recycled steam may significantly increase gas yields and is easily removed from the raw gas product. If hot gas is not recycled, then superheated steam may be used to promote char gasification.

The adiabatic flame temperatures for several conditions of interest in gasification are plotted in Fig. 6-17 (a). Abscissa values are equivalence ratios and range from 0 to 0.30. In all cases, oxygen is used in the blast; thus, for example, ER = 0.2 refers to

$$0.20 \times 1.476 = 0.295 \text{ g } O_2/\text{g dry wood} .$$

Curves a, b, c, d [Fig. 6-17 (a)] are all calculated for 1 atm of pressure. The first curve is for dry wood, while increasing amounts of water are added in cases b, c, d. The water is added either as liquid water at ambient temperature or as steam at 1000 K (1340 F). The large separation between curves a and b illustrates the effect of moisture on the adiabatic flame temperature. Further water addition was made as steam to minimize temperature differences to isolate the effect of moisture on gas composition and char consumption. Curves e, f are for elevated pressure. The quantities of water and steam listed in the tables below each figure are in g/g dry wood.

The influence of water/steam addition on gasification of char is illustrated in Fig. 6-17 (b,c). As more char is consumed, more chemical energy is stored in the gas [Fig. 6-17 (b)]. The breaks in the gas energy curves coincide with the stability limits of carbon. It appears from comparing curves c, e and d, f in Fig. 6-17 (c) that increasing pressure promotes char takeup. Most of this effect is a result of a temperature increase, however. When excess water is present, only the methanation reaction is pressure-dependent. The CO produced in the water gas reaction ($C + H_2O = CO + H_2$) is shifted by steam to $CO_2 + H_2$, resulting in the net reaction

$$2C + 2\,H_2O = CH_4 + CO_2 ,$$

which would exhibit a negligible pressure dependence.

Whenever a steam blast is used in gasification, it is important to know under what conditions the steam can be considered inert: whether it acts as a diluent or as a reactive species. Net water formation is plotted in Fig. 6-17 (d).

$$\text{Net } H_2O \text{ formation} = \frac{\text{weight } H_2O \text{ in product} - \text{weight } H_2O \text{ in feed}}{\text{weight dry wood in feed}} .$$

Since curves b through f are below 0, it is apparent that water is not just a diluent under gasification conditions. All the curves turn upward after their break points, the latter coinciding with the limiting equivalence ratios for carbon stability. This upward trend at higher ER values agrees with experience since water is a diluent (thermodynamically, not kinetically) under combustion conditions.

The dry gas composition for these six cases is plotted in Figs. 6-17 (e) through 6-17 (h). In all these figures, the curves for case (a), for dry wood, always lie well separated from the others. The effect on the CO shift reaction of increasing water addition is evident in Fig. 6-17 (g). The partitioning of these conditions into three sets a; b, c, d; and e, f is especially evident in the AFT plot [Fig. 6-17 (a)] and in the LHV plot [Fig. 6-17 (i)].

Stoichiometric reactions for several conditions at ER = 0.2032 are listed in Table 6-5.

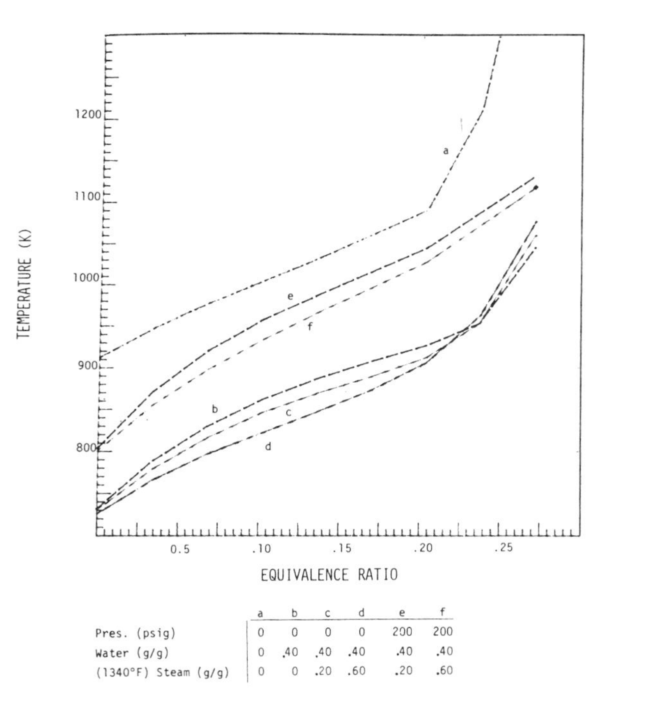

(a) Temperature Vs. Equivalence Ratio

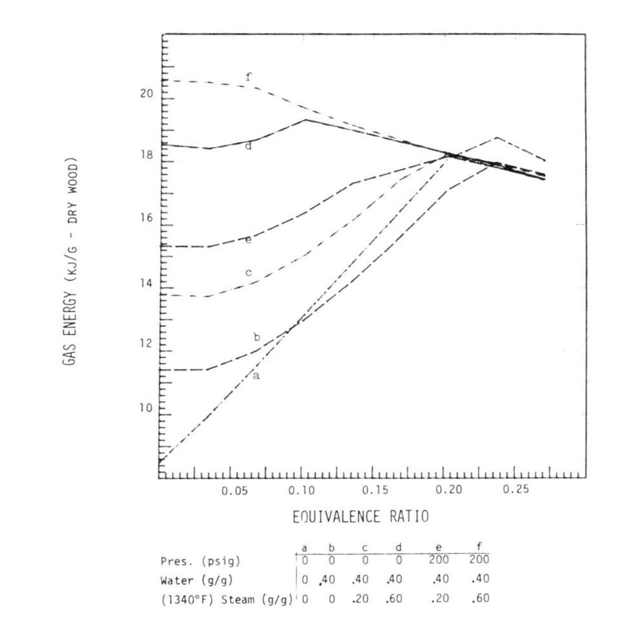

(b) Energy Distribution Vs. Equivalence Ratio

Figure 6-17. Water Addition to Gasification

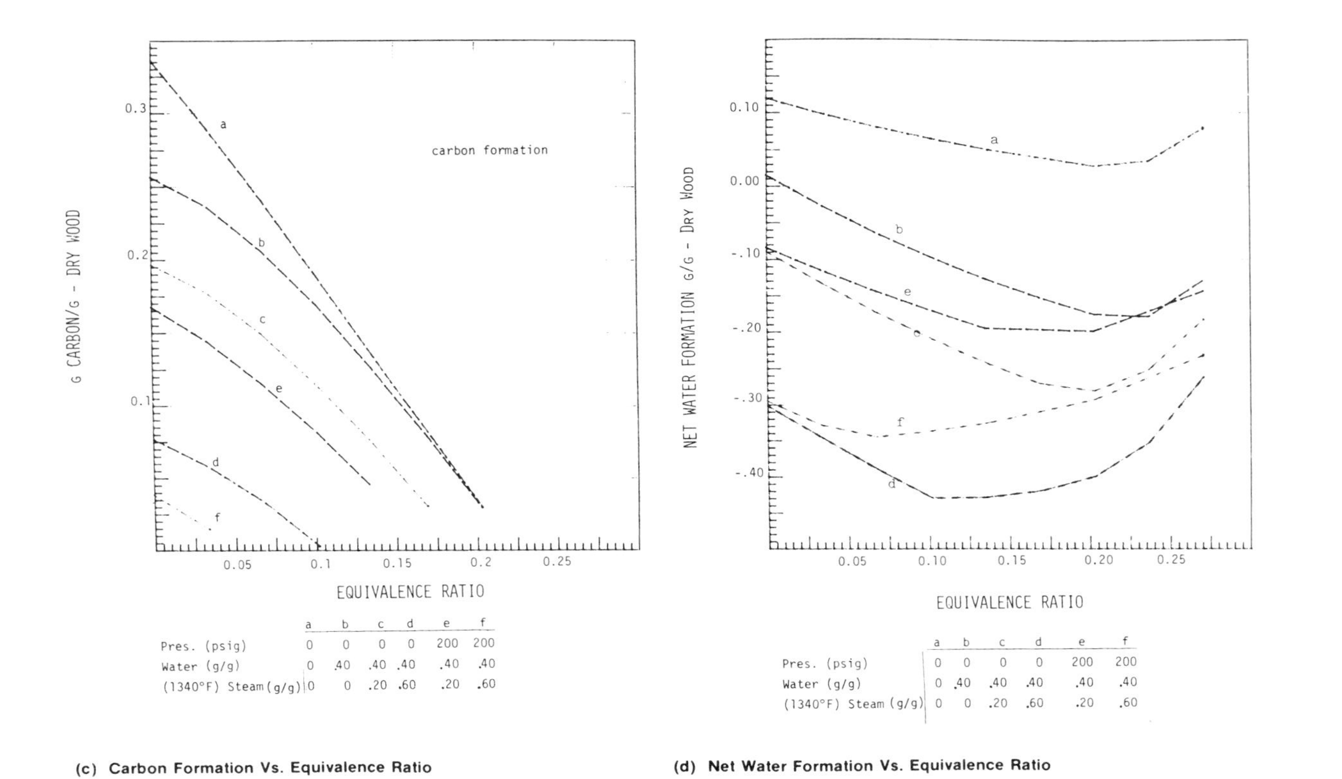

(c) Carbon Formation Vs. Equivalence Ratio

(d) Net Water Formation Vs. Equivalence Ratio

Figure 6-17. Water Addition to Gasification

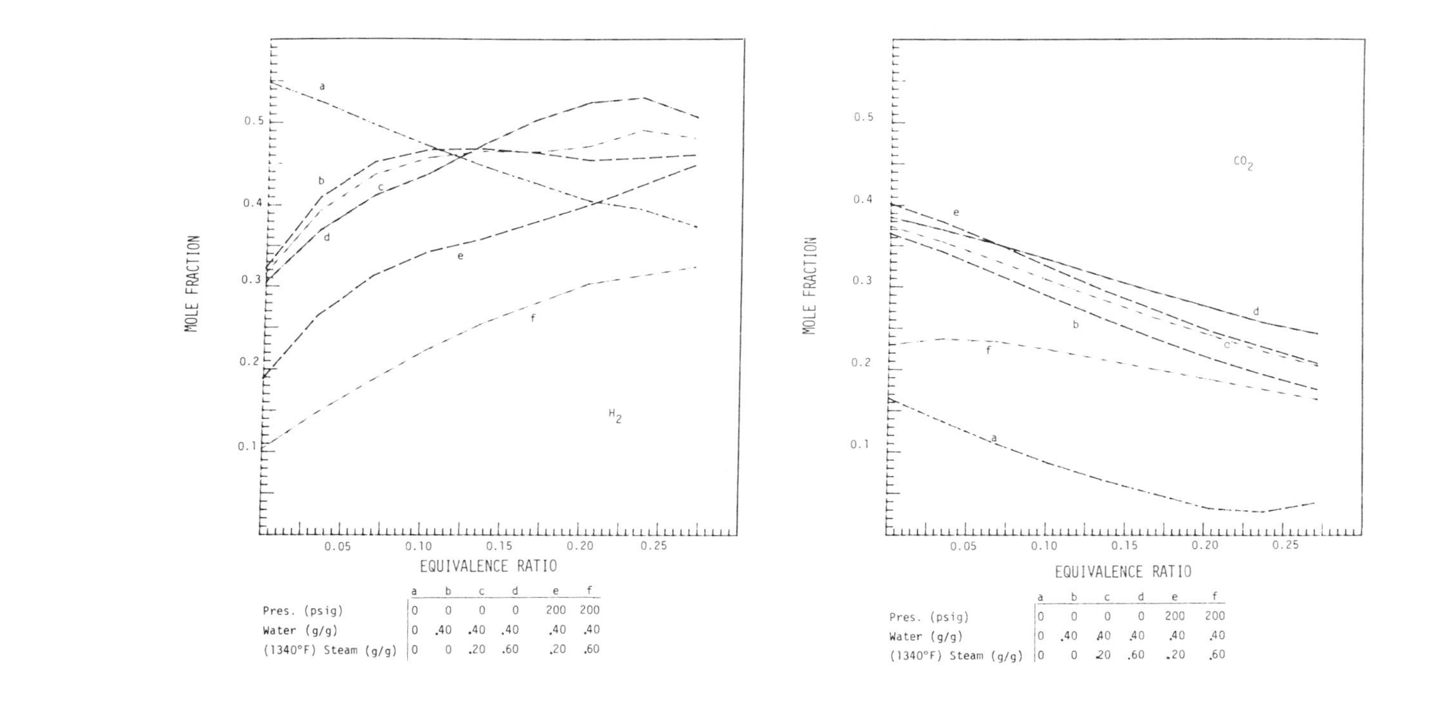

(e) Gas Composition (H_2) Vs. Equivalence Ratio

(f) Gas Composition (CO_2) Vs. Equivalence Ratio

Figure 6-17. Water Addition to Gasification

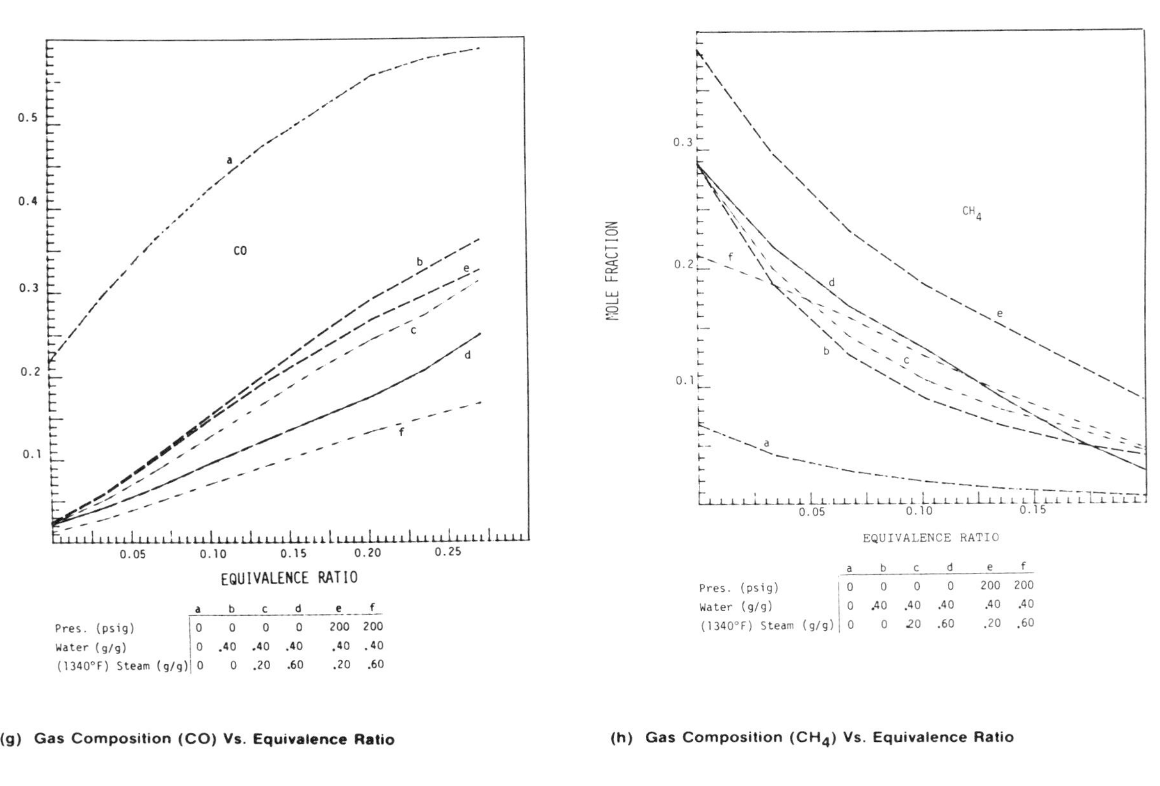

	a	b	c	d	e	f
Pres. (psig)	0	0	0	0	200	200
Water (g/g)	0	.40	.40	.40	.40	.40
(1340°F) Steam (g/g)	0	0	.20	.60	.20	.60

(g) Gas Composition (CO) Vs. Equivalence Ratio

	a	b	c	d	e	f
Pres. (psig)	0	0	0	0	200	200
Water (g/g)	0	.40	.40	.40	.40	.40
(1340°F) Steam (g/g)	0	0	.20	.60	.20	.60

(h) Gas Composition (CH_4) Vs. Equivalence Ratio

Figure 6-17. Water Addition to Gasification

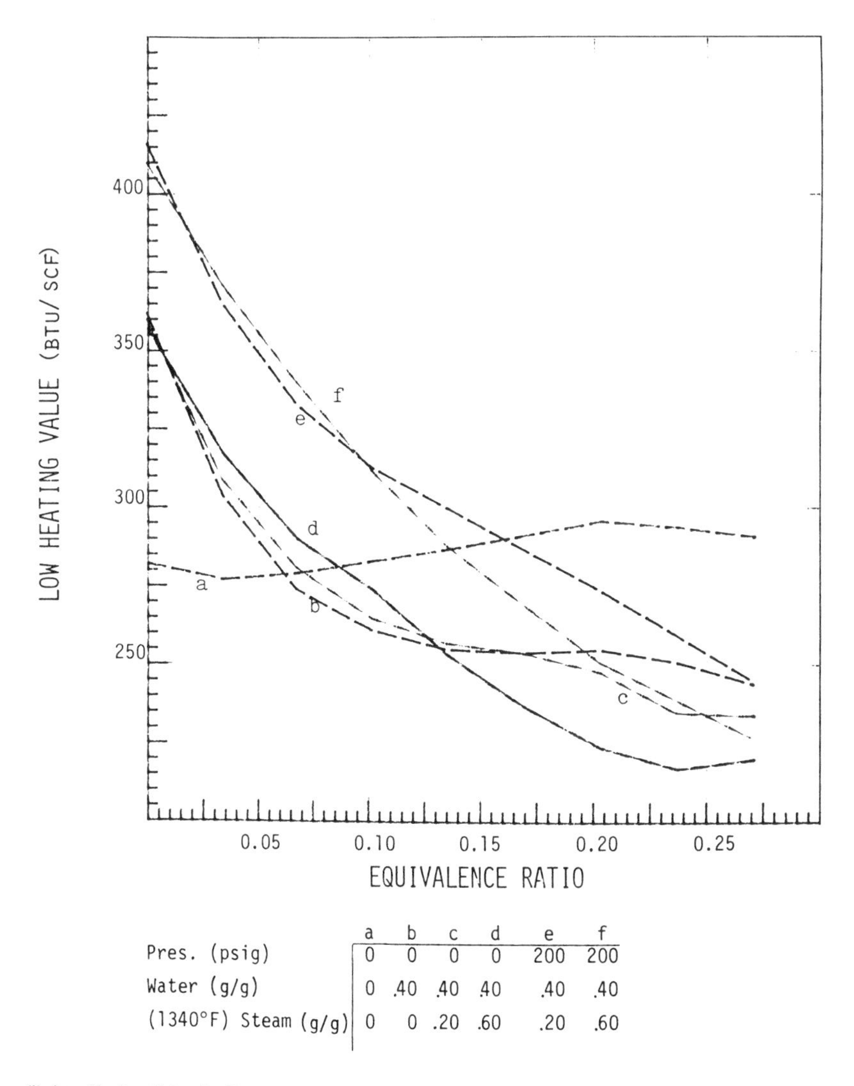

	a	b	c	d	e	f
Pres. (psig)	0	0	0	0	200	200
Water (g/g)	0	.40	.40	.40	.40	.40
(1340°F) Steam (g/g)	0	0	.20	.60	.20	.60

(i) Low Heating Value Vs. Equivalence Ratio

Figure 6-17. Water Addition to Gasification

Table 6-5. ADIABATIC OXYGEN GASIFICATION REACTIONS OF WOOD UNDER STEAM (1000 K, 1340 F)

$$CH_{1.4}O_{0.59} + yO_2 + wH_2O = x_1C_s + x_2H_2 + x_3H_2O + x_4CO + x_5CO_2 + x_6CH_4$$

Pressure (psig)	Temperature (K)	Water (g/g-dry wood)	Steam (g/g-dry wood)	ER	y	w	x_1	x_2	x_3	x_4	x_5	x_6
0	1091	0	0	0.2032	0.21	0	0.06	0.64	0.036	0.88	0.05	0.01
0	927	0.40	0	0.2032	0.21	0.51	0.06	0.78	0.29	0.50	0.37	0.07
0	912	0.40	0.20	0.2032	0.21	0.76	0	0.89	0.41	0.46	0.46	0.08
0	906	0.40	0.60	0.2032	0.21	1.27	0	1.10	0.76	0.37	0.58	0.05
200	1045	0.40	0.20	0.2032	0.21	0.76	0	0.67	0.51	0.45	0.41	0.14
200	1028	0.40	0.60	0.2032	0.21	1.27	0	0.83	0.90	0.36	0.51	0.12

6.4.4 Series 4 — Steam Addition to Pyrolysis

Pyrolysis offers the advantage of producing a medium energy gas without the requirement of an oxygen plant. The results of pyrolysis for the conditions considered to this point are all represented by points on the left hand axis at ER = 0. In this series of plots we consider the effect of steam addition [steam temperature 1000 K, (1340 F)] on pyrolysis. In Figs. 6-18 (c) and 6-18 (d), dry gas compositions are plotted for three different pressures (0, 400, and 1000 psig) as a function of steam addition. The feed is wet wood (40% moisture, dry basis). The values on the x-axis of each figure represent the weight of steam added per unit of dry wood. Thus the system at a point with x-coordinate 0.6 is composed of 100 g of wood, 40 g of water, and 60 g of steam. As observed previously, pressure strongly affects the concentration of H_2 and CH_4 but has little effect on CO or CO_2. Carbon formation is superimposed on each figure. Steam addition affects the quantity of gas produced but the composition is fixed by the water-gas shift and methanation reactions.

The AFT and LHV curves are plotted in Fig. 6-18 (a) and net water formation in Fig. 6-18 (b).

A higher steam temperature (1144 K, 1600 F) has a negligible effect on the results, as can be seen by comparing Figs. 6-18 (e) and 6-18 (f) with Figs. 6-18 (a) through 6-18 (d). [Notice that steam addition is plotted on an expanded scale in Figs. 6-18 (e) and 6-18 (f).] This observation highlights the difficulty of introducing sufficient sensible heat in a pyrolysis scheme to significantly affect the system composition. Some designs rely on recirculating hot solids to fluidized or entrained beds. The solid is usually char or sand that is withdrawn from the bed and heated externally. The thermal duty of the process is supplied easily by providing a sufficiently high solids/gas ratio; design problems arise only in the solids handling area. Pyrolysis schemes which rely on heating a recycled gas stream, however, have more stringent equipment limitations because of the volume of gas that must be handled. Consider the example illustrated in Fig. 6-19, in which 100 g wood, 40 g water, and 67 g steam (at 1600 F) are fed to a pyrolysis reactor equipped with a gas recycle stream which is reheated from the adiabatic reactor temperature (T_{aft}) to a reheat temperature (T_{rh}) of 1300 K (1880 F). The computer program used in these calculations includes the option of introducing an enthalpy adjustment for heat losses or additions. For several enthalpy additions (ΔH) in kcal, the system composition and T_{aft} were calculated. Then, using the heat capacity of the product gas stream, the quantity of gas which would have to be recycled to provide the stipulated heat input within the temperature rise $T_{rh} - T_{aft}$ was determined. The results of a series of such calculations for the system pictured are listed in Table 6-6. For the level of reheat considered, over 72% of the reactor effluent would have to be recycled in order to effect a change in temperature of 100 K.

The stoichiometry of the pressure-steam pyrolysis reactions considered in this section are listed in Table 6-7. The methane concentrations achieved under steam pyrolysis can be as high as 48%. The Wright-Malta process operates under conditions similar to those described here and offers a very attractive route to SNG and a medium energy boiler fuel.

6.4.5 Series 5 — Pyrolysis Equilibria Versus Pressure

Although pressure has been considered as a parameter in several cases discussed in previous sections, the effect of pressure has not been presented from a global perspective. The major advantage of high pressure operation in gasification is the diminished compression required for downstream gas processing. A second benefit is an enhanced rate of reaction. The effect on equilibrium conversion is very small, as is demonstrated in Figs. 6-20 (a) and 6-20 (b). In these figures the calculated equilibrium properties for two systems are plotted: wood and water (0.4 g/g) and wood with steam (1 g/g, at 1600 F).

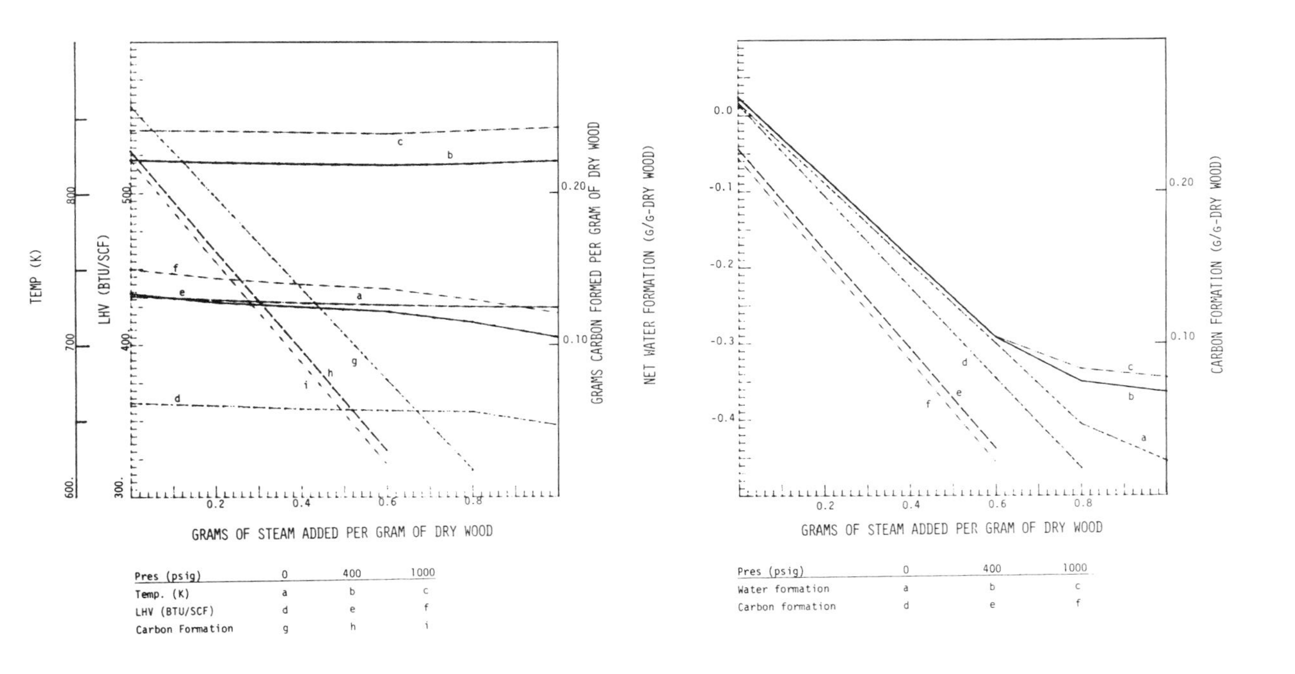

(a) Adiabatic Flame Temperature, Low Heating Value and Carbon Formation

(b) Net Water Formation and Carbon Formation

Figure 6-18. Steam Addition to Pyrolysis

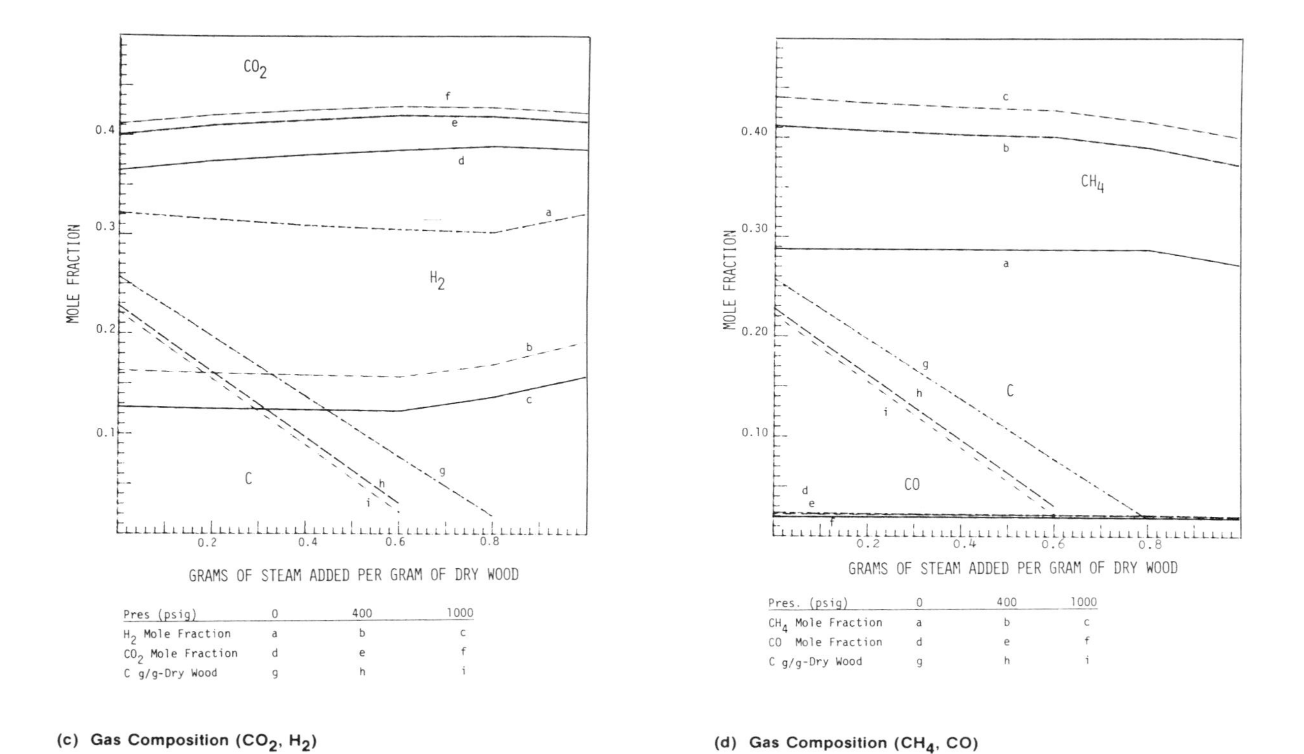

Pres (psig)	0	400	1000
H_2 Mole Fraction	a	b	c
CO_2 Mole Fraction	d	e	f
C g/g-Dry Wood	g	h	i

(c) Gas Composition (CO_2, H_2)

Pres. (psig)	0	400	1000
CH_4 Mole Fraction	a	b	c
CO Mole Fraction	d	e	f
C g/g-Dry Wood	g	h	i

(d) Gas Composition (CH_4, CO)

Figure 6-18. Steam Addition to Pyrolysis

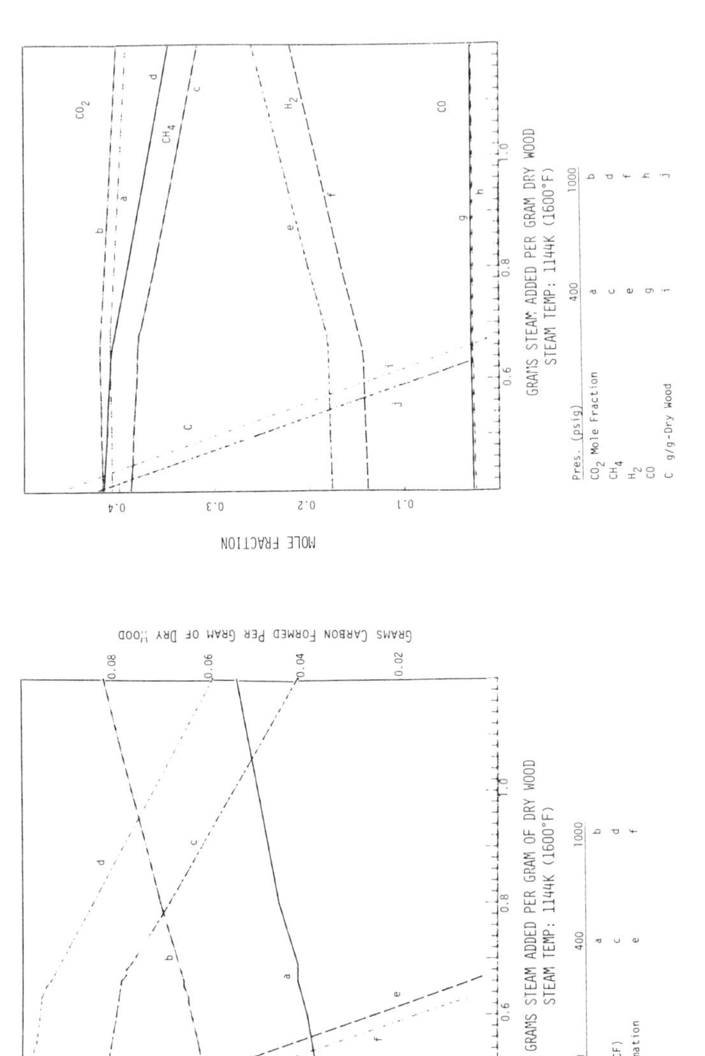

(e) Adiabatic Flame Temperature, Low Heating Value, and Carbon Formation, Steam at 1144 K

(f) Gas Composition, Steam at 1144 K

Figure 6-18. Steam Addition to Pyrolysis

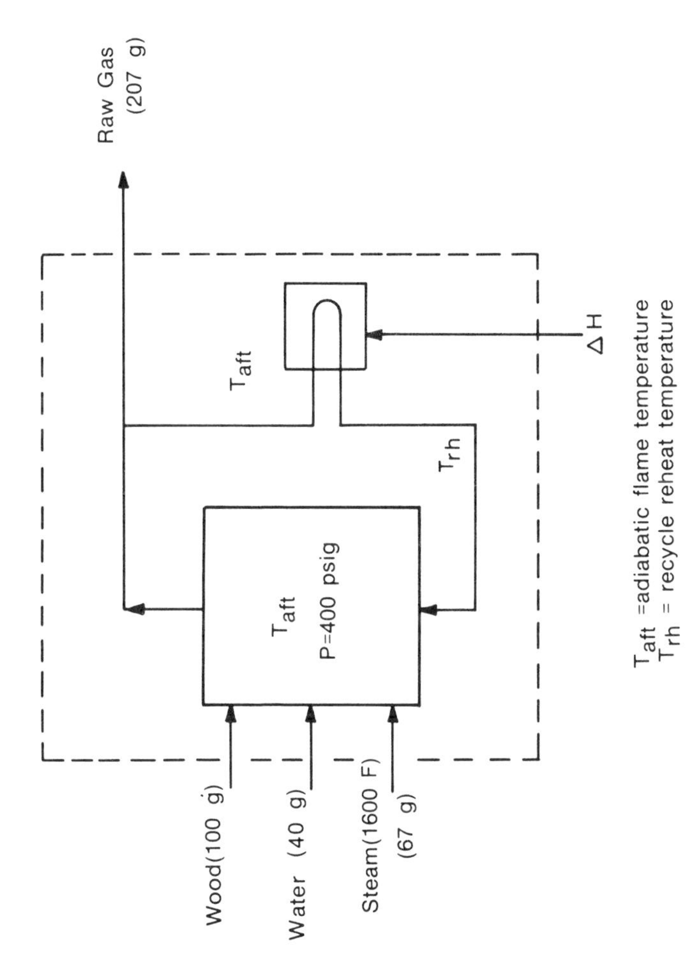

Figure 6-19. Pyrolysis with Recycle Reheat

Table 6-6. HEAT ADDITION IN PYROLYSIS

$$h = \bar{c}\,(T_{rh})\,(T_{rh} - 298) - \bar{c}\,(T_{aft})\,(T_{aft} - 298)$$

$$\text{Recycle \%} = \Delta H \times 10^5/h \times 207$$

ΔH	T_{aft}		Products (mole %)				H_2O	LHV	$\bar{c}(T_{aft})$	$\bar{c}(T_{rh})$	h	Recycle
(kcal)	(K)	(F)	H_2	CO	CO_2	CH_4	(g)	(MJ/Nm^3)	(cal/g C)		(cal/g)	(%)
0	842	1056	18.05	3.03	40.94	37.91	72.27	15.96	—	—	—	0.00
5	862	1092	20.24	3.92	39.79	35.98	70.95	15.62	0.445	0.493	243	9.94
10	881	1126	22.47	4.89	38.57	34.01	70.08	15.27	0.449	0.495	234	20.65
20	914	1186	26.58	6.98	36.08	30.29	68.50	14.65	0.457	0.497	216	44.73
30	943	1238	30.29	9.20	33.60	26.86	67.10	14.10	0.463	0.499	201	72.10

Table 6-7. ADIABATIC PYROLYSIS REACTIONS OF WOOD[a] UNDER STEAM (1000 K, 1340 F)

$$CH_{1.4}O_{0.59} + wH_2O = x_1C_s + x_2H_2 + x_3H_2O + x_4CO + x_5CO_2 + x_6CH_4$$

Pressure (psig)	Temperature (K)	Steam (g/g-dry wood)	w	x_1	x_2	x_3	x_4	x_5	x_6
0	732	0	0.51	0.49	0.24	0.53	0.018	0.28	0.22
	728	0.40	1.02	0.26	0.33	0.77	0.024	0.41	0.31
	725	0.80	1.52	0.03	0.42	1.01	0.029	0.54	0.40
400	822	0	0.51	0.43	0.11	0.54	0.015	0.27	0.28
	819	0.40	1.02	0.18	0.15	0.78	0.021	0.40	0.39
	819	0.80	1.52	0	0.20	1.08	0.026	0.50	0.47
1000	842	0	0.51	0.42	0.08	0.54	0.013	0.27	0.29
	840	0.40	1.02	0.17	0.12	0.78	0.019	0.40	0.41
	841	0.80	1.52	0	0.16	1.10	0.022	0.50	0.48

[a]Wood at 40% moisture, dry basis.

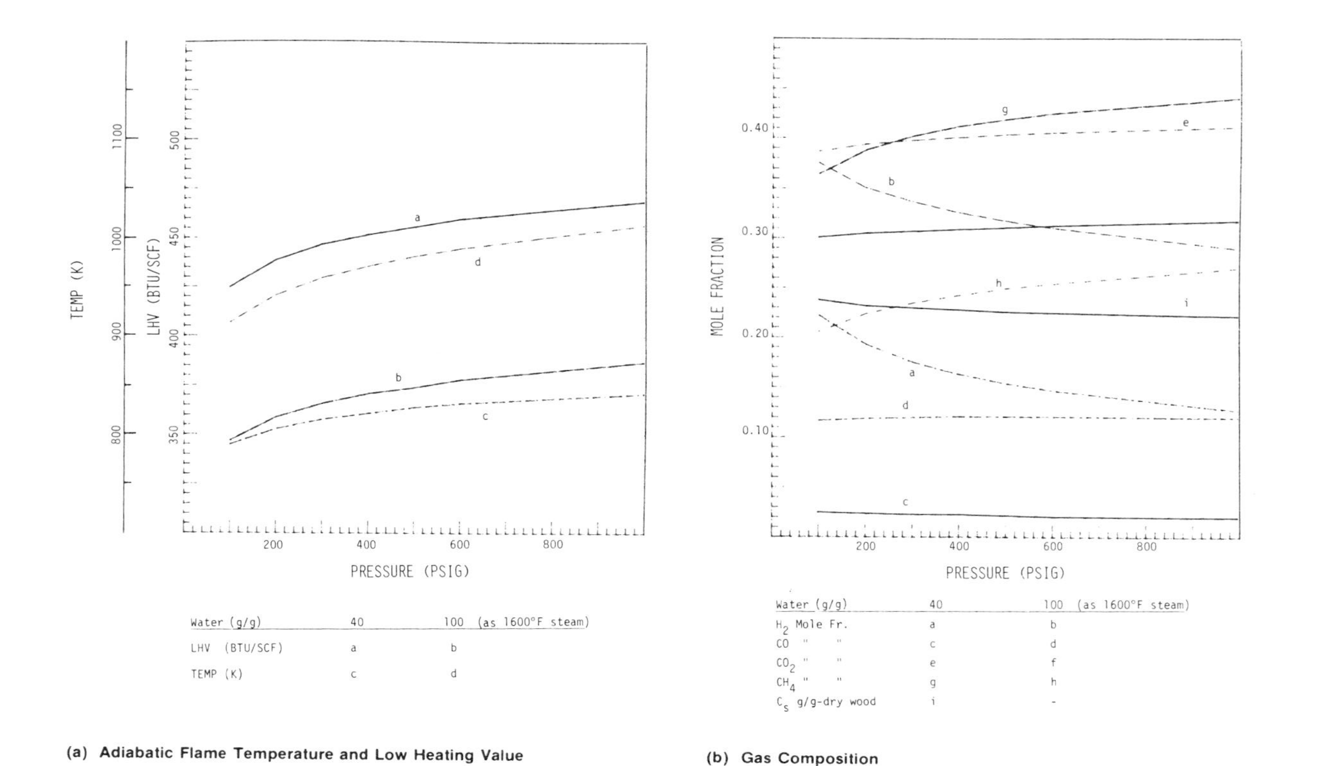

(a) Adiabatic Flame Temperature and Low Heating Value

(b) Gas Composition

Figure 6-20. Pyrolysis Equilibria Vs. Pressure

6.5 REFERENCES

Cruise, D. R. 1964. "Notes on the Rapid Computation of Chemical Equilibrium." J. Physical Chemistry. Vol. 68: p. 3797.

Desrosiers, R. E. 1977. "Computer Techniques for Determining Flame Temperature and Composition." ERDA report FE 2205-5; 20 Mar.

Stull, D. R.; Prophet, H. 1971. "JANAF Thermochemical Tables." NBS 37; June, and supplements.

Chapter 7

Kinetics of Char Gasification Reactions

M. Graboski
Colorado School of Mines

Biomass chars, like coal chars, are composed principally of carbon. Chars produced by pyrolytic reactions can be oxidized to synthesis gas through heterogeneous reactions with carbon dioxide, steam, oxygen, and hydrogen. The reactivity of chars in gaseous atmospheres is a complicated function of temperature, particle structure, carbon source, and thermal history of the char. The overall rate of char gasification may be affected not only by chemical kinetics but also by intraparticle and external mass transfer resistances. Additionally, the gasification reactions have large associated heat effects, making possible a significant temperature gradient within the particle. Therefore, any discussion of kinetics must include the effects of mass and heat transfer. Many early studies on gasification kinetics have been invalidated because they ignored these effects.

This chapter introduces some of the fundamental concepts of heat and mass transfer in chemical reactions. The true kinetics of the important gasification reactions are then summarized in terms of mechanisms and the effects of pertinent variables on the observed rates of gasification.

7.1 CHEMICAL REACTION SCHEMES

The principal objective in char gasification is to produce from the carbon-containing char a mixture of gases containing a substantial quantity of carbon monoxide and hydrogen. The reaction scheme usually involves the coupling of the exothermic oxygen combustion reaction with a number of endothermic gasification reactions to produce the synthesis gas. In order for the overall process to occur at a significant rate, temperatures in excess of 500 C are required, and more likely the process will operate at temperatures on the order of 800 C. The energy content of the synthesis gas will depend on the mode of heat addition to the gasification reactions. Oxygen gasification will yield a medium energy gas of 300 Btu/SCF or higher depending on the amount of methane produced during gasification. Air gasification, on the other hand, will yield a low energy gas of 150 Btu/SCF or less with little or no methane in the product. The lower energy content is due mainly to the nitrogen content of the air. In either case, the principal reactions are:

$$C + O_2 \rightleftharpoons 2CO, \qquad \Delta H_R = -26 \text{ kcal/mole} \tag{7-1}$$

$$C + H_2O \rightleftharpoons CO + H_2, \qquad \Delta H_R = +31 \text{ kcal/mole} \tag{7-2}$$

$$C + CO_2 \rightleftharpoons 2CO, \qquad \Delta H_R = +41 \text{ kcal/mole} \tag{7-3}$$

In many gasification systems, the gasifying medium is a mixture of air or oxygen and steam. Equation 7-2, termed the carbon-steam reaction, is the principal endothermic step in such systems. In the absence of steam, as in partial oxidation, Eq. 7-3, termed the Boudouard reaction, converts CO_2 produced by oxidation to CO. The rates of reactions 7-2 and 7-3 are similar (the carbon steam reaction being several times faster) for most carbons. Thus reaction 7-3 can serve as an indicator of the activities of different chars. It is much easier to study the kinetics of reaction 7-3 in comparison to reaction 7-2, since in reaction 7-2 parallel competing reactions can occur when the hydrogen generated reacts with other species.

Reaction 7-3 is very important in downstream heat transfer equipment and piping, where steel surfaces can promote the reverse reaction and deposit soot.

Since the combustion and gasification reactions produce carbon monoxide, the water-gas shift reaction can take place in the presence of steam:

$$CO + H_2O \rightleftarrows CO_2 + H_2, \Delta H_R = -10 \text{ kcal/mole} \quad . \tag{7-4}$$

This reaction is thought to occur as a result of heterogeneous catalysis by the carbon surface at temperatures below about 2000 F. At higher temperatures it may occur as a homogeneous reaction.

Methane formation by the hydrogasification reaction,

$$C + 2H_2 \rightleftarrows CH_4, \Delta H_R = -17 \text{ kcal/mole} \quad , \tag{7-5}$$

is important in oxygen gasification for two reasons: (1) The energy content of the synthesis gas is increased; (2) the oxygen required is reduced because of the heat released in methane generation. For coal chars, two methane-forming processes have been observed (for example, Johnson 1974). Freshly devolatilized char is highly reactive and forms methane at a high rate in the first seconds of its existence. After that time, the char becomes graphitized (or stabilized) to some degree and, subsequently, methane is formed at a very slow rate. To have significant production rates of methane relative to the rate of carbon gasification by the steam carbon reaction, rapid heating and high pressure operation are necessary since the kinetics of reaction 7-5 are strongly dependent on hydrogen partial pressure. Air gasification is usually carried out at low pressure. This fact, coupled with the dilution effect of the nitrogen in the air, all but eliminates methane production from char by reaction 7-5.

In fixed bed gasifiers, there are different kinetic regions depending on whether the gasifier is operated in the updraft or downdraft mode (Fig. 7-1). In downdraft gasifiers, the steam and oxidant are fed directly to the gasifier with the fresh biomass. Pyrolysis and combustion occur simultaneously; tars are gasified to CO, CO_2, and H_2. The hot gases are swept downward over the remaining char to yield a relatively hydrocarbon-free, low energy gas at the gasifier outlet.

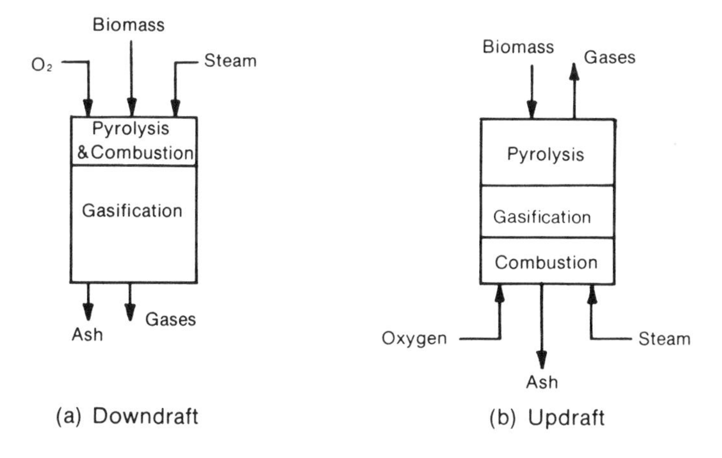

Figure 7-1. Modes of Gasifier Operation

In the updraft mode, steam and oxygen contact spent char. Combustion occurs at the base of the gasifier, and above the combustion zone the slower gasification reactions take place. In the top zone, the biomass is devolatilized to produce a synthesis gas containing substantial quantities of hydrocarbons.

The operating mode depends on the use of the synthesis gas. The downdraft method is especially useful for conversion of biomass materials to methanol and ammonia synthesis gas. Updraft gasification yields a fuel gas suitable as a boiler fuel or feedstock for manufacture of synthetic natural gas.

7.2 EFFECT OF MASS TRANSFER ON REACTION RATE

Figure 7-2 shows a porous char particle typical of biomass materials, such as wood, which contain negligible quantities of ash. In char gasification, the following reaction steps are considered to occur in series:

- diffusion of reactants across the stagnant film to the external char surface;
- diffusion of gas down the pore toward the center of the particle;
- adsorption, surface reaction, and desorption on the pore wall;
- diffusion of products out of the pore; and
- diffusion of products across the stagnant film to the gaseous reaction environment.

Depending on the temperature, pressure, gas composition, and extent of reaction, any or all of these steps may be important.

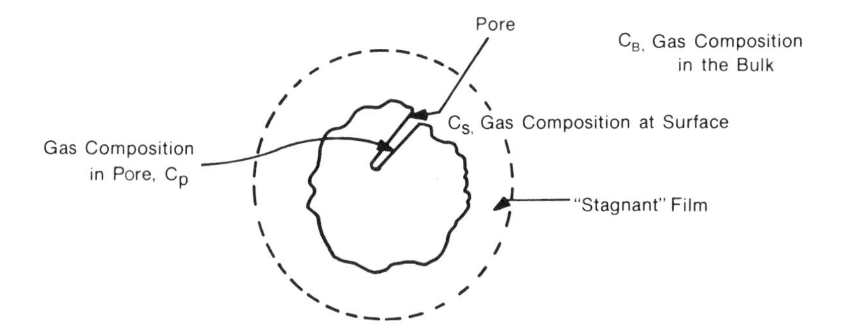

Figure 7-2. Model of Porous Char Particle

7.2.1 External Mass Transport and Heat Transfer

Diffusion across the film is termed external mass transport. At steady state, the rate of transport to the surface is given by the standard mass transfer expression:

$$W = k_m A_p C(Y_B - Y_S) = k_m A_p (C_B - C_S) \quad , \tag{7-6}$$

where

W = transfer rate, moles/time/weight of solid;

k_m = mass transfer coefficient, length/time;

A_p = external surface area per weight of solid;

Y_B = bulk gas concentration, mole fraction units;

Y_S = concentration of gas adjacent to surface, mole fraction units;

C = total gas concentration, moles/volume;

C_B = concentration of component in the bulk, moles/volume; and

C_S = concentration of component adjacent to surface, moles/volume.

The mass transfer coefficient is a weak function of absolute temperature and velocity, as is shown later in this section. The total concentration C is given approximately by the ideal gas law:

$$C = P/RT \ ,$$

where

P = absolute pressure,

R = gas constant, and

T = absolute temperature

The external heat transfer process by convection across the film is described by the following equation:

$$Q = hA_p\,(T_B - T_S) \ , \qquad (7\text{-}7)$$

where

h = heat transfer coefficient, energy/area/temperature;

A_p = external surface area per mass of solid;

T_B = absolute bulk temperature, R;

T_S = absolute particle surface temperature, R; and

Q = total heat flow to the gas per mass of particle.

Equation 7-7 assumes that there is minimal heat transfer by conduction between particles and negligible radiation exchange. The total heat flow Q is given by the following expression:

$$Q = (W)(\Delta H_R) \ , \qquad (7\text{-}8)$$

where

H_R = heat of reaction, energy/mole.

In fixed-bed operation, Satterfield (1970) recommends correlations for heat and mass transfer coefficients based on the Colburn j factor defined as follows:

$$j = \frac{k_m}{V}\,(Sc)^{2/3} = \frac{h}{C_p(\rho^* V)}\,(Pr)^{2/3} \ , \qquad (7\text{-}9)$$

where

j = Colburn j factor, dimensionless;

k_m = mass transfer coefficient, length/time;

h = heat transfer coefficient, energy/time;

C_p = heat capacity, energy/mole temperature;

ρ^* = molar density, moles/volume;

V = superficial velocity, length/time based on empty tube;

Sc = Schmidt number, $\mu/\rho D$, dimensionless;

Pr = Prandtl number, $C_p\mu/k_T$, dimensionless;

μ = viscosity, mass/length/time;

ρ = mass density, mass/volume;

D = diffusivity through the film, $(\text{length})^2/\text{time}$; and

k_T = thermal conductivity of the film, energy/length/time/temperature.

The j factor depends on the external bed porosity ϵ, and the Reynolds number, $Re = D_p V\rho/\mu$, where D_p is the particle diameter as follows:

$$j = \frac{0.357}{Re^{0.359}} \qquad 3 \leq Re \leq 2000 \quad . \tag{7-10}$$

The appropriate particle diameter is given as:

$$D_p = \frac{6V_{ex}}{S_{ex}} \quad , \tag{7-11}$$

where

V_{ex} = volume of particle, and
S_{ex} = surface area of particle.

In fluidized beds, Chu et al. (1953) recommend the j factor as follows for mass and heat transfer:

$$j = 5.7 \left(\frac{Re}{1-\epsilon}\right)^{-0.78} \qquad 0 < \frac{Re}{1-\epsilon} < 30 \quad , \tag{7-12a}$$

$$j = 1.77 \left(\frac{Re}{1-\epsilon}\right)^{-0.44} \qquad 30 < \frac{Re}{1-\epsilon} < 10{,}000 \quad . \tag{7-12b}$$

Equations 7-9 and 7-12 show that the heat and mass transfer coefficients vary with velocity, gas density, and particle size due to the Reynolds number dependency.

For fixed beds:

$$k_m h = f(V^{0.641})$$

For fluid beds:

$$k_m h = f(V^{0.22}) \text{ low Reynolds numbers}$$

$$k_m h = f(V^{0.56}) \text{ high Reynolds numbers}$$

By definition, the j factor for mass transfer and for heat transfer are identical. In terms of temperature dependency, the mass transfer coefficient behaves like a diffusion coefficient. Therefore:

$$\frac{k_m(2)}{k_m(1)} = \left(\frac{T_2}{T_1}\right)^{1.75} \tag{7-13}$$

For a temperature change from 800 C to 1000 C, $k_m(2)/k_m(1) = 1.35$. If an Arrhenius behavior were assigned to the mass transfer coefficient:

$$k = k_o \exp(-E/RT) \ , \tag{7-14}$$

a value of about 4 kcal is obtained for the temperature change from 800 C to 1000 C. Thus the mass transfer process has a very low activation energy; that is, the rate of mass transfer is not affected significantly by temperature.

External mass transfer reduces the concentration of reactant gas close to the particle surface and thus reduces the overall process rate. To demonstrate this phenomenon, consider gasification to be a first order reaction. Then at steady state the rate of gasification equals the rate of mass transfer:

$$kC_S = k_m A_p (C_B - C_S) \ . \tag{7-15}$$

Solving for the surface concentration yields:

$$C_S = \frac{k_m A_p C_B}{k + k_m A_p} \ . \tag{7-16}$$

The process rate is given by $-r_c = kC_S$:

$$-r_c = \frac{k k_m A_p C_B}{k + k_m A_p} \ . \tag{7-17}$$

If the mass transfer rate constant k_m is large, $k_m > k$, the rate reduces to:

$$-r_c = KC_B; \tag{7-18}$$

that is, the true kinetic rate is based directly on the bulk concentration. At high temperatures, $k \geq k_m$ since the activation energy for k is typically 50 kcal and the process becomes controlled by mass transfer. Due to the low activation energy for the mass transfer coefficient, the process rate becomes almost independent of temperature at high temperatures.

External mass transfer effects can be minimized by increasing the velocity (v) or mass flux (ρ v) and decreasing the particle size (D_p). The mass transfer coefficient increases with an increase in particle size according to Eqs. 7-9 through 7-12 as $k \propto D_p^n$ with $0.22 \leqq n \leqq 0.641$. Since the particle external surface area per unit weight is inversely proportional to D_p, decreasing the particle size increases the $k_m A_p$ product by D_p^m with $0.359 \leqq m \leqq 0.78$.

Similarly, for heat transfer:

$$T_S = T_B - \frac{(-r_c)(\Delta H_r)}{hA_p} \ . \tag{7-19}$$

For large reaction rates, high heats of reaction, large particles, or low velocities the solid temperature may be significantly different from the bulk gas temperature. For endothermic reactions the particle is cooler whereas for exothermic reactions the particle temperature is higher. The effect of velocity and particle size on h are the same as for k_m.

7.2.2 Pore Diffusion

The gasification reaction occurs principally within the particle. Except at very high temperatures, reactants must diffuse into the pore to the reacting surface. The average reaction rate within the particle may be related to the rate based on the surface concentration in terms of the effectiveness factor (for example, Satterfield 1970) defined as follows:

$$\eta = \frac{(r_{avg})}{r_{surface}} \quad . \tag{7-20}$$

The effectiveness factor is a function of a dimensionless group termed the Thiele modulus, which depends on the diffusivity in the pore, the rate constant for reaction, pore dimension, and external surface concentration C_S.

The effectiveness factor for a wide range of reaction kinetic models differs little from the first order case. For an isothermal particle, the first order reaction effectiveness factor is given as follows:

$$\eta = \frac{\text{Tanh } \phi}{\phi} \quad , \tag{7-21}$$

where ϕ is the Thiele modulus,

$$\phi = L_p \left(\frac{k C_s^{m-1}}{V_p D} \right)^{1/2} \quad , \tag{7-22}$$

and

L_p = effective pore length, cm = R/3 for spheres (R = particle radius);
k = reaction rate constant, $(cc/mole)^{m-1} s^{-1}$;
C_S = external surface concentration, moles/cc;
m = reaction order;
V_p = pore volume, cc/g; and
D = diffusivity, cm^2/s.

When diffusion is fast relative to surface kinetics, $\phi \to 0$, $\eta \to 1$, and $r_{avg} = r_{surface}$. Under these conditions all of the pore area is accessible and effective for reaction. When $\phi \to \infty$, that is, diffusion is slow relative to kinetics, the reaction occurs exclusively at the particle external surface; reactant gas does not penetrate into the pores.

For the process controlled by pore diffusion, the apparent reaction rate constant k_{app} is given as follows based on the Thiele modulus:

$$k_{app} \propto (kD)^{1/2} \quad . \tag{7-23}$$

Therefore, the apparent activation energy is given by Eq. (7-24), under the assumption that the activation energy for diffusion is much less than that for reaction.

$$E_{app} = \frac{E_{diff} + E_{act}}{2} \cong \frac{E_{act}}{2} \tag{7-24}$$

The effect of pore diffusion is to halve the activation energy for the process.

7.2.3 Surface Kinetics

The surface kinetics depend on the reaction and carbon species under consideration. Kinetic models are presented in the following sections. The activation energies for gasification reactions are on the order of 50 kcal/mole.

7.2.4 Global Kinetics

The global kinetic expression combines the effects of mass transfer, pore diffusion, and kinetics. Thus at steady state:

$$W = (-r_{surface}) = \text{overall process rate.} \tag{7-25}$$

At low temperatures, the kinetic rate constant approaches zero. Thus the pore diffusion and mass transfer processes are very fast relative to the kinetics; the kinetic step is rate-limiting. As the reaction temperature increases, pore diffusion tends to be important and, at sufficiently high temperatures, external mass transfer dominates. The effects of these processes on the activation energy are shown in Figure 7-3.

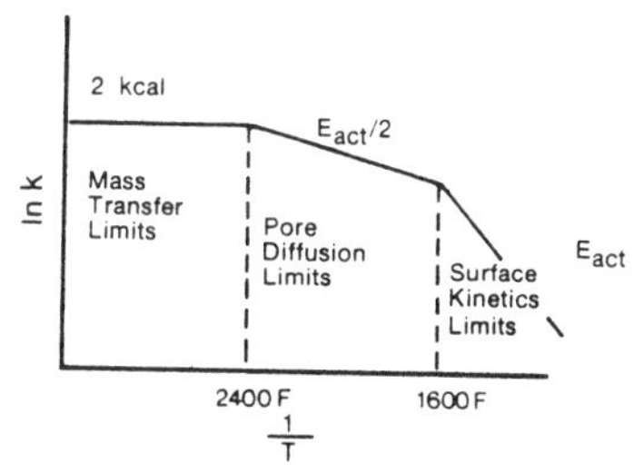

Figure 7-3. The Effect of Temperature on Reaction Rate of Heterogeneous Processes

7.2.5 Estimates of Pore Diffusion Effects

As an example, consider a biomass char with a pore volume and surface area of 0.5 cc/g (0.008 ft^3/lb) and 200 m^2/g, respectively. This translates to a mean pore radius of 50 Å. The Knudsen diffusivity controls the diffusional transport process at low pressure, and a value of the diffusion coefficient is estimated as 0.06 cm^2/s (0.004 ft^2/min). For a steam concentration of 1.1 x 10^{-5} gmole/cc (6.86 x 10^{-7} lbmole/ft^3) at 1 atm, and pore length of one-third the particle radius, the data presented in Table 7-1 were generated.

Table 7-1 shows that for biomass chars with the assumed properties, particles smaller than 20 mesh will be free of pore diffusion effects at gasification temperatures. Chars undergoing combustion may be diffusion-limited.

Table 7-1. EFFECTIVENESS FACTORS FOR BIOMASS CHARS

	Kinetic Rate (g char/g-min)			
Particle Diameter	0.01 (700 C)[a]	0.10 (800 C)[a]	1.00 (900 C)[a]	240 Combustion (900 C)[a]
1/4 in.	0.413	0.133	0.042	0.0027
20 mesh	0.968	0.765	0.317	0.021
200 mesh	0.999	0.998	0.975	0.233

[a]Approximate temperature at which the rate will be observed for biomass chars for gasification with CO_2 or steam.

7.2.5.1 Gasification Reactions

For coal chars, the rate of steam gasification is on the order of 0.01 g/g-min to 0.05 g/g-min at 900 C (1650 F). Thus for coal chars having the same properties used to construct Table 7-1, pore diffusion limitations appear to become important for particles larger than 20 mesh at 900 C.

For a variety of bituminous coal chars, Dutta et al. (1975) found that pore diffusion became important above 980 C for particles of -35 + 60 mesh size and larger for CO_2 gasification.

Hedden and Lowe (1965) claim that pore diffusion is unimportant for graphite particles smaller than 35 mesh at 900 C.

Walker and Hippo (1975) examined the effect of particle size on gasification rate for lignite and bituminous chars. At 900 C (1650 F), pore diffusion mass transfer appeared to be important at particle sizes of 325 mesh, at least during the initial stages of gasification. Since coal chars contain some very fine pores, elimination of pore diffusion is probably not possible but smaller particle sizes tend to minimize the effect.

For biomass materials, the reactivity is such that comparable gasification rates for wood chars are obtained at temperatures 100 C to 200 C lower than those required for coal. Thus, diffusion mass transfer may be important for 20 mesh particles at temperatures on the order of 700 C to 800 C (1300 F to 1475 F). For fluidized-bed gasification of biomass chars, smaller particles are used and thus mass transfer within the particles does not tend to affect the process rate. However, for the large particles (1/4 in.) used in fixed bed operations, intraparticle mass transfer may decrease the process rate significantly even at 700 C (1300 F). At present there is insufficient information to quantify more fully the diffusional effect; data on the structure, effect of particle size, and rate for a variety of biomass chars are limited and none are available for pelletized materials.

7.2.5.2 Combustion Reactions

The combustion reaction occurs at a much more rapid rate than gasification. For Saran chars, Tomita et al. (1977) show that the rate of combustion in air at 550 C is comparable to the rate of CO_2 gasification at 900 C. The gasification rate of the char was 1 g/g-min, which is typical of biomass chars.

Thring and Essenhigh (in Lowry [1963]) report an activation energy of about 30 kcal for the carbon combustion reaction. At 900 C, the approximate ratio of the kinetic combustion rate to gasification is estimated to be:

$$\left(\frac{r_{combustion}}{r_{gas}}\right)_{900C} = 240 \ . \qquad (7\text{-}26)$$

Based on the relative rate data, an estimated intrinsic combustion rate would be 240 g/g-min for biomass chars at 900 C. The combustion reaction rates for various particle sizes, including the effects of pore diffusion, are given in Table 7-2.

Table 7-2. THE EFFECT OF PORE DIFFUSION ON GASIFICATION AND COMBUSTION RATES FOR BIOMASS CHAR

Particle Diameter	Pore Diffusion Corrected Rate, (g/g-min at 900 C)		
	Combustion	Gasification	$\frac{r_{combustion}}{r_{gas}}$
1/4 in.	0.648	0.042	15.4
20 mesh	5.04	0.317	15.9
200 mesh	55.9	0.975	57.3

The effect of pore diffusion is to bring the rates closer together. The combustion reaction, however, is still more than an order of magnitude faster than the gasification reaction.

7.2.6 Estimates of External Mass Transfer Effects

External mass transport generally becomes dominant at temperatures higher than that at which pore diffusion limits the gasification rate. For small particles, $D \leq 20$ mesh, mass transfer limitations generally are not important because these particles have external surface areas that are large compared to their unit volume. Furthermore, mass transfer coefficients are greater in fluid bed operations due to the motion of the solid particles. Thus in fluid bed operations, external mass transfer limitation in the temperature region below 900 C to 1100 C is never important. For fixed bed operation, mass transfer to large particles can be important.

In fixed beds, the mass transfer coefficient is given by the following expression resulting from Eq. (7-9) and (7-10):

$$k = \frac{j\ (\rho *V)}{(Sc)^{2/3}} = \frac{0.357\ Re^{-0.359}\ (\rho *V)}{(Sc)^{2/3}} \qquad (7\text{-}27)$$

Table 7-3 presents mass transfer calculations for a fixed-bed gasifier operating at 1 atm and 1 ft/s gas velocity with 1/4-in. particles. For a kinetic gasification rate of 1.0

Table 7-3. EFFECT OF MASS TRANSPORT ON THE OVERALL GASIFICATION RATE IN A FIXED BED GASIFIER

Reaction Rate	= 0.014 g/g-min = 0.014 lb/lb-min
Particle Size, D_p	= 1/4-in. cylinders
Gas Superficial velocity, v	= 1 ft/s
Pressure	= 1 atm CO_2 or steam
Sc (Schmidt Number)	= 0.7, dimensionless
Gas Density, ρ	= 5.54×10^{-4} lbmoles/ft^3
Viscosity of gas, μ	= 0.035 cP
Bed Voidage, ϵ	= 0.5 ft^3 of voids/ft^3 of bed
Re (Reynolds Number)	$= \frac{\rho D_p V}{\mu} = 21.6$
Particle Density (ρ_p)	= 30 lb/ft^3
j_D (Colburn j Factor)	= 0.237, dimensionless
$\bar{k}_m = k_m C$	$= 0.01 \frac{\text{moles}}{ft^2\text{-min}} \times 12\ \text{lb/mole} = 0.12 \frac{\text{lb carbon}}{ft^2\text{-min}}$
A_p, Specific External Particle Surface Area,	$= \frac{S_p}{V_p} \cdot \frac{1}{\rho_p} = \frac{\left[2 R_p^2 + 2 R_p (2 R_p)\right]}{R_p^2 (2R_p)} \times \frac{1}{\rho_p}$
	$= \frac{3}{R_p \rho_p} = 9.60\ ft^2/lb$
$k_m A_p$	= 1.15 lb/lb-min = 1.15 g/g-min

g/g-min (see Table 7-1), the observed rate, limited by pore diffusion mass transfer, would be 0.014 g/g-min. Using the external mass transfer rate constant from Table 7-3, the mole fraction drop, Δy, across the external film would be:

$$\Delta y = \frac{0.014 \text{ g/g-min}}{1.15 \text{ g/g-min}} = 0.012 \quad .$$

If relatively pure steam or CO_2 were being used in gasification, external mass transfer again would not limit the process.

From this analysis, it may be concluded tentatively that particle size is important in terms of diffusional limitations during gasification but may not be in terms of external mass transfer effects. This conclusion is dependent on the structural properties of the feedstock. For biomass chars, the process rate can be increased significantly by using small particles that tend to eliminate pore diffusion mass transfer.

For combustion, the mole fraction drop at 900 C is more significant:

$$\Delta y = \frac{0.648 \text{ g/g-min}}{1.15 \text{ g/g-min}} = 0.56 \quad .$$

The combustion reaction may become limited by external mass transfer for larger particles at 900 C. Thus, combustion is predominantly a surface phenomenon at the higher gasification temperatures whereas gasification occurs more uniformly throughout the particle. For smaller particles, as in a fluidized bed, external mass transport has a lesser effect on the combustion rate.

7.3 Mechanistic Considerations for CO_2 and Steam Gasification

Considerable information is available concerning the mechanisms of the gasification reactions. The C-CO_2 reaction has been the most extensively studied of the gasification reactions because its products do not enter into side reactions. The steam-C reaction is technically the more important of the two reactions; it has been found that both reactions are similar in their kinetics.

7.3.1 Gasification with CO_2

Mentser and Ergun (1973) recently reviewed the literature on the C-CO_2 reaction and performed a number of isotope experiments on spheron carbon to learn about the reaction mechanism. The mechanistic studies indicate:

- The exchange of oxygen by Eq. (7-28) occurs reversibly at all temperatures investigated, including those below that required for gasification:

$$C_f + CO_2 \underset{k_1'}{\overset{k_1}{\rightleftharpoons}} C(O) + CO \quad . \qquad (7\text{-}28)$$

  C(O) represents a surface oxide, not adsorbed oxygen. C_f are free carbon sites.

- Interchange of carbon between CO and solid carbon occurs only at temperatures on the order of 1500 C.

- Deposition of carbon on the surface by decomposition of CO_2 occurs at an insignificant rate.

Bonner and Turkevich (1951) found that during the initial stages of reaction 7-28, 95% of the radioactive CO_2 charged was converted to radioactive CO with no increase in system pressure. This result further substantiates reaction 7-28. During later stages of the reaction, the gas pressure increased, suggesting surface decomposition of the oxide species:

$$C(O) \underset{k_2'}{\overset{k_2}{\rightleftharpoons}} CO + nC_f \tag{7-29}$$

In several studies (Bonner and Turkevich 1951; Orning and Sterling 1954), it was found that the oxygen exchange reaction (Eq. 7-28) was potentially faster than the oxide decomposition reaction (Eq. 7-29) at low temperatures, suggesting that the oxide decomposition was rate controlling.

Assuming that Eqs. 7-28 and 7-29 in the forward direction apply, the reaction rate is given by the following kinetic expression if it is assumed that the reactions are far from equilibrium:

$$-r_c = \frac{k_1 C_{tot} \; P_{CO_2}}{1 + \frac{k_1'}{k_2} P_{CO} + \frac{k_1}{k_2} P_{CO_2}} \; . \tag{7-30}$$

According to the rate equation, CO_2 and CO may suppress the reaction. CO can decrease the gasification rate by reversibly removing the surface complex C(O) by Eq. 7-28. At high CO_2 partial pressures, the reaction becomes independent of CO_2 pressure because the surface sites become saturated with CO_2. The oxygen exchange reaction (Eq. 7-28) limits the rate at high temperatures due to its lower activation energy. The surface oxide decomposition reaction (Eq. 29) has a large activation energy and is much more temperature sensitive than the exchange reaction. Grabke (1966) estimated that in pure CO_2 gaseous environments, and at temperatures above 1000 C, reaction 7-28 becomes rate controlling; in environments with equal CO and CO_2, reaction 7-28 does not become limiting until about 1200 C for the carbon he used. Mentser and Ergun (1973) suggest that the forward exchange reaction has an activation energy of about 53 kcal; the reverse exchange reaction 36 kcal; and the oxide decomposition reaction 58 kcal. In Eq. 7-30, the temperature dependence of the ratios k_1'/k_2 and k_1/k_2 are such that these terms become negligible at the high temperatures at which the rate equation goes to a first order form:

$$-r_c = k_1 \, C_{tot} \, P_{CO_2} \; . \tag{7-31}$$

At the lowest temperatures, in pure CO_2 atmospheres, the gasification rate varies from zero order to first order:

$$-r_c = \frac{k_1 \, C_{tot} \, P_{CO_2}}{1 + \frac{k_2}{k_1} P_{CO_2}} \; . \tag{7-32}$$

Equation 7-32 explains the observation that at moderate pressures and temperatures the reaction varies by about $(P_{CO_2})^{1/2}$.

Equation (7-30) also states that the rate is proportional to C_{tot}, the total number of active carbon sites available. C_{tot} is not the total surface area of the carbon; it has been proposed that only edge carbon atoms, atoms present along crystal defects, and atoms adjacent to mineral matter deposits (particularly CaO, MgO, and FeO_4) are sufficiently reactive to be gasified. Thus C_{tot} is only a small fraction of the surface atoms.

In studies where oxygen, hydrogen, and carbon monoxide chemisorption areas and total surface areas of chars were compared (Laine et al. 1963a, b; Menster and Ergun 1973), it was found that the chemisorption area amounted typically to only several percent of the total surface area. Thus the number of exposed reactive carbon sites is small compared to the total number of exposed sites. Further, Ergun (1956) and Menster and Ergun (1973) showed that the activation energies of the elementary steps are independent of char type. Thus the reaction rate at a given temperature is dependent on the number of active sites only.

The mechanistic equation (7-32) can be extended to consider the reversible approach to equilibrium by adding in the reverse reaction from Eq. 7-29. Under these conditions, the gasification rate equation becomes:

$$-r_c = \frac{k_1 C_{tot} (P_{CO_2} - P^2_{CO}/k_{eq})}{1 + \frac{k_1 + k_2'}{k_2} P_{CO} + \frac{k_1}{k_2} P_{CO_2}} \quad . \tag{7-33}$$

At pressures substantially above 1 atm Blackwood and Ingeme (1960) report that the rate varies with a CO_2 partial pressure order somewhat greater than unity.

7.3.2 Kinetics of Carbon-Steam Reaction

The kinetics of the carbon-steam reaction are in many respects analogous to those for the carbon-CO_2 reaction. The carbon-steam reaction is made more difficult to analyze by possibly competitive reactions resulting from the generation of H_2 and CO_2 by side reactions.

Lowry (1963) reports general agreement that the products of the carbon-steam reaction at low pressures are CO and H_2:

$$C + H_2O \longrightarrow CO + H_2 \quad . \tag{7-34}$$

Carbon dioxide is produced through the water-gas shift reaction which is catalyzed by the carbon surface.

As with CO_2, water vapor deposits oxygen on the carbon surface at temperatures below gasification temperatures. This oxygen may be removed by reaction with either carbon monoxide or hydrogen at temperatures below gasification temperatures. At gasification temperatures the surface oxide readily decomposes to liberate carbon monoxide.

A plausible mechanistic model is the following:

$$C_f + H_2O \underset{k_6'}{\overset{k_6}{\rightleftharpoons}} C(O) + H_2 \tag{7-35}$$

$$C(O) \underset{k_7'}{\overset{k_7}{\rightleftharpoons}} CO + nC_f . \tag{7-36}$$

In Eq. 7-35, C_f represents free carbon sites for reaction while C(O) is the surface oxide. Assuming this two-step mechanism, at steady state the rate expression is:

$$-r_c = \frac{k_6 C_{tot} P_{H_2O}}{1 + \frac{k_6'}{k_7} P_{H_2} + \frac{k_6}{k_7} P_{H_2O}} \quad . \tag{7-37}$$

According to the model, steam and hydrogen may suppress the reaction. Hydrogen can reduce the reaction rate by removing the surface oxide by the reverse reaction in Eq. 7-35. Thus the number of complexes available to decompose to CO is reduced. At sufficiently high steam partial pressures, the surface becomes saturated with the oxide complex and the decomposition of the oxide by Eq. 7-36 becomes rate controlling.

The rate is also dependent on C_{tot}, the number of carbon sites available on the surface for reaction. These sites are the same as those that are capable of reacting with CO_2.

The rate constant (k_6) exhibits an Arrhenius dependency and thus increases with temperature. The ratios of rate constants in the denominator of Eq. 7-37 can exhibit a positive or negative temperature dependence depending on the activation energies (E) of the individual rate constants. Since the surface oxide is identical in both CO_2 and steam gasification reactions, the activation energy for k_7 should be on the order of 58 kcal. For graphite tubes, Lowry reports E_6 = 32.7 kcal, E_6' = 14.2 kcal, and E_7 = 46.6 kcal. While the activation energy E_7 is lower than for the CO_2 reaction (E_2 = 58 kcal) the difference is probably within experimental error.

An appropriate extension of the rate expression as Eqs. 7-36 and 7-37 approach equilibrium is:

$$-r_c = \frac{k_6\, C_{tot} \left(P_{H_2O} - P_{CO} P_{H_2} / k_{eq}\right)}{1 + \frac{k_6'}{k_7} P_{H_2} + \frac{k_6}{k_7} P_{H_2O}} \quad . \tag{7-38}$$

The reactivities of the chars are affected by thermal annealing or graphite formation; the pretreatment and thermal history of the char are important. If freshly prepared char and char thermally stabilized (annealed) at the reaction temperature are reacted under the same conditions, the fresh char will have a higher initial reactivity than the stabilized char. The rates will tend to become the same at longer reaction times, after the fresh char has stabilized. Thermal annealing becomes important at temperatures above 700 to 1100 C. During thermal annealing, carbon active sites (edges and dislocations) are lost due to surface reorganization, and the char structure becomes more graphitic. Additionally, thermal annealing causes a decrease in porosity of the char that reduces the accessibility of the internal surface to reacting gases. The overall effect of pretreatment on yield is not extremely significant since the carbon-steam reaction is relatively slow. If the pretreatment is conducted at a temperature higher than the reaction temperature, the reactivity of the char will be lower than that of the char prepared at the reaction temperature. Table 7-4 shows the specific rate data for a coal devolatilized at two different temperatures. The rate of gasification was reduced by a factor of two to three by thermal annealing.

Table 7.4. RATE OF GASIFICATION OF COAL CHAR IN STEAM AT 850 C (Jolley et al. 1953)

	Rate (g/g-min)	
	Pretreatment Temperature	
Burnoff (%)	850 C	1000 C
0	0.0041	0.0014
30	0.0059	0.0026
60	0.0087	0.0038

At high temperatures, the thermal stabilization becomes rapid enough to interfere with the rate of gasification. Figure 7-4 from Yang and Steinberg (1977) shows that above about 1200 C thermal annealing tends to make the rate insensitive to temperature for a variety of carbons; above 1400 C, the rate decreases with increasing temperature due to the very rapid loss of active sites.

The water-gas shift reaction is considered to occur catalytically on the carbon or ash surface at sites not undergoing gasification and always occurs during gasification.

It has been found experimentally that the rate equation (7-39) correlates catalytic water-gas shift data:

$$-r_{CO} = \frac{k_8 \; P_{CO} P_{H_2O} - P_{H_2} P_{CO_2}/k_{eq}}{(1 + k_9 \; P_{CO_2} + k_{10} \, P_{CO}} \quad . \tag{7-39}$$

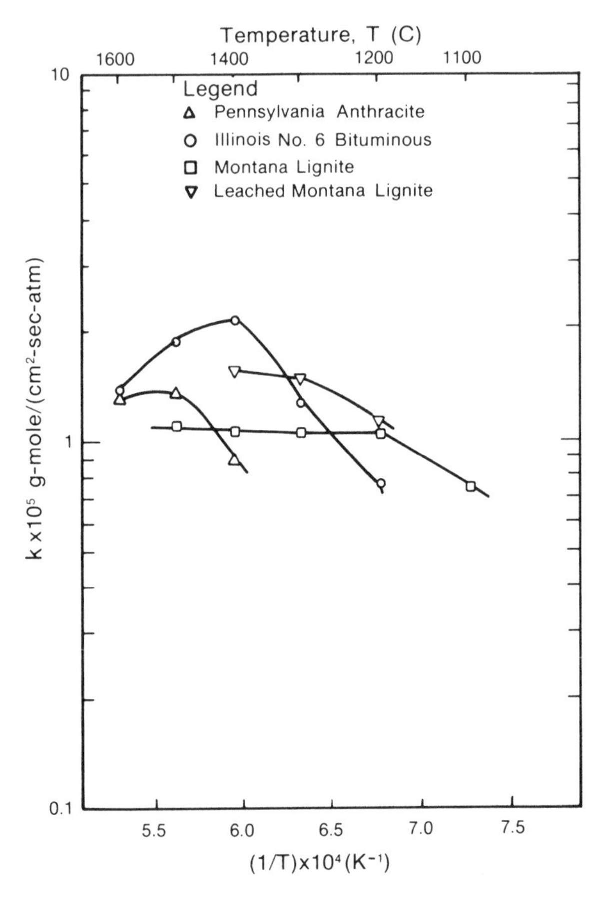

Figure 7-4. Rate Constant as a Function of Temperature for Coal Chars

7.4 RELATIVE REACTIVITIES OF CARBONS DURING GASIFICATION

A number of recent studies have examined the relative reactivities of carbons to carbon dioxide and steam. All of the studies discussed in this section used chars that were devolatilized by slow heating in nitrogen to a temperature from 900 C to 1000 C; the chars were held at the higher temperature for at least 30 min to stabilize (graphitize) the char.

7.4.1 Reactivity in CO_2 at 1 atm, 900 C

Considerable data have been reported on the reactivity of chars with CO_2 at 900 C and 1 atm pressure. Most runs were made for -40 mesh particles. The reactivity data (Table 7-5) show clearly the effect of carbon rank (degree of aromatization) on gasification reactivity. In carbon dioxide lignites are, on the average, ten times more reactive than the bituminous chars at the same conditions.

Table 7-5. COMPARISON OF CHAR REACTIVITY IN C–CO_2 REACTION AT 900 C

	Reference	Reactivity (g/g-min)	Method
Montana Lignite	Walker and Hippo 1975	0.058	TGA
ND Lignite (1)	Walker and Hippo 1975	0.060	TGA
ND Lignite (2)	Walker and Hippo 1975	0.045	TGA
Wyoming Sub C	Walker and Hippo 1975	0.055	TGA
Wyoming Sub A	Walker and Hippo 1975	0.028	TGA
Illinois HVC	Walker and Hippo 1975	0.011	TGA
IGT No. 155 (ILL 6)	Dutta et al. 1975	0.0107	TGA
Hydrane No. 49 (ILL 6)	Dutta et al. 1975	0.0117	TGA
Synthane No. 122 (ILL 6)	Dutta et al. 1975	0.0129	TGA
Hydrane Char (ILL 6)	Fuchs and Yavorsky 1975	0.011[a]	Fluid bed
Hydrane Char (HVA)	Dutta et al. 1975	0.0063	TGA
High Vol A (KY)	Walker and Hippo 1975	0.0032	TGA
Graphite From Pitch	Peterson and Wright 1955	0.002	TGA
Coconut Shell Charcoal	Lowry 1963 (See Table 7-13)	0.01	—

[a]Corrected to 1 atm assuming rate is related by $(P_{CO_2})^{1/2}$.

Baird et al. (1976) report relative reactivities for biomass and coal chars based on the approach to equilibrium of the C-CO_2 reaction. In their fixed bed reactor using pure CO_2, the ratio of exit CO to CO_2 is a measure of reactivity at steady state. The materials studied included paper board, wood chips, lignite, and subbituminous coals. Each was pyrolyzed at reaction temperature prior to reaction. The wood chips consisted of chips and branches of 1/8 in. by 3/4 in. size. The paper board, composed of compressed paper and plastic, had a density of 60 lb/ft^3 and was charged in 1-in. squares. No particle size was reported for the coals. Because of the variability in particle sizes, it is not possible to rate the feedstocks quantitatively. Table 7-6 shows the qualitative rating of data taken at the lowest temperatures of the study, where mass transport is least important.

Table 7-6. QUALITATIVE COMPARISON OF THE REACTIVITY OF BIOMASS AND COAL CHARS IN CARBON DIOXIDE

	Activity Ratio $\left(\frac{CO}{CO + CO_2}\right)$		
Feedstock	500 C	600 C	700 C
Wood chips	8.2	11.8	21.2
Lignite	2.3	6.4	16.6
Pressed paper	2.3	1.5	7.9
Subbituminous coal	1.7	4.9	8.8
Equilibrium	9.1	34.0	71

According to the study, the wood chips exhibit a reactivity greater than lignite, but a quantitative ranking in terms of weight loss of carbon is not possible. The paper exhibits an activity close to that of the subbituminous char, a result which is probably unreasonable. No physical property data for the chars or explanation of the reactivity is given.

7.4.2 Reactivity in Steam

A number of char reactivity studies have been conducted in steam. The same general pretreatment policy was followed; that is, slow devolatilization of the char to be tested at the reaction temperature. Table 7-7 summarizes a number of these gasification studies. Data are available for several biomass chars from Rensfelt et al. (1978). The rates have been adjusted to 900 C and 1 atm of steam with the assumptions that the reported activation energy is correct where applicable and that the rate is proportional to $(P_{H_2O})^{1/2}$. Table 7-7 shows that the data for the bituminous coal chars from Friedman (1975) and Lowry (1963) are consistent. The rates for biomass gasification appear to be four to ten times greater than those for the lignite chars, which are the most reactive coal chars available.

A comparison of Tables 7-5 and 7-7 shows that the gasification rate in pure steam is greater than that in pure CO_2 by a factor of about 3 to 5. Thus for rating chars on a relative basis, kinetic studies in CO_2 are satisfactory.

In coal chars, rank is a measure of the degree of aromatization or graphitization of the carbon. As rank increases, reactivity decreases. Since biomass chars are much more reactive than even lignite, the degree of aromatization resulting from pyrolysis may be less than that in the coal chars. Such a hypothesis is consistent with the monoaromatic structure of lignin. The difference in reactivity also may be partly explained by differing pore structures; biomass pore structures are much more open than coal pore structures.

Table 7-7. REACTIVITY OF CHARS IN STEAM AT 900 C

	Reactivity (g/g-min at 900 C)		
	Reference	Reported	Corrected to 1 atm steam
Illinois COED	Friedman 1975	0.0027	0.0027
Utah COED	Friedman 1975	0.0054	0.0054
WKY COED	Friedman 1975	0.0069	0.0069
Pittsburgh	Friedman 1975	0.0093	0.0093
Pittsburgh oxidized[a]	Friedman 1975	0.0037	0.0037
Lignite 247	Nandi et al. 1975	0.020	0.132
Lignite LLL	Nandi et al. 1975	0.0159	0.105
Montana Lignite[b]	Linares et al. 1977	0.045	0.296
ND Lignite	Linares et al. 1977	0.044	0.290
Wy. Sub C	Linares et al. 1977	0.0390	0.257
Wy. Sub A	Linares et al. 1977	0.0156	0.103
Sub BC 2A8	Nandi et al. 1975	0.005	0.033
Ill HVC	Linares et al. 1977	0.0070	0.046
MVB 274	Linares et al. 1977	0.0004	0.0027
Coconut Shell	Linares et al. 1977	0.070	0.070
Solid Waste[b]	Rensfelt et al. 1978	0.318	0.372
Poplar Wood[b]	Rensfelt et al. 1978	0.942	1.102
Straw[b]	Rensfelt et al. 1978	0.463	0.542
Bark[b]	Rensfelt et al. 1978	0.725	0.849
High Moor Peat[b]	Rensfelt et al. 1978	0.152	0.178

[a]Oxidized during pyrolysis; corrected to 900 C using 18 kcal activation energy.
[b]Measured at 45% burnout, 0.73 atm steam.

7.5 EFFECT OF BURNOFF AND SURFACE AREA

Numerous studies have been conducted on the effect of burnoff (% weight loss, sometimes called burnout) on gasification rate in carbon dioxide and steam atmospheres; they show that the rate is dependent on burnoff in a complicated and somewhat confusing manner. Some materials show little change of rate with burnoff while others show marked changes, either positive or negative. Burnoff alters the pore size distribution, pore volume, and hence surface area available for reaction. It is logical, therefore, to believe that the rate of reaction should depend on the burnoff to the extent that burnoff alters the porous structure of the solid. In studies by Hedden and Lowe (1965), the chars to be gasified were produced by slow heating (20 C to 50 C per minute) to a temperature of 900 C. The graphites used to produce the chars were commercial products, and it is assumed that at some point during their preparation they were exposed to temperatures higher than 900 C. The maximum temperature and timing of exposure are important in that the stabilization or graphitization process is sensitive to the highest temperature experienced by the char. Any rate studies carried out at or below the temperature of manufacture will not be affected by major changes in the char structure due to thermal annealing.

Hedden and Lowe (1965) examined the rate of gasification of two graphites at 1030 C and 1 atm of CO_2 as a function of burnoff. The BET surface area based on liquid nitrogen was also determined at each burnoff level. Table 7-8 summarizes their results.

Table 7-8. COMPARISON OF BURNOFF SURFACE AREA AND RELATIVE RATES FOR BURNOFF OF GRAPHITES[a]
(Nitrogen BET)

% Burnoff (X)	Specific Surface (S)	Surface Area Ratio [S(O)/S(X)]	Reactivity Ratio [R(O)/R(X)]
Graphite G-S			
0	1.7	1.00	1.00
2	4.2	2.47	—
5	7.4	4.35	2.41
12	9.3	5.47	3.88
20	10.0	5.88	4.49
38	9.8	5.76	4.97
60	9.6	5.65	4.65
Graphite G-9			
0	1.6	1.00	1.0
5	7.3	4.6	2.5
9	10.5	6.6	3.4
21	14.5	9.1	3.8

[a]From Hedden and Lowe 1965.

There is a direct relationship between BET surface area and reaction rate. However, the rate is not directly proportional to the measured area of the particle, indicating that the low temperature BET surface might not be the "correct" area for correlating reaction rates.

Turkdogan and Vinters (1969) report N_2 BET areas and kinetic rates for the gasification of a coconut shell charcoal and a graphite at 900 C. The surface areas and rates were reported for a 10% burnoff. The data are presented in Table 7-9.

For the two very different carbons, the ratio of gasification rates and surface areas is essentially the same, indicating a proportionality between area and rate. The gasification rates reported are unusually low for the carbons investigated.

Table 7-9. GASIFICATION RATES AND SURFACE AREAS AT 10% BURNOFF FOR TWO CARBONS[a]
(Nitrogen BET)

Carbon	Surface Area (S) (m^2/g)	Gasification Rate (R) (g/g-min)
Graphite	4	8×10^{-5}
Coconut shell	850	0.014
Ratio	200	175

[a]From Turkdogan and Vinters (1969).

Dutta et al. (1975) investigated the relationship between N_2 BET area and reactivity at 900 C for four different coal chars. Their results are given in Table 7-10.

Table 7-10. RELATIONSHIP BETWEEN SURFACE AREA AND GASIFICATION RATE FOR COAL CHARS[a]
(Nitrogen BET)

Coal Char	Total Surface Area(S) (m^2/g)	Surface Area r_p 27.5 Å	Rate (R) (g/g-min)	$\frac{R}{S(27.5)}$
Hydrane No.150	18.75	18.75	0.067	3.57×10^{-3}
IGT No. HT155	423.87	25.43	0.113	4.44×10^{-3}
Hydrane No. 49	171.69	34.42	0.123	3.57×10^{-3}
Synthane No. 122	280.87	38.06	0.136	3.57×10^{-3}

[a]From Dutta et al. (1975).

There was no correlation between the reactivity and total BET surface area. Based on the pore size distribution, Dutta et al. found a strong correlation between the area for pores greater than 27.5 Å and the reactivity. At 900 C and 1 atm the bulk diffusion coefficient is about 1.7 cm^2/s while the Knudsen diffusion coefficient is 0.014 cm^2/s. Thus the diffusion process is pure Knudsen diffusion. Since the Knudsen diffusion coefficient is directly proportional to the pore radius, diffusion limitations in the fine pores might be responsible for making part of the surface completely inaccessible.

Wen et al. (1977) investigated the effect of burnoff on CO_2 BET surface area and on reactivity, on the assumption that the CO_2 surface area would be a better parameter against which to correlate reactivities. Table 7-11 shows the results of this study based on a lignite char. The char was prepared by devolatilization at 1000 C for 30 min. The reactivity was determined at 900 C.

Table 7-11. EFFECT OF BURNOFF ON SURFACE AREA AND REACTIVITY FOR LIGNITE CHARS
(Carbon Dioxide BET)

% Burnoff (X)	CO_2 Surface Area (S) (m^2/g)	Rate Ratio [R(O)/R(X)]	Surface Area Ratio [S(O)/S(X)]
0	137	1.0	1.0
21.4	186	1.4	1.34
45.0	281	2.03	2.05
56.3	306	2.35	2.24
71.4	404	2.50	2.95

These data indicate a strong correlation between the CO_2 reactivity and CO_2 surface area for the lignite char.

Rensfelt et al. (1978) studied the gasification of biomass and coal chars at various temperatures and burnoffs. Figure 7-5 shows the influence of burnoff on the gasification rate at constant temperature for various chars. Biomass materials such as bark and wood exhibit a rate that is strongly dependent on burnoff. This suggests that gasification is generating major increases in surface area in these materials. Figure 7-6 shows that the temperature also affects the rate dependency on burnoff for poplar wood. This might suggest that activated diffusion in molecular size pores contributes significantly to the gasification rate at higher temperatures. Activated diffusion is a transport process that occurs in molecular size pores. The diffusion coefficient is very small and highly temperature-dependent due to molecular interactions between the gas and surface. No surface area or pore size data were reported.

A number of conclusions can be drawn from these studies. Surface area and reaction rate are related during gasification. Furthermore, it is suggested that the percentage of active sites on the total surface remains constant during gasification. Most of the studies suggest that the N_2 BET surface is not a correct measure of the reactive surface. For low area solids (solids with a large mean pore radius), the nitrogen areas correlate reasonably well with reactivity. For high area solids where the bulk of the area is associated with fine micropores, the N_2 BET area is not related to reactivity. Nitrogen cannot penetrate micropores readily; the CO_2 can more readily diffuse into the micropores due to polar interactions with the char surface. The only study not consistent with this conclusion is that of Turkdogan and Vinters (1969), which shows a direct correlation between areas and rates for two solids, one of which has a significant area tied up in micropores.

For correlating reactivities, the CO_2 surface area is probably a better measure of reactive surface area than the N_2 surface area. No comprehensive studies of this type have been done for biomass materials. Since the kinetic rates for the C-H_2O reaction and C-CO_2 reaction are similar, it is expected that similar conclusions could be drawn for the carbon steam reaction. Since steam is a smaller molecule and is more polar than CO_2, it can penetrate more micropores, a fact which may explain the difference in reactivity.

No data are available in the literature which relate surface area to gasification rate for biomass chars.

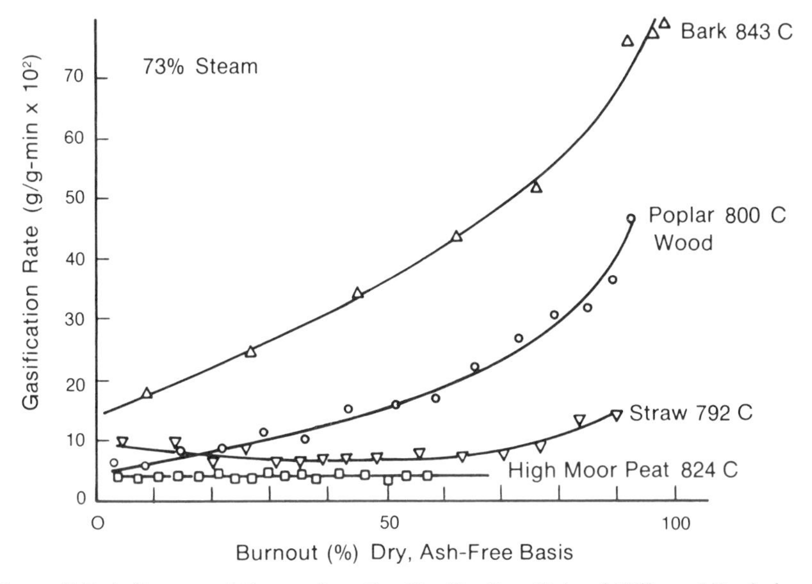

Figure 7-5. Influence of Burnout on the Gasification Rate of Different Fuels in a Steam-Argon Mixture

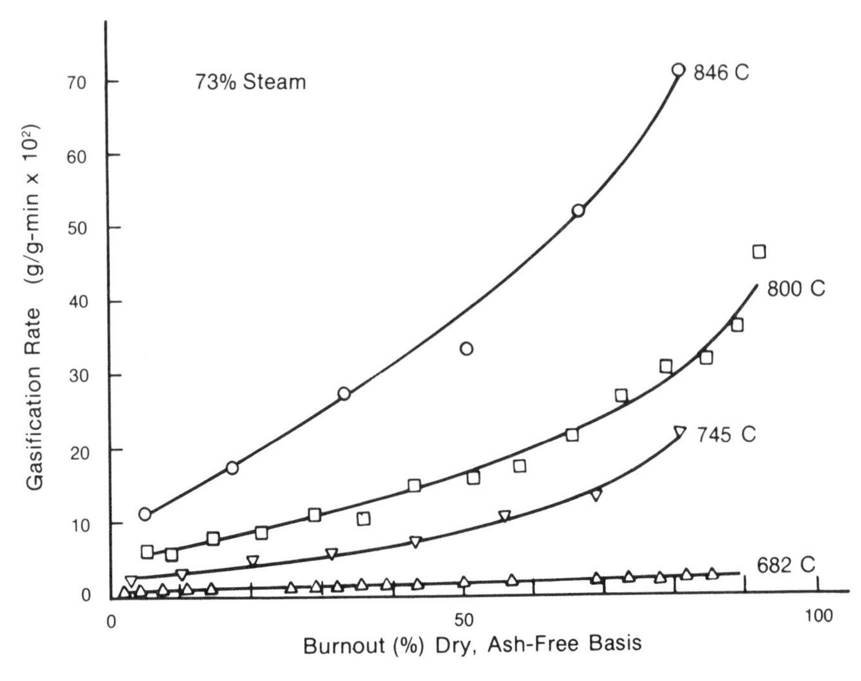

Figure 7-6. Influence of Burnout on the Gasification Rate of Poplar Wood in Steam-Argon Mixture

7.6 RATE CONSTANTS FOR BIOMASS CHARS

Most of the available studies ranking gasification reactivity are based on a constant gas composition. Researchers infer an activation energy based on the rate of carbon gasification from such data. These activation energies are apparent values in as much as there is an effect of temperature on the composition factor as discussed in Section 7-3. The only data available for biomass chars were reported by Rensfelt (1978). Table 7-12 shows apparent activation energies and frequency factors for biomass chars gasified in 0.73 atm of steam.

Table 7-12. KINETIC PARAMETERS FOR BIOMASS AND COAL CHARS
$[k=k_o \exp(-E_{act}/RT)]$

Fuel	Activation Energy (E_{act}) (kcal/mol)	Frequency Factor (k_o) (g/g-min)
Solid waste	59.5	3.9×10^{10}
Poplar wood	43.4	1.2×10^{8}
Straw	43.4	5.9×10^{7}
Bark	37.0	5.8×10^{6}
Peat	40.4	5.1×10^{6}
Bituminous coal	48.8	5.9×10^{7}

Except for the solid waste, the apparent activation energy is essentially constant at 42.6 ± 3.1 kcal/mole. The constancy of activation energy for a variety of charcoals and graphites has been observed by a number of researchers, particularly Ergun as cited earlier. The nature of the carbon is apparent in the frequency factor, which generally decreases with rank for the carbons investigated. The frequency factor is related to the number of active sites, while chemical activity of the sites is related to the activation energy. Thus there appear to be more energetically similar reactive sites available in biomass chars than in coal chars, suggesting either more available surface area or a less ordered structure.

No data are available for the effect of the ambient gas composition on the reactivity of biomass chars. Hedden and Lowe (1965) found that their data and data from other are studies could be fit to Eq. 7-30 over a wide range of conditions. Table 7-13 presents rate constant data for several studies from Lowry (1963). Paralleling the apparent rate data, the frequency factor and activation energy data show the appropriate trends with rank; that is, E_{act} is constant and k_1 decreases with increasing rank. The constant k_1/k_2, which accounts for the retardation of CO_2, is essentially independent of the carbon type. The constant k_1'/k_2 indicates that retardation by CO is strong and more variable than for CO_2. Using the rate constant data, the following equation applies at 800 C for coconut shell charcoal:

$$-r_c = \frac{0.00796\ P_{CO_2}}{1 + 23.37\ P_{CO} + 2.34\ P_{CO_2}} \quad . \qquad (7\text{-}40)$$

Figure 7-7 shows the effect of gas composition on rate for the coconut shell carbons relative to the rate in pure CO_2 at 1 atm. The rate is relatively independent of CO_2 pressure above 1 atm and is strongly inhibited by carbon monoxide. Thus, at 1 atm pressure, the rate in the gaseous environment of one third CO_2 is about one fifth that in a CO_2-free gaseous environment. For more active chars like the biomass materials, the behavior of coconut shell char should be typical. Thus, for biomass chars, changing the pressure should have little effect on gasification rate with CO_2.

Rate data were determined by Long and Sykes (1948) for coconut shell charcoal in steam.

At 800 C,

$$-r_c = \frac{0.0387\ P_{H_2O}}{1 + 33\ P_{H_2} + 2.54\ P_{H_2O}} \quad . \qquad (7\text{-}41)$$

This equation is also plotted in Figure 7-7. In a gaseous environment of one third hydrogen, the rate is depressed by a factor of five as compared to a H_2-free atmosphere. The rate in water is roughly five times as great in steam as in CO_2. Data for coal chars show a similar behavior.

From these data, it is evident that the rate is extremely sensitive to the partial pressure of the products CO and H_2. In practice, the gaseous environment may contain appreciable amounts of both CO and H_2.

7.7 CATALYTIC EFFECTS

A number of investigations have studied the effects of catalysis on gasification. Most metals, their oxides, and salts are more or less catalytic. Tingly and Morrey (1973) report that iron, calcium, and magnesium have the greatest potential effect on reactivity. Surface impurities can also affect the water-gas shift reaction. Biomass is essentially ash-free. Therefore, any catalyst to promote reaction would have to be added from an external source, increasing the ash disposal problem.

As an example related to biomass chars, Rensfelt (1978) investigated the C-steam gasification of peat char with and without a 2% K_2CO_3 catalyst. The alkali tripled the rate of C-steam gasification.

Table 7-13. ARRHENIUS CONSTANTS FOR THE CARBON-CARBON DIOXIDE REACTION

	Temperature Range (C)	k_1 $\left(\frac{\text{g-mole}}{\text{g-min-atm}}\right)$	E_{act} (kcal)	k_1'/k_2 (1/atm)	E_1-E_2 (kcal)	k_1/k_2 (1/atm)	E_1-E_2 (kcal)	Value at 800 C	
								k_1'/k_2	k_1/k_2
Coconut shell charcoal	734-830	6.3×10^8	58.8	1.26×10^{-8}	-45.5	3.16×10^6	30.1	23.37	2.34
New England coke	800-1090	6.9×10^5	47.6	1.4×10^{-2}	-15.0	0.21	-6.3	15.91	4.03
New England coke	--	3.16×10^7	61.7	4.0×10^6	-40.3	3.16×10^{-2}	-6.1	647.3	0.55
Electrode carbon	--	1.0×10^6	50.1	3.16×10^9	-60.6	0.16	-6.6	697.9	3.54
Pitch coke	926-1150	1.05×10^7	40.1	2.0×10^9	-55.1	--	--	33.5	--

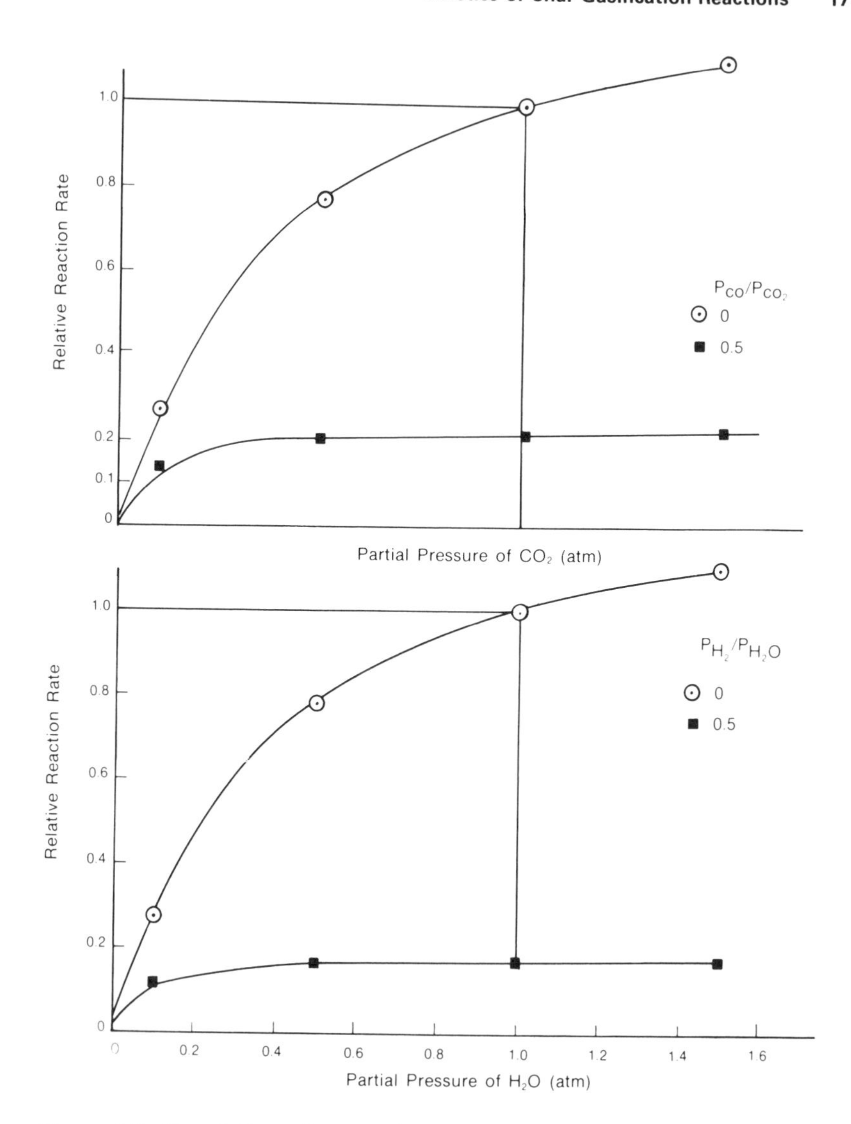

Figure 7-7. Reactivity of Coconut Shell Charcoal in H_2O and CO_2 at 800 C

7.8 MECHANISM AND KINETICS OF COMBUSTION

It is generally believed that the combustion mechanism involves attack of the same active sites as in gasification. Also as in gasification, the reaction is thought to proceed through an intermediate surface oxide that decomposes at a characteristic rate. At low temperatures, the surface oxides are stable and essentially cover the surface. The reaction is zero order under these conditions. At high temperatures, the rate of decomposition is so fast that the rate becomes limited by the formation of surface oxides and the reaction order approaches unity.

At intermediate temperatures, the rate is typically 1/2 order. It also has been determined that both CO and CO_2 are primary products of combustion. However, at all temperatures of interest, CO production is the dominant reaction, with CO_2 being produced in the film surrounding the particle.

A possible mechanism summarized by Laurendeau (1978) is the following:

$$2C_f + O_2 \underset{k_{11}}{\overset{k_{10}}{\rightleftharpoons}} 2\,C'(O) \quad , \tag{7-42}$$

$$C'(O) \xrightarrow{k_{11}} C(O) \quad , \tag{7-43}$$

$$C(O) \xrightarrow{k_{12}} CO + nC_f \quad , \tag{7-44}$$

$$C'(O) \xrightarrow{k_{13}} CO + nC_f \quad , \tag{7-45}$$

$$2\,C'(O) \xrightarrow{k_{14}} CO_2 + C_f \tag{7-46}$$

In this mechanism, C'(O) are primary mobile surface oxides while C(O) are immobile complexes. Such a two-site adsorption helps explain the rapidity of the combustion process since the most active sites are constantly regenerated while, as reaction proceeds, the less reactive sites are removed by the oxide decomposition. Equation 7-46 accounts for primary CO_2 production. Since the number of active sites is small compared to the total surface, the probability that reaction 7-46 will proceed, to produce significant quantities of CO_2, is very small. Equations 7-42 to 7-46 lead to a rate expression of the form:

$$-r_c = k' \; C_{tot} \; P^m_{O_2} \tag{7-47}$$

where m can be 0, 1/2, or 1 with the appropriate simplifications. At combustion temperatures of interest in biomass gasification, m of 1/2 should apply.

Experimental studies have shown that thermal annealing of the carbon is important at temperatures above about 1300 K. Below that point, k' exhibits an Arrhenius behavior with an activation energy of 30-40 kcal. Above 1300 K, the rate becomes independent of temperature and, at very high temperatures, falls to a low value. The annealing or graphitization phenomenon is a stabilizing reaction that eliminates the necessary defects or surface active sites for combustion.

It is generally observed that the primary products of combustion are CO and CO_2. Arthur (1951) has shown that the product mixture is a function of temperature but probably not of carbon type. Between 460 C and 900 C, the primary product distribution is given as follows:

$$\frac{[CO]}{[CO_2]} = 10^{3.4} \exp\,[-12{,}400/RT]\,, \; T \text{ in K} \quad . \tag{7-48}$$

7.9 HYDROGASIFICATION

Pyrolysis is normally carried out in an inert gaseous environment. When the pyrolysis is conducted in hydrogen, with rapid heating (>1000 C/s), it is possible to increase the devolatilization of the feedstock and enhance the hydrocarbon yield. Recently, Anthony and Howard (1976) reviewed the state of the art for hydrogasification of coals.

The hydrogasification reaction takes place in two stages. If the char is prepared and stabilized in an inert atmosphere, the rate of hydrogasification is very low. On the other hand, if hydrogen is in direct contact with the freshly formed char during pyrolysis, the rate of gasification is several orders of magnitude greater. Figure 7-8 from Gray et al. (1975) shows the effect of heating on the hydrogasification rate. For rapid heating, the gasification rate is almost 100 times faster than for the slow heating case.

The rate of hydrogasification depends on temperature, hydrogen pressure, and time. As an example of a rate model, Gray et al. (1975) propose that a parallel sequence occurs.

$$\text{Char} + H_2 \rightarrow CH_4 + \text{Oil} + \text{Light hydrocarbons} \qquad (7\text{-}49)$$
$$\text{Char} \rightarrow \text{Stabilized char} .$$

The rate of reaction of stabilized char with hydrogen is negligible during normal residence times. The stabilization reaction is assumed to consume only a fraction of the char available for hydrogasification. The proposed rate equation then, is:

$$-r_c = \frac{1}{W}\frac{dW}{dT} = KP_{H_2}\left(\alpha - \frac{W}{W_o}\right) . \qquad (7\text{-}50)$$

The parameter α is a function of temperature, coal type, and heating rate. The rate constants plotted in Figure 7-8 are based on the model given in Eqs. 7-49 and 7-50. Hydrocarbon production in atmospheric gasifiers by direct hydrogenation of char is slow, even in comparison to the steam gasification reaction.

The only published hydrogasification studies on biomass have been conducted for a naturally occurring peat. Punwani et al. (1978) and Weil et al. (1978) recently investigated the hydrogasification of peat with a volatile content of 63.2%. Figure 7-9 shows the effect of hydrogen pressure on the rapid heating gasification of peat. At 60 atm of hydrogen in 4 s to 7 s at 1400 F, the amount of carbon gasified increased by roughly 40% over that for pyrolysis in an inert atmosphere.

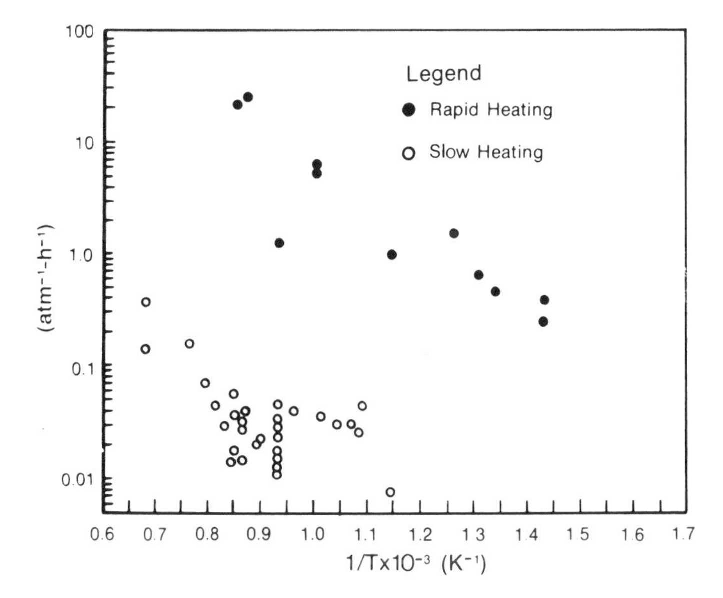

Figure 7-8. Temperature Dependence of Hydrogasification Rate Constants for Coal and Char

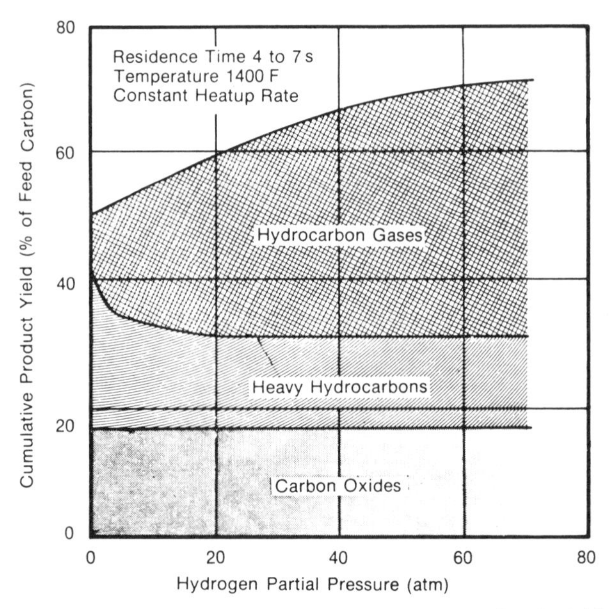

Figure 7-9. Effect of Hydrogen Partial Pressure on Product Yields Obtained During Peat Gasification

Figure 7-10 compares the effect of pretreatment on conversion. The base carbon is the fixed carbon, as determined from the proximate analysis. With char devolatilized and stabilized in nitrogen, the additional gasification in steam and hydrogen at 1500 F is minimal even for long residence times. For the raw peat, the initial extra gasification is significant, representing 70% of the fixed carbon after 10-min residence time.

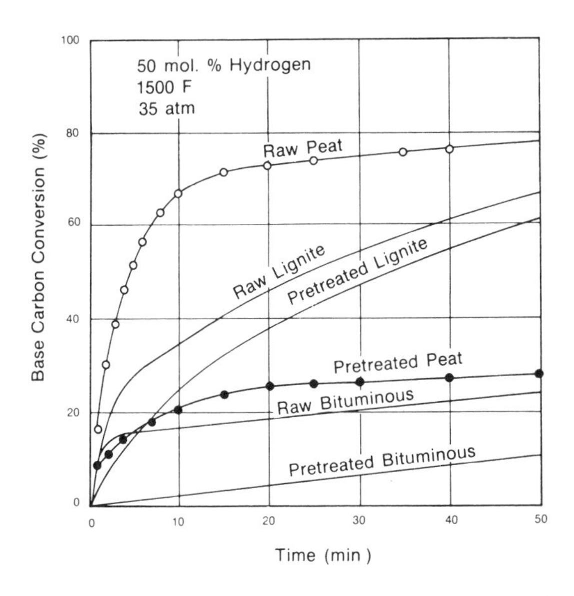

Figure 7-10. Comparison of Carbon Conversions for Peat, Lignite, and Bituminous Coal During Gasification with a Steam-Hydrogen Mixture

Hydrogasification of other biomass feedstocks is presently being investigated (Babu 1979), and it is believed that a considerable quantity of hydrocarbons can be derived from such materials under high hydrogen pressures and rapid heating.

7.10 REFERENCES

Anthony, D. B.; Howard, J. B. 1976. AICHE J. Vol. 22: p.625.

Arthur, J. R. 1951. Transactions Faraday Society. Vol. 47: p.164.

Babu, S. P. 1979. Private communication from IGT.

Baird, M. J.; Reimers, G. W.; Khalafalla, S. 1976. "Reactivity and Gasification Characteristics of Low Ranking Coals and Potentially Reducing Waste Materials." PERC/RI-76/2.

Blackwood, J. D.; Ingeme, A. J. 1960. Austrian J. Chemistry. Vol. 13: p. 194.

Bonner, F.; Turkevich, J. 1951. J. American Chemical Society. Vol. 73: p. 561.

Chu, J. C.; Kalil, J.; Wetteroth, W. A. 1953. Chemical Engineering Progress. Vol. 49: p. 141.

Dutta, S.; Wen, C. Y.; Belt, R. J. 1975. "Reactivity of Coal and Char in CO_2 Atmosphere." ACS Div. Fuel Chemistry Preprint. Vol. 20 (No. 3): p. 163.

Ergun, S. 1956. "Kinetics of the Reaction of Carbon Dioxide with Carbon." J. Physical Chemistry. Vol. 60: p. 480.

Flandrois, S.; Tinga, A. 1972. "Etude Cinetique De La Graphitisation: Determination Du Coefficient D'Activation." Carbon. Vol. 10: p.1.

Friedman, L. D., 1975. "Development of a Fluidized Bench-Scale Reactor for Kinetic Studies." ACS Division Fuel Chemistry Preprint. Vol. 20 (No. 3): p. 35.

Fuchs, W.; Yavorsky, P. 1975. "Gasification of Hydrane Char in Reactions with Carbon Dioxide and Steam." ACS Division Fuel Chemistry Preprint. Vol. 20 (No. 3): p. 115.

Grabke, H.J. 1966. Bunsengesellschaft Phys. Chem. Vol. 70: p. 66A.

Gray, J.; Donatelli, P.; Yavorsky, P. 1975. "Hydrogasification Kinetics of Bituminous Coal and Coal Char." ACS Division Fuel Chemistry Preprint. Vol. 20 (No. 4): p. 103.

Hedden, K.; Lowe, A. 1965. "Rate Constants of the Reaction of Carbon with Carbon Dioxide." AEC Winfrith (England): CONF - 641031-4.

Johnson, J. 1974. "Kinetics of Bituminous Coal Char Gasification with Gases Containing Steam and Hydrogen." Advances in Chemistry. Vol. 131: p. 145.

Johnson, J. 1975. "Relationship Between the Gasification Reactivities of Coal Char and the Physical and Chemical Properties of Coal and Coal Char." ACS Division Fuel Chemistry Preprint. Vol. 20 (No. 4): p. 85.

Jolly, L. J.; Pohl, A. J. 1953. Institute of Fuel. Vol. 26: p.33.

Laine, N. R.; Vastola. F. J.; Walker, P. L. 1963a. J. Physical Chemistry. Vol. 67: p. 2030.

Laine, N. R.; Vastola, F. J.; Walker, P. L. 1963b. Proceedings of the Fifth Carbon Conference. Vol. 2. New York: Pergamon Press; p. 211.

Laurendeau, N. 1978. "Heterogeneous Kinetics of Coal Char Gasification and Kinetics." Progress in Energy Combustion Science. Vol. 4: p. 221.

Linares, A.; Mahajan, O. P.; Walker, P. L. Jr. 1977. "Reactivities of Heat-Treated Coals in Steam." ACS Division Fuel Chemistry Preprint. Vol. 22 (No. 1) p. 1.

Long F. J.; Sykes, K. W. 1948. Proceedings Royal Society (London). Vol. A 193: p. 377.

Lowry, H. H. 1963. Chemistry of Coal Utilization, Supplementary Volume. Wiley.

Mentser, M.; Ergun, S. 1973. "A Study of the Carbon-Dioxide Carbon Reaction by Oxygen Exchange." USBM Bull. 664.

Nandi, S.; Lo, R.; Fisher, J. 1975. "Rate of Reaction of Oxygen and Steam with Char/Coke." ACS Division Fuel Chemistry Preprint. Vol. 20 (No. 3): p. 88.

Orning, A. A.; Sterling, E. 1954. J. Physical Chemistry. Vol. 58: p. 1044.

Peterson, E. E.; Wright, C. C. 1955. Industrial and Engineering Chemistry. Vol. 47: p. 1624.

Punwani, P. V.; Nandi, S. P.; Gavin, L. W.; Johnson, J. L. 1978. "Peat Gasification-An Experimental Study." Presented at 85th AIChE Meeting, Philadelphia, PA; June.

Rensfelt, E.; Blomkvist, G.; Eastrom, C.; Engstrom, S.; Esperas, B. G.; Liinanki, L. 1978. "Basic Gasification Studies for Development of Biomass Medium-Btu Gasification Processes." Energy From Biomass Wastes. IGT; 14 Aug.

Satterfield, C. N. 1970. Mass Transfer in Heterogeneous Catalysis. Cambridge, MA: MIT Press.

Tien, R. H.; Turkdogan, E. T. 1970. "Incomplete Pore Diffusion Effect on Internal Burning of Carbon." Carbon. Vol. 8: p. 607.

Tingley, G. L.; Morrey, J. R. 1973. Coal Structure and Reactivity. Richland, WA: Pacific Northwest Laboratories; Battelle Energy Program Report.

Tomita, A.; Mahajan, O. P.; Walker, P. L. Jr. 1977. "Catalysis of Char Gasification By Minerals." ACS Division Fuel Chemistry Preprint. Vol. 22 (No. 1): p. 4.

Turkdogan, E. T.; Vinters, J. R. 1969. "Kinetics of Oxidation of Graphite and Charcoal in Carbon Dioxide." Carbon. Vol. 7: p. 101.

Walker, P. L.; Hippo, E. 1975. "Factors Affecting Reactivity of Coal Chars." ACS Division Fuel Chemistry Preprint. Vol. 20 (No. 3): p. 45.

Walkup. 1978. "Investigation of Gasification of Biomass in the Presence of Multiple Catalysts." Fuels from Biomass. p. 301; June 20.

Wen, C. Y.; Sears, J. T.; Galli, A. F. 1977. The Role of the C-CO_2 Reaction in Gasification of Coal and Char. DOE Contract EF-76-C-01-0497; Dec.

Wiel, S.; Nandi, S; Punwani, D.; Kopstein, M. 1978. Peat Hydrogasification. Presented at 176th Meeting, American Chemical Society. Miami, FL; Sept.

Wilks, K.; Gardner, N.; Angus, J. 1975. "Catalyzed Gasification of Coals and Chars." ACS Division Fuel Chemistry Preprint. Vol. 20 (No. 3): p. 52.

Yang, R.; Steinberg, M. 1977. "The Reactivity of Coal Chars With CO_2 at 1100-1600 C." ACS Division Fuel Chemistry Preprint. Vol. 22 (No. 1): p. 12.

Part III

Technology and Research

The information in Part III is from *A Survey of Biomass Gasification. Volume III–Current Technology and Research* (SERI/TR-33-239), edited by T.B. Reed of the Solar Energy Research Institute, prepared for the U.S. Department of Energy, April 1980.

Chapter 8

Types of Gasifiers and Gasifier Design Considerations

T. B. Reed
SERI

8.1 INTRODUCTION

Gasifiers come in a seemingly bewildering variety. The principal types are shown in Fig. 8-1. This chapter explains why the various types exist and delineates the factors needed to choose among them or to design a new one. Later chapters give a comprehensive list of biomass and other gasifiers and discuss in some depth the work of a number of groups engaged in gasifier research or development.

8.2 GENERAL CONSIDERATIONS FOR GASIFIER DESIGN

The development of gasifiers has been and continues to be largely empirical. Inventors study existing gasifiers and design improvements to fit specific concepts and needs. Initial models generally do not work well and require a great deal of effort and learning to become operational. Many problems are mechanical and can be solved by trial and error. Other problems are conceptual or chemical, or involve nonobvious heat transfer problems that remain unidentified—yet which fundamentally determine allowable conditions for practical operation. It would be presumptuous to claim that all the areas that must be considered in designing or choosing a gasifier are identified in Chapter 8, but it does offer a framework in which to consider the most important factors contributing to successful operation of gasifiers.

8.2.1 Chemistry of Biomass Gasification

The central problem in gasification is the conversion of a solid fuel (biomass, MSW, coal, peat, lignite, etc.) to a gaseous fuel, as can be seen from studying Figs. 8-1, 8-2, and 8-3.

The chemical composition of solid and gaseous fuels, along with the various processes of converting solid fuels to gaseous fuels, are shown in the ternary diagram of Figure 8-2. The atomic compositions of the biomass, coal, and char samples from Tables 3-4 and 3-7 (Chapter 3, Part II) are plotted, and they define the practical range of variation of these solid fuels. It is interesting to note that the composition of biomass ranges between that of lignin (L) and that of cellulose (C). The average composition of the biomass used in the calculations of Chapter 6 (Part II) is shown with the larger point marked B (biomass) with composition $CH_{1.4}O_{0.6}$. The chart also shows the wide variation of char compositions, overlapping the composition (but not the physical structure) of coals. These compositions are especially arbitrary. Chars formed at low temperatures (between say 500°C and 800°C) have a surprisingly high H and O content. The compositions of three peats have been included (Punwani 1979), and it is seen that peats are very close in composition to lignin.

In this diagram, fuel gases lie to the right of the line defined by the composition CO and C_2H_4. At high temperatures, only CO and H_2 are stable, defining the gas fuel range to be to the right of the H-CO line. However, at lower temperatures, CH_4 becomes stable and CO becomes unstable, so there is no exact position for the line separating gas fuels from solid fuels unless thermodynamic and kinetic conditions are specified. Finally, the products of complete combustion are CO_2 and H_2O, so that this line defines the low

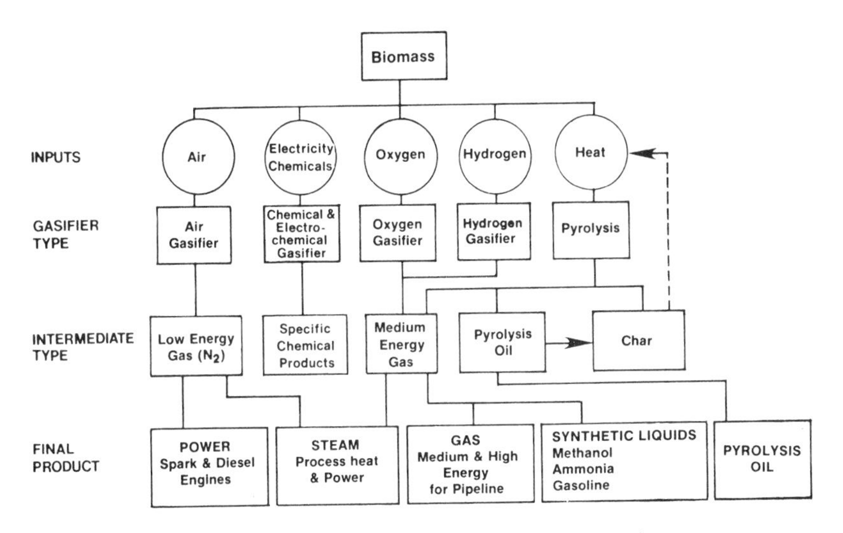

Figure 8-1. Gasification Processes and Their Products

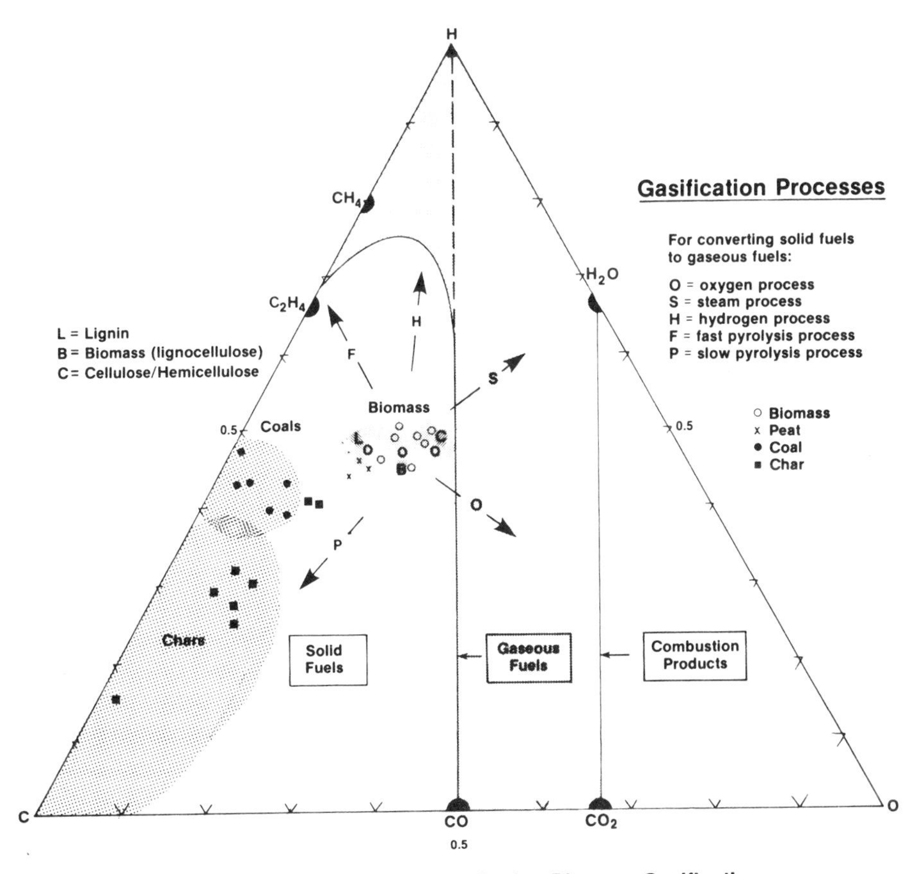

Figure 8-2. Chemical Changes During Biomass Gasification

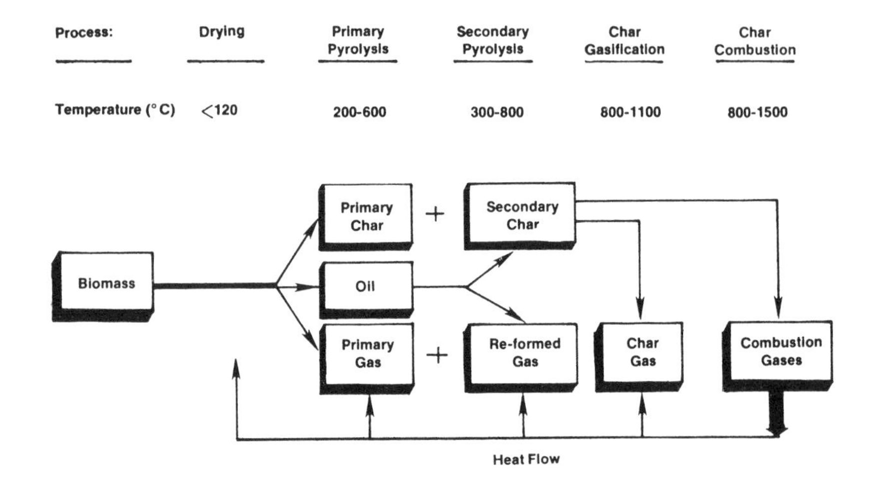

Figure 8-3. Heat and Mass Flows in Pyrolysis and Gasification Processes

energy limit of gaseous fuels. Compositions to the right of this line represent combustion with excess air or oxygen.

Thus, the problem of gasification becomes the problem of shifting the composition of the solid fuels of the left side of Fig. 8-2 along one or more of the arrows to a gaseous composition. A simple equation for high-temperature biomass pyrolysis to gas and char,

$$CH_{1.4}O_{0.6} \longrightarrow 0.7\ H_2 + 0.6\ CO + 0.4\ C\ \text{(solid)},$$

suggests that with heat alone char must result and that there must be a change in composition if biomass is to be completely gasified (with the possible exception of flash pyrolysis). The arrows of Fig. 8-3 show the various methods of accomplishing this. Pyrolysis is the disproportionation of biomass to yield some gases (typically methane, CO, and H_2) and the arrow P shows that in addition there will be a char formed. Oxygen/air gasification is mechanically the simplest method of producing gas because the initial reaction is exothermic (arrow O), and by far the largest number of the gasifiers of Chapters 9 and 10 use this method. Oxygen gasification is quite exothermic; in many cases, steam is used in conjunction with the oxygen to conserve energy and produce a fuel higher in hydrogen (arrow S). Steam can be used alone for biomass gasification, producing a gas high in methane, but the temperature of operation must be kept relatively low (see the Wright-Malta process and Fig. 6-10b in Part II).

Hydrogen has been used in the past for the liquifaction and gasification of coal, and it can be seen from the arrow H in Fig. 8-2 that this shifts the composition of solid fuels toward high-methane and high-energy content fuels. However, the reaction with hydrogen requires high pressures, high temperatures, and a source of hydrogen—a fuel in its own right. Furthermore, at the low temperatures at which biomass volatilizes (200-500 C) it is not clear that there is any primary reaction between the biomass and the hydrogen, while the high temperatures required for coal volatilization make primary reactions more likely. Several groups are working on hydrogen gasification, but the processes are not ready for commercial demonstration.

A new area of biomass gasification involves the production of ethylene and higher olefins such as ethylene, propylene and butylene. These molecules are relatively unstable compared to methane or CO at high temperatures, yet their decomposition is slow, so that they can be formed in high yields by the flash pyrolysis of hydrocarbon feedstocks at temperatures of 750-1000 C. Recent experiments have shown that the rapid pyrolysis of biomass also gives high yields of these olefins with correspondingly low char yields (see Table 5-6, Section 5.3.2, Part II). Fast pyrolysis of biomass to ethylene is shown diagrammatically by the arrow F on Fig. 8-2.

8.2.2 Energetics of Biomass Gasification

The thermodynamics of gasification was discussed in Chapters 3 (Heats of Combustion and Formation) and 6 (Thermodynamics of Gas-Char Reactions), both in Part II. The energy requirements for idealized cellulose reactions to form gases, liquids, and chars are shown in Table 8-1. The increase or decrease in energy content of the products is illustrated diagrammatically in Fig. 8-4 (Reed 1978). These reactions show that the energy involved in conversion to gas, liquid, or solid products runs from -5 to 5 kJ/g (-5 to 5 MBtu/ton), which is small compared to the heat released on combustion (18 kJ/g or 16 MBtu/ton). In any practical gasifier, however, it is necessary to heat the biomass to the required reaction temperature (typically 500-1100 C) and then add the necessary energy for reaction, if any.

There is little reliable experimental data on the amount of energy necessary for the gasification reactions. Laboratory work on pyrolysis (see Figure 5.5, Part II) suggests that during very slow reactions with high char formation (Fig. 8-4) pyrolysis can be mesothermic or exothermic, but faster pyrolysis, producing a higher proportion of gases, is endothermic.

As mentioned previously, the gasification of biomass and char with oxygen or air is exothermic, while the oxidation by decomposition of steam is highly endothermic. Thus, practical gasifiers sometimes use mixtures of oxygen/steam to maintain proper reaction temperature. Similarly, the production of methane and CO_2 at lower temperatures can be exothermic, but it proceeds relatively slowly and may require a catalyst.

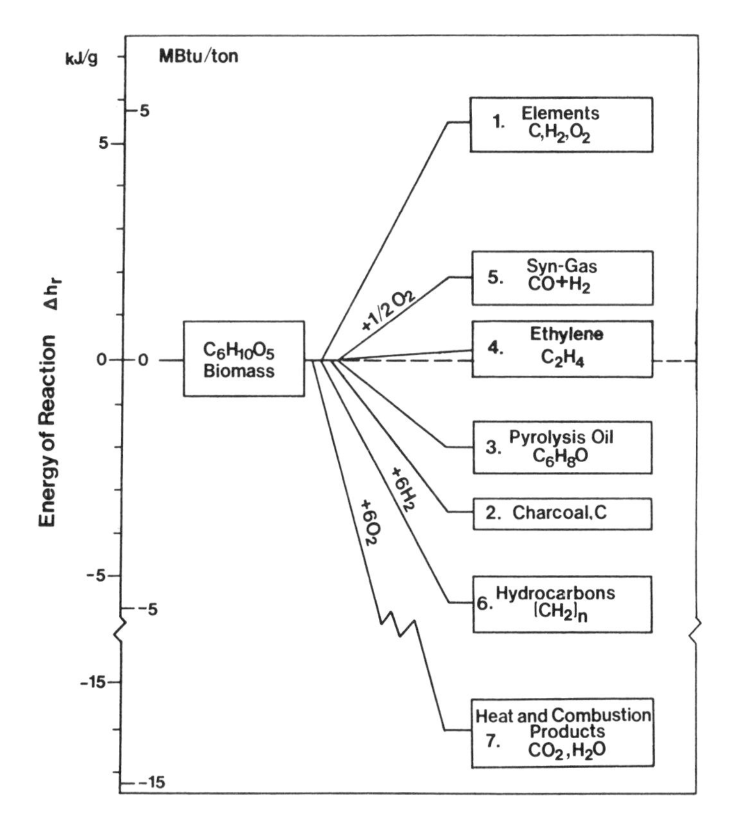

Figure 8-4. Energy Change for the Cellulose Thermal Conversion Reactions Shown in Table 8-1

Table 8-1. ENERGY CHANGE FOR IDEALIZED CELLULOSE THERMAL CONVERSION REACTIONS

Chemical Reaction				Energy consumed[a]		Products	Process
				ΔH_r(kcal/m)	Δh_r(kJ/g)		
1. $C_6H_{10}O_5$		→	$6\ C + 5\ H_2 + 5/2\ O_2$	+229.9[b]	+5.94	Elements	Dissociation
2. "		→	$6\ C + 5\ H_2O$ (g)	-110.6	-2.86	Charcoal	Charring
3. "		→	$0.8\ C_6H_8O + 1.8\ H_2O$ (g) $+ 1.2\ CO_2$	-80.3[c]	-2.07	Pyrolysis oil	Pyrolysis
4. "		→	$2\ C_2H_4 + 2\ CO_2 + H_2O$ (g)	+6.2	+0.16	Ethylene	Fast Pyrolysis
5. "	$+ 1/2\ O_2$	→	$6\ CO + 5\ H_2$	+71.5	+1.85	Synthesis gas	Gasification
6. "	$+ 6\ H_2$	→	6 "CH_2" $+ 5\ H_2O$ (g)	-188.0[d]	-4.86	Hydrocarbons	Hydrogenation
7. "	$+ 6\ O_2$	→	$6\ CO_2 + 5\ H_2O$ (g)	-677.0	-17.48	Heat	Combustion

[a] 1 kJ/g = 0.239 kcal/g = 430 Btu/lb = 0.860 MBtu/ton.

[b] The negative of the conventional heat of formation calculated for cellulose from the heat of combustion of starch.

[c] Calculated from the data for the idealized pyrolysis oil C_6H_8O (ΔH_c = -745.9 kcal/mol, ΔH_f = -149.6 kcal/g).

[d] Calculated for an idealized hydrocarbon with ΔH_c = -149.6 kcal/mol. Note H_2 consumed.

Each gasification process has its own energy requirements—some are exothermic, some endothermic, and all have process heat losses that have to be accounted for. The adiabatic reaction temperature (ART) is, of course, a measure of the degree of energy production in any process, and Fig. 8-5 shows the ART for pyrolysis, air, and oxygen gasification as a function of the amount of air or oxygen added relative to that required for combustion (the equivalence ratio). These results were calculated assuming equilibrium among the products, a fairly good assumption for downdraft gasification. Results of calculations for other conditions are given in Chapter 6, Part II. In many other processes, the products are far from equilibrium (see Chapters 6 and 7, Part II).

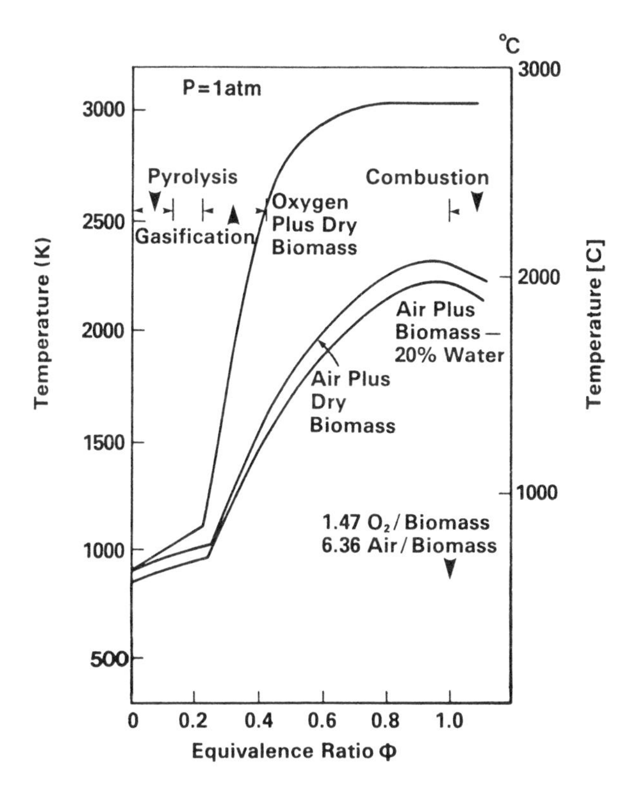

Figure 8-5. Biomass Adiabatic Reaction Temperatures

8.2.3 Pyrolysis and Char Gasification Reactions

Although the mechanics of gasification vary widely in different processes, each particle of biomass must undergo some or all of the stages shown in Fig. 8-3.

The first stage, drying, occurs below about 110 C, and locally the temperature cannot rise above this until all physical water has been driven off. Due to the low thermal conductivity of biomass (0.05-0.1 Btu/ft-h-F) and the even lower conductivity of char (0.03 Btu/ft-h-F), larger pieces can be burning on the outside while there is still moisture inside.

Once dry, pyrolysis converts the biomass to oil vapors, primary char, and primary gas (cellulose typically produces very little primary char, whereas the lignin and hemicellulose components produce higher char yields).

For small particles, the oil vapors are generated near the surface and can escape into the gas phase before being cracked to secondary char. These oil vapors can be condensed for use, burned with the gases, or cracked at higher temperatures to form re-formed gas. Recent experiments show that this re-forming only occurs at temperatures over 600 C (see Section 5.3, Part II). For larger particles, the longer escape path provides more time for cracking the oil vapors, thus resulting in higher char production.

Char gasification and combustion are the most difficult tasks in gasifier design. The gasification of char proceeds relatively slowly even at temperatures high enough to require special construction materials. Combustion of char, on the other hand, is rapid and exothermic. Again, materials of construction will set the upper temperature limits. In some biomass gasifier designs, these reactions are carried out in the same vessel as the pyrolysis and drying; in other processes, the char may be gasified or burned separately. In coal gasification, the char reactions are even more difficult, because there is a higher proportion of char and it is much less reactive.

8.2.4 Heat Transport and Heat Transfer in Gasification

The micro- and macroscopic paths of heat transfer to the biomass suggested in Fig. 8-3 must be a central consideration in the design of any gasifier. Both the heat flow within the biomass particle and the heat flow to its surface from other parts of the gasifier must be examined.

In general, heat will flow within the biomass particle by conduction from its externally heated surface. The thermal conductivity of wood, and even more so of char, is especially low relative to most other materials (see Section 3.4.1, Part II); for large pieces of biomass, it may require minutes, even hours, before the pyrolysis is complete at the core, despite the outer surface being maintained at 1000 C. Early charcoal manufacture required reaction times in excess of weeks ! Thus, it is necessary to consider that there can be very steep temperature gradients inside the particle, with microzones of drying, pyrolysis, and char gasification from the center to the surface of the biomass particle undergoing external heating.

It is also possible to conceive of a biomass particle being uniformly heated throughout, either by solar radiation for small particles or by microwave radiation, but this is not likely to be important at higher temperatures after char coats the surface. Heat is transported and transferred, in general, by conduction, natural and forced convection, radiation, and change of state (as in a heat pipe); all of these mechanisms are active in gasification.

Conduction through a solid metal wall was used in early gas generators to heat a retort, producing gas, char, and oil. This has the advantage of yielding a relatively high Btu gas, since there is no dilution by air. However, it also produces the maximum char yield, because of the slow transfer of heat through the biomass volume. Indirect heating can be made more efficient by increasing the surface to volume ratio, for instance by using a multitude of small heat exchange tubes as is done for fast pyrolysis.

Forced convection gas heating of biomass is accomplished by passing a hot gas through the interstices in a fixed bed or around fluidized or suspended biomass particles in most gasifiers. In addition to the obvious forced flow of gases caused by the passage of the gas, currents due to natural convection and aspiration can occur unexpectedly and greatly alter the gasifier behavior.

"Solid convection" may seem like a contradiction in terms, but a fluidized bed accomplishes rapid, even, heat transfer by the movement of biomass or inert particles in a rising gas column. In a true fluidized bed, the temperature is considered to be uniform throughout, but in spouted beds and other forms there may be different temperature zones. Solid convection of particles can even be used to transfer heat alone from one vessel to another, permitting combustion of char with air in one vessel to provide heat for pyrolysis in a second vessel.

Liquid conduction and convection can provide much faster heating rates than gas convection, and baths of molten salts or metals have been used to heat biomass very rapidly. Solid and liquid convection are used in a number of the processes discussed in Chapter 10.

Radiation in gasifiers is an important heat transfer mechanism at higher temperatures between particles or with the wall, but only over short distances, since charred biomass is opaque to most radiation.

Finally, friction can be used to generate intense heat at the biomass surface. Change of state is an important mechanism of heat transfer that is generally overlooked in operating gasifiers. Oil and water vapors are generated in higher-temperature zones of reactors; if they pass to low-temperature areas, they can condense, releasing very large quantities of heat directly at the condensing surface. This is a very important heat transfer mechanism, comparable to that found in "heat pipes," and it must be considered in understanding any practical gasifier.

8.2.5 Mass Transport in Gasification

Both micro- and macroscopic aspects of mass flow are important in gasifier design as suggested by Fig. 8-3.

In a particle of biomass undergoing pyrolysis, there must be a continuous flow of gases and oil vapors to the surface and into the surrounding gas stream. This flow of gas tends to produce a boundary layer of cooler gases around the particle. As the gases pass through the char layer, there can be cracking reactions of the larger molecules, and this is probably one reason why char production is higher in larger particles.

In addition, in a fixed bed there will be a macroscopic flow of the solids (generally down), of the ash produced, of the oil vapors, and of the gas, all of which must be accounted for in the design of any gasifier.

8.2.6 Fuel and Ash Handling

A major consideration in gasifier design is the type of fuel or fuels to be used. Fixed bed gasifiers are most suitable for fuels of larger sizes (more than 1/4 in.); fluidized beds can operate with a range of sizes; suspended fuel gasifiers operate with smaller sizes (less than 1/4 in.), whereas fast pyrolysis may require very small particles to maximize heating rate and minimize internal vapor cracking.

Fuel feeding is often a major difficulty in gasifier operation, as is bridging inside the reactor. These problems can be minimized by densification (pelleting or briquetting), if this is economically justified (see Section 4.1.3, Part II). The strength of the char is often important in the successful operation of fixed bed gasifiers since a weak char is likely to have high losses to the ash pit. Densification of the biomass increases the density and strength of the char.

Ash production is usually very low for wood fuels, higher for agricultural and aquatic biomass, and higher still for municipal wastes. In fixed bed gasifiers, provision must be made to either keep the ashes below about 1100 C to prevent aglomeration ("dry ash" operation) or heat the ash above 1300 C so the ash can be removed as a liquid ("slagging" operation). In fluidized and suspended bed gasifier operation, the ashes are typically removed after gasification by a cyclone.

8.2.7 Gasifier Pressure

In most cases, gasifiers will be operated close to 1 atmosphere of pressure in order to minimize sealing difficulties. Gasifiers designed for engine operation generally operate under slightly negative pressure and are called "suction" gasifiers. Those used to provide gaseous fuel for boilers typically operate slightly above atmospheric pressure.

Pressurized gasifiers require sturdy construction, lock hoppers, and pressurized feed gas. Nevertheless, these added requirements may be justified if the gas is subsequently to be used in a turbine, pipeline, or for chemical synthesis (to make ammonia or methanol), because of the elimination of compression costs, and commercial coal gasifiers are operated at pressures as high as 100 atmospheres.

8.3 GASIFIER TYPES

In designing, buying, or building a gasifier, one must make the following choices (discussed in the previous sections):

- Chemical change: air, oxygen, hydrogen, and slow or fast pyrolysis (5 types).
- Method of heat and mass contact—direct: updraft (counter-flow), downdraft (co-flow), fluidized bed, and suspended; and indirect: solids (fluidized bed), liquids, and gaseous recirculation (7 types).
- Fuel type and form: biomass, MSW, and pellets, powder, etc. (4 types).
- Ash type: dry ash and slagging (2 types).
- Pressure: suction, low pressure, and high pressure (3 types).

Gasifiers could also be categorized by products (gas, gas/oil, gas/oil/char, gas/char); by purpose (for power, for making steam, for pipeline distribution, for synthetic liquids); and in many other ways.

The world of gasifiers is potentially rich and varied. The possible combinations of the above five categories give over 500 types; only a few dozen are listed in Chapters 9 and 10. Figure 8-1 shows one possible simple breakdown of the major processes, and some important characteristics of the most common varieties are discussed below. Chapter 9 lists manufacturers and research groups working in these major areas, Chapter 10 gives more detail on certain specific research and development projects.

8.3.1 Air Gasification

The simplest way to produce gas is by air gasification, according to the (oversimplified) formula

$$CH_{1.4}O_{0.6} + \underbrace{0.2\ O_2 + 0.8\ N_2}_{\text{"Air"}} \rightarrow CO + 0.7\ H_2 + 0.8\ N_2\ .$$

Unfortunately, this reaction is slightly endothermic and in practice somewhat more air must be added and some CO_2 and H_2O produced to provide the process energy. The nitrogen results in a dilute "low energy gas" of 120-200 Btu/SCF.

8.3.1.1 Updraft

The simplest air gasifier is the updraft (counterflow) gasifier shown in Fig. 8-6, in which air is introduced to the biomass through grates in the bottom of the shaft furnace. Rather high temperatures are generated initially where the air first contacts the char, but the combustion gases immediately enter a zone of excess char, where any CO_2 or H_2O present is reduced to CO and H_2 by the excess carbon. As the gases rise to lower-temperature zones, they meet the descending biomass and pyrolyze the mass in the range of 200 C to 500 C. Continuing to rise, they contact wet, incoming biomass and dry it. The counterflow of gas and biomass exchanges heat so that the gases exit at low temperatures.

A simpler arrangement can hardly be imagined, but the updraft gasifier has several drawbacks. A wide variety of chemicals, tars, and oils is produced in the pyrolysis zone and, if allowed, will condense in cooler regions. For this reason, this gas is generally used in the "close-coupled" mode in which it is mixed immediately with air and burned completely to CO_2 and H_2O. The close-coupled mode is quite suitable for supplying a biomass gas to existing oil or gas furnaces for process heat (see Chapter 11). The high temperature at the grate may melt the ash and produce slagging on the grates with feedstocks such as rice hulls and corn cobs. Indeed, in the Andco-Torrax solid municipal waste (SMW) gasifier, the incoming air is preheated to give slagging temperatures on the grate, which then convert the high mineral content of SMW to a clean glass frit that can be used in road building. The Purox process uses oxygen to achieve high temperatures to melt minerals.

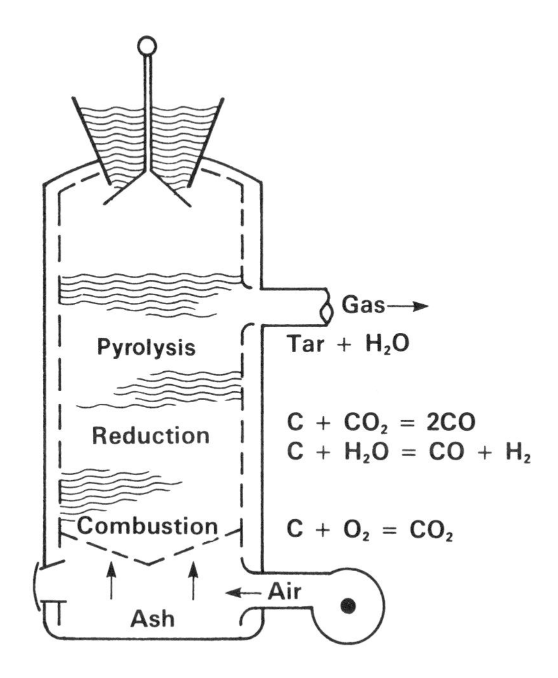

Figure 8-6. Schematic Diagram of Updraft Gasifier

8.3.1.2 Downdraft

The downdraft (co-flow) gasifier shown in Fig. 8-7, is designed specifically to eliminate the tars and oils from the gas. Air is introduced to the gasifier through a set of nozzles called "tuyeres" and the products of combustion are reduced as they pass through a bed of hot charcoal extending some distance down to the grate. Continuing operation pyrolyzes descending biomass, but the oil vapors also pass through the bed of hot charcoal, where they are cracked to simpler gases or char. An important result of this cracking is an effect called "flame stabilization" in which the temperature is maintained in the range from 800 C to 1000 C by these cracking reactions. If the temperature tends to rise, the endothermic reactions predominate, thus cooling the gas. If the temperature drops below this range, the exothermic reactions predominate, keeping the gas hot.

The tars and oils are reduced to less than 10% of the value produced in updraft gasifiers, and these gases can then be used with minimal filtering to fuel spark and diesel engines, the principal use of downdraft air gasifiers. Typically, the gas velocities are low in updraft and downdraft gasifiers, and the ash settles through the grate, so that very little is carried over with the gas.

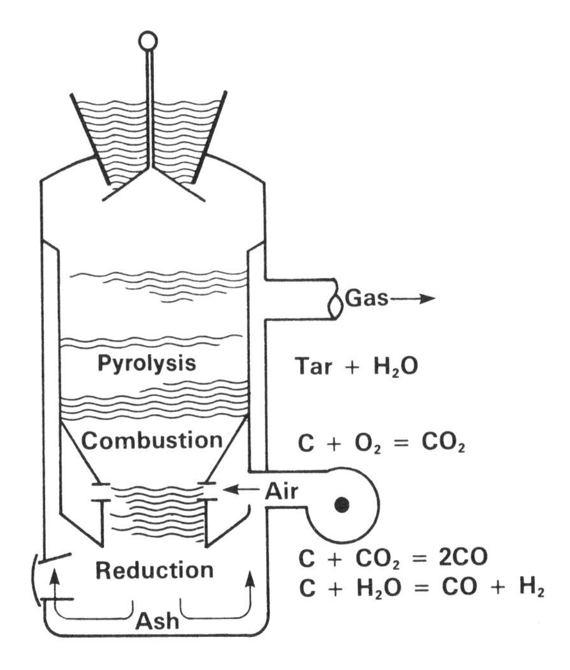

Figure 8-7. Schematic Diagram of Downdraft Gasifier

8.3.1.3 Fluidized Beds

Fluidized beds have been developed over the last few decades to provide uniform temperatures and efficient contact between gases and solids in process industries. A typical fluidized bed is shown in Fig. 8-8. Because of its higher throughput, it is more compact (Section 8.4.2), but the higher velocities carry the ash and char out with the gas and they must be separated in cyclones or bag houses.

Fluidized beds usually contain either inert material (such as sand) or reactive material (such as limestone or catalysts). These aid in heat transfer and provide catalytic or gas-cleaning action. The material is kept in suspension, simulating a "fluid," by a rising column of gas. In a true fluidized bed, the solids mix very rapidly and provide high heat transfer between all parts of the bed. In "spouted" beds and other modified gasifiers, there may be temperature gradients established and less mass exchange between the lower and upper parts.

Since fluidized bed gasifiers are a newer development than updraft and downdraft, their characteristics are not as well known. It is claimed that they can produce very low tars and char with recirculation, but to date this remains to be proven. A number of fluidized beds are under development and are discussed in Chapters 9 and 10.

8.3.1.4 Suspended Gasification

Suspended combustion is quite common for coal and fine biomass, utilizing a vortex action to obtain sufficient gas-solid contact to ensure complete combustion. Smaller particles such as sawdust can also be gasified in suspension. Only one suspended gasifier has been tested to date (Fig. 8-9).

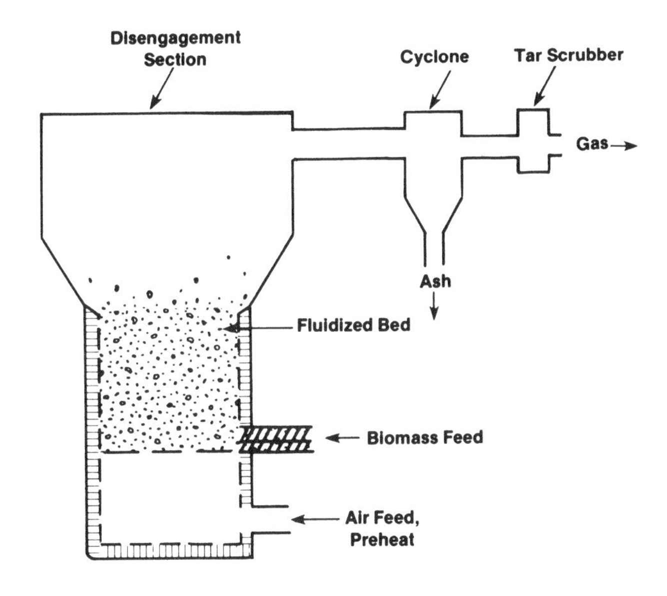

Figure 8-8. Schematic Diagram of Fluidized Bed Gasifier

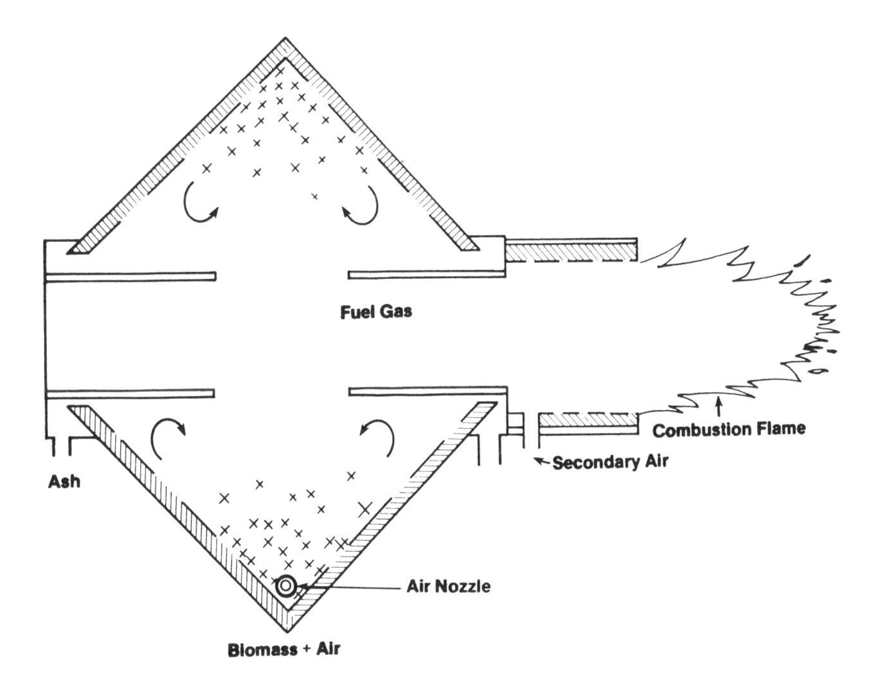

Figure 8-9. Schematic Diagram of Suspended Fuel Gasifier (After Frederick/Morback Design)

8.3.2 Oxygen Gasification

Oxygen can also be used for gasification of biomass; it has the advantage that it produces a medium energy (300-400 Btu/SCF) gas that can be used in pipelines or for chemical synthesis to make methanol, ammonia, gasoline, or methane. Reaction rates are higher and velocities are lower than with air, resulting in easier gas cleanup and handling.

Oxygen production is the second largest of that of any chemical produced in the United States (after that of sulfuric acid), and it presently sells for $20-$60/ton in bulk. Since it requires about 1/3 of a ton of oxygen to gasify a ton of biomass, this will add $0.40-$1.20/MBtu of biomass to the cost of gasification. Bulk oxygen is available in most U.S. cities.

At present, no gasifiers have been designed specifically for biomass, but the Union Carbide PUROX process (see Section 10.2.2) processes 300 ton/day of solid municipal waste using oxygen in a updraft slagging gasifier. The mineral content of the waste is converted to a clean frit, and the tars and oils are scrubbed and reinjected into the hot zone for conversion to gas. An extended analysis of a gasifier that was designed using PUROX data to work on biomass is given in Desrosiers (1979, Section 5.4). Oxygen has been tested recently in an air downdraft gasifier with biomass. The temperatures observed were surprisingly low, which suggested that downdraft gasifiers for biomass may be simpler than updraft (Solar Energy Research Institute 1979). Oxygen has not yet been demonstrated for fluidized bed or suspended operation with biomass or SMW, but it has been used with coal in these modes; there is no obvious hindrance to its use with biomass and SMW.

8.3.3 Pyrolysis and Pyrolysis Gasification

8.3.3.1 Pyrolysis Processes for Gas/Oil/Char

Air gasification has the disadvantage that it produces a low energy gas; oxygen, that it uses high-cost input (oxygen) to achieve a medium energy gas. Biomass has a high content of volatile gas relative to coal and can be pyrolyzed to form a medium energy gas containing methane and higher hydrocarbons. Unfortunately, there is also a moderate amount of char and oil produced; these are assets if they can be sold but are disposal problems if they cannot. A number of pyrolysis processes are described in Chapter 9 and 10.

8.3.3.2 Slow Pyrolysis Gasification

Pyrolysis gasification uses many ingenious schemes to recycle the energy contained in char and oil into gas energy. This recycle results in a process of greater complexity, but one which yields a medium energy gas with no other products. Various pyrolysis gasification processes are listed in Chapters 9 and 10.

The char energy can be recycled in a variety of ways. The char can be burned in a fluidized bed with sand. The resulting hot sand is transferred to a second bed in which the biomass is pyrolyzed. In other variations, char is burned to heat pyrolysis gas, which is then recycled to a fluidized fixed-bed pyrolysis unit, or external heat is fed to a slurry of wet biomass at high pressure.

8.3.3.3 Fast Pyrolysis Gasification

Many experiments have demonstrated that the degree of char and oil formation during pyrolysis increased with particle size, with reaction time, and with lignin content. Sufficiently rapid heating of finely divided biomass, on the other hand, need produce no char at all (see Chapter 4, Part I).

Even more recently, it has been found that the vapor molecules generated during pyrolysis can be cracked at high temperatures to yield olefins (especially ethylene), and these products can be preserved if the gas is quenched before further reactions can occur. Since olefins form the basis of much of our chemical synthesis today, and since they can be easily converted to either gasoline-type hydrocarbons or, through hydration, to alcohols, it is understandable that there is a good deal of interest in "fast pyrolysis." Several such processes are discussed in Chapter 10.

8.3.4 Hydrogen Gasification

Hydrogen can be used at very high pressure to change the composition of biomass as shown in Fig. 8-2; this results in the formation of liquids or gasses, depending on the reaction conditions. This approach will be most attractive where hydrogen is readily available. Several projects in hydrogen gasification are described in Chapter 10.

8.3.5 Chemical and Electrochemical Gasification

A number of innovative approaches to gasification are being explored in which specific chemical reactions are induced to produce specific products. Examples include reaction of biomass with Br_2 to produce HBr and CO_2. The HBr is then electrolyzed to produce H_2 (Darnell 1979). As a second example, one might envision an electrochemical scheme for H_2 from biomass analagous to the recent proposed method for coal (Coughlin and Farooque 1979).

8.4 FIGURES OF MERIT FOR GASIFICATION AND COMBUSTION PROCESSES

"Figures of merit" useful in comparing gasification and combustion processes are discussed in this section.

8.4.1 Volumetric Energy Content of Fuel Gases

The "volumetric energy content," typically quoted in Btu/SCF in the U.S., is a "figure of merit" for gases.

Caution must be used in reporting or reading energy contents of gases, as they can be misleading. The measurement of the volumetric energy content is straightforward for cold, clean gases. However, if gases are produced and used hot and containing combustible tars, the "equivalent volumetric energy content" released on combustion may be as much as 50% higher than that for cold, clean gas.

The energy contents of gases are seen in Table 8-2 to vary from less than 100 Btu/ft^3 for blast furnace gas to 1000 Btu/ft^3 for natural gas (methane). The volumetric energy content is indeed important in the distribution and storage of gases. Pipelines are expensive; at present, only natural gas can be distributed economically over long distances. Before the transcontinental pipelines were built during the 1940s, however, medium energy gas was regularly distributed city-wide and presumably this could be done again for industrial parks or city use. (The presence of carbon monoxide may rule out distribution to homes, although this was done prior to 1940.)

The volumetric energy content is not of prime importance in determining the suitability of gases for combustion applications, except for gases below about 200 Btu/ft^3, where flame temperature and heat transfer may be affected (see Fig. 11-1). Low Btu gases may also require special burner designs.

8.4.2 Energy Conversion Rates in Various Processes

Two other figures of merit often used in combustion and conversion processes are the heat released or converted per unit area and the heat released or converted per unit volume. These figures in turn dictate the size and cost of equipment. Typical combustion processes for solid fuels release 400 Btu/ft^2-h and 30 Btu/ft^3-h. In contrast,

Table 8-2. ENERGY CONTENT OF FUEL GASES AND THEIR USES

Name	Source	Energy Range (Btu/SCF)	Use
Low Energy Gas (LEG) [Producer Gas, Low Btu Gas]	Blast Furnace, Water Gas Process	80-100	On-site industrial heat and power, process heat
Low Energy Gas (LEG) [Generator Gas]	Air Gasification	150-200	Close-coupled to gas/oil boilers Operation of diesel and spark engines Crop drying
Medium Energy Gas (MEG) [Town Gas; Syngas]	Oxygen Gasification Pyrolysis Gasification	300-500	Regional industrial pipelines Synthesis of fuels and ammonia
Biogas	Anaerobic Digestion	600-700	Process heat, pipeline (with scrubbing)
High Energy Gas (HEG) [Natural Gas]	Oil/Gas Wells	1000	Long-distance pipelines for general heat, power, and city use
Synthetic Natural Gas (SNG)	Further Processing of MEG and Biogas	1000	Long-distance pipelines for general heat, power, and city use

combustion of gas or oil typically releases 100 Btu/ft^3-h in process heat burners and up to 5000 Btu/ft^3-h in automobile engines and turbines—hence the necessity of using gas or liquid fuels in these important applications.

Gasification processes typically convert 500-1000 Btu/ft^2-h in updraft and downdraft air gasifiers (50-100 Btu/ft^3-h) while fluidized beds convert 100-500 Btu/ft^3-h. Operation on oxygen and/or at high pressure can increase these rates three- to tenfold. Thus, it is apparent that gasification processes have a high thruput relative to their combustion counterparts. This is due to the fact that most of the energy is not actually converted to heat in the gasifier, but only converted to another form.

Char conversion to gas is the most difficult stage of gasification and accounts for most of the dwell time of biomass in the gasifier. Pyrolysis systems, producing char, oil, and gas, therefore have even higher throughputs than gasifiers: typically 500 Btu/ft^3-h. Again, this is due to the fact that pyrolysis makes a minimal change in the feedstock at quite low temperatures, and the char is not gasified.

8.4.3 Turndown Ratio

A figure of merit that is likely to become widely used in evaluation of gasifiers is the "turndown ratio":

$$R = \text{maximum gasification rate/minimum gasification rate.}$$

The turndown ratio is an inherent property of most common processes. As an illustration, a light bulb typically has a turndown ratio of 1; that is, it can only operate at full rated power. Recently, solid-state devices have been used in low-cost switch controls that give a turndown ratio of more than 10 for dimming the lights, and many homes now have several of these devices in selected rooms. An automobile has an infinite turndown ratio, since it will go all speeds including zero.

On the other hand, many devices have no turndown capability (a ratio of $R = 1$), and in many cases such capability would be very desirable. An oil-fired furnace is either on or off, and though the heating rate is made variable by cycling, the efficiency suffers in comparison to that which could be achieved by a continuous lower-level operation.

The recent advent of airtight woodstoves is an attempt to get a high turndown ratio for wood heat, since it is difficult to operate wood heat on an on/off basis. However, operation at low air input involves the problem of creosote generation, air pollution, and chimney fires.

Fixed bed air gasifiers have a high turndown ratio, typically at least five. This property is very useful in situations where the gas is required on an intermittent or varying-load basis, such as operation of engines, drying, and heating.

On the other hand, fluidized bed gasifiers have a narrower range of operation ($R = 2$) and must operate close to their design limit at all times or be started and stopped. Unfortunately, the field of gasification is so new that very little reliable data on established systems is available. We hope that the turndown ratio will be recognized as an important parameter of gasifiers and will be included in measurements and specifications of gasifiers.

8.5 REFERENCES

Coughlin, R. W. and Farooque, M. 1979. Nature. Vol. 279: May 24; p. 301.

Darnell, A. J. 1979. "Production of Hydrogen from Biomass." T. Veziroglu, ed. Proceedings of the 2nd Miami International Conference on Alternative Energy Sources; Miami Beach, FL; Dec. 10, 1979. Coral Gables, FL: University of Miami, Clean Energy Research Institute.

Desrosiers, R. 1979. Process Designs and Cost Estimates for a Medium Btu Gasification Plant Using a Wood Feedstock. Golden, CO: Solar Energy Research Institute; SERI/TR-33-151.

Punwani, D. U.; Weil, S. A.; Paganessi, J. E. 1979. "Synthetic Fuels from Peat Gasification." Proceedings of the 14th Intersociety Energy Conversion Engineering Conference (IECEC). Boston; Aug. 5-10, 1979. New York: American Institute of Chemical Engineers (AIChE).

Reed, T. B. 1978. "Survey of Pyroconversion Processes for Biomass." Nystrom, J. M.; Barnett, S. M., eds. Biochemical Engineering: Renewable Sources of Energy and Chemical Feedstocks. American Institute of Chemical Engineers (AIChE) Symposium Series. New York: AIChE. Vol. 74 (No. 181).

Solar Energy Research Institute. 1979. Retrofit '79: Proceedings of a Workshop on Air Gasification. Seattle, WA; Feb. 2, 1979. Golden, CO: Solar Energy Research Institute; SERI/TP-49-183.

Chapter 9

Directory of Current Gasifier Research and Manufacturers

T. B. Reed and D. Jantzen
SERI

9.1 INTRODUCTION

The first part of this chapter is a summary, in tabular form, of industrial and institutional facilities performing biomass and municipal waste gasifier research and development or manufacturing biomass gasifiers. Information presented includes gasifier type (air, oxygen, pyrolysis, etc.) contact mode (updraft, downdraft, or fluidized bed), primary fuel products, number of operating units, and size of units. For comparison, a summary of major coal gasification processes is included.

Questionnaires were sent to the manufacturers and researchers listed in Section 9-2; their detailed responses are given as a directory listing characteristics of existing gasifiers.

Attempts have been made to make this list as complete as possible, but the rate at which this field is developing makes it very difficult to maintain a completely current list.

9.2. SURVEY OF GASIFIER RESEARCH, DEVELOPMENT, AND MANUFACTURE *

NOTATION: (by columns)

Input: A = air gasifier; O = oxygen gasifier; P = pyrolysis process; PG = pyrolysis gasifier; S = steam; H = hydrogasification; C = char combustion.

Contact Mode: U = updraft; D = downdraft; O = other (sloping bed, moving grate); Fl = fluidized bed; S = suspended flow; MS = molten salt; MH = multiple hearth.

Fuel Products: LEG = low energy gas (about 150-200 Btu/SCF) produced in air gasification; MEG = medium energy gas produced in oxygen and pyrolysis gasification (350-500 Btu/SCF); PO = pyrolysis oil, typically 12,000 Btu/lb; C = char, typically 12,000 Btu/lb.

Operating Units: R = research; P = pilot; C = commercial size; CI = commercial installation; D = demonstration.

Size: Gasifiers are rated in a variety of units. Listed here are Btu/h derived from feedstock throughput on the basis of biomass containing 16 MBtu/ton or 8000 Btu/lb, SMW with 9 MBtu/ton. ( ) indicates planned or under construction.

	Gasifier Type					
Organization	Input	Contact Mode	Fuel Products	Operating Units	Size (Btu/h)	Comments
9.2.1 Air Gasification of Biomass						
Alberta Industrial Dev. Edmonton, Alb., Can.	A	Fl	LEG	1	30M	
Applied Engineering Co. Orangeburg, SC 29115	A	U	LEG	1	5M	
Battelle-Northwest Richland, WA 99352	A	U	LEG	1-D	—	

*Unless otherwise noted, the gasifiers listed here produce dry ash (T less than 1100 C) and operate at 1 atm pressure. (Coal gasifiers and future biomass gasifiers may operate at much higher pressures.)

B.C. Research Vancouver, B.C., Can. VC5 262	A	Fl	LEG	2	1-4M
Biomass Corp. Yuba City, CA 95991	A	D	LEG	1	2M
Bio-Solar Research & Development Corp. Eugene, OR 97401	A	U	LEG	1	—
Century Research, Inc. Gardena, CA 90247	A	U	LEG	1	80M
Davy Powergas, Inc. Houston, TX 77036	A	U	LEG-Syngas	20	—
Deere & Co. Moline, IL 61265	A	D	LEG	1	100 kW
Eco-Research Ltd. Willodale, Ont., Can. N2N 558	A	Fl	LEG	1	16M
Environmental Energy Eng., Inc. Morgantown, WV 26505	A	Fl	?	1	3M
Environmental Energy Eng., Inc. Morgantown, WV 26505	A	D	LEG	1	0.1-0.5M
Environmental Energy Eng., Inc. Morgantown, WV 26505	A	Fl	MEG	1	—
Forest Fuels, Inc. Keene, NH 03431	A	U	LEG	4	1.5-30M
Foster Wheeler Energy Corp. Livingston, NH 07309	A	U	LEG	1	—

Organization	Gasifier Type		Fuel Products	Operating Units	Size (Btu/h)	Comments
	Input	Contact Mode				
Georgia Institute of Tech. Eng. Exp. Station Atlanta, GA 30332	A	U	LEG	1	0.5M	
Halcyon Assoc., Inc. East Andover, NH 03231	A	U	LEG	4	6-50M	
Imbert Air Gasifier 5760 Arnsberg Z, Germany	A	D	LEG	500,000	34k-34M	
Industrial Development & Procurement, Inc. Carle Place, NY 11514	A	D	LEG	Many	100-750 kW	
Lamb-Cargate Industries, Ltd. New Westminster, B.C., Can.	A	U/Fl	LEG	?	4M	
Lamb-Cargate Industries, Ltd. New Westminster, B.C., Can.	A	U	LEG	2	25M	
Pioneer Hi-Bred International, Inc. Johnston, IA	A	D	LEG	—	9M	
Pulp & Paper Research Inst.,* Pointe Claire, Quebec, Can. H9R 3J9	A	D	LEG	—	—	
Purdue Univ. Agricultural Eng. Dept. W. Lafayette, IN 47907	A	D	LEG	1	0.25M	
Saskatchewan Power Corp. Regina, Sask., Can. S4P-0S1	A	Fl	LEG	1 or 2	25M	

*Operates at 1-3 atm pressure.

Texas Tech Univ. Dept. of Chem. Eng. Lubbock, TX 79409	A	Fl	LEG	1	0.4M
Texas Tech Univ. Dept. of Chem. Eng. Lubbock, TX 79409	A	U	LEG	1	—
Univ. of California Dept. of Agricultural Eng. Davis, CA 95616	A	D	LEG	1	64,000
Univ. of California Dept. of Agricultural Eng. Davis, CA 95616	A	D	LEG	1	6M
Univ. of Missouri at Rolla Rolla, MO	A	—	—	1P	—
Vermont Wood Energy Corp. Stowe, VT 05672	A	D	LEG	1	0.08M
Westwood Polygas Vancouver, B.C., Can. V6G 2Z4	A	U	LEG	1	—
9.2.2 Oxygen Gasification of Biomass					
Battelle-Northwest Richland, WA 99352	O,A-S	U	—	1	—
Davy Powergas, Inc. Houston, TX 77036	—	—	—	—	—
Environmental Energy Eng., Inc. Morgantown, WV	O	D	MEG	1P	0.5

Organization	Gasifier Type		Fuel Products	Operating Units	Size (Btu/h)	Comments
	Input	Contact Mode				
IGT-Renugas Chicago, IL	O,S	Fl	MEG	—	—	
Rockwell Int. Canoga Park, CA 91304	O,A	—	—	—	—	
9.2.3 Pyrolysis Gasification of Biomass						
A&P Coop (Angelo Industries) Jonesboro, AR	P	O	MEG (C)	1C	—	
Arizona State Univ. Tempe, AR	PG	Fl	MEG	1	—	
Battelle–Northwest Richland, WA 99352	P	Fl	MEG	1	—	
ENERCO Langham, PA	P	—	MEG, PO, C	1P, 1C	—	
ERCO Cambridge, MA	P	Fl	PO, C	1P, (1C)	16, (20)	
Garrett Energy Research & Engineering Ojai, CA	MH	—	MEG	1P	—	
Gilbert Associates Reading, PA 19603	P	Fl	—	1R	—	
Princeton Univ. Princeton, NJ 08544	PG	O	MEG,C	IR	—	
M. Rensfelt Sweden	PG	O	MEG,C	IR	—	

Tech Air Corp. Atlanta, GA 30341	P	U	MEG, PO, C	4P, 1C	33
Texas Tech Univ. Lubbock, TX	PG	Fl	MEG	1P	—
Univ. of Arkansas Fayetteville, AR	P	O	MEG (C)	IR	—
Wright-Malta Ballston Spa, NY*	PG	O	MEG (C)	1R, 1P	4
9.2.4 Biomass Hydrogasification					
Battelle-Columbus** Columbus, OH 43201	H	Fl,U,S	PG,PO,C	1-Res	—
9.2.5 Air Gasification of Solid Municipal Waste (CSMW)					
Andco-Torrax*** Buffalo, NY	A	U	LEG	4C	100M
Battelle-Northwest Richmond, VA 99352	—	—	—	—	—
9.2.6 Oxygen Gasification of SMW					
Calorican Murray Hill, NJ	O	U	—	—	9M
Union Carbide Corp. (Linde) Tonowanda, NY***	O	U	MEG	1	100M

*Operates at 1-3 atm pressure.

**Operates at less than 70 atm pressure.

***These gasifiers produce slagging (T greater than 1300 C) instead of dry ash.

Organization	Gasifier Type: Input	Gasifier Type: Contact Mode	Fuel Products	Operating Units	Size (Btu/h)	Comments
9.2.7 Pyrolysis Gasification of SMW						
Envirotech Concord, CA	P	MH	LEG	1 P	—	
ERCO Cambridge, MA	P	Fl	MEG	1P	16	
Garrett Energy Research & Eng. Hanford, CA	P	MH	MEG	1P	—	
Michigan Tech Houghton, MI	P	ML	MEG	—	—	
Monsanto Enviro-chem. Systems Baltimore, MD	P, C	K	LEG, O, C	1 D	20 (375)	
Nichols Engineering Belle Mead, NJ	P	—	MEG, C	—	—	
Occidental Research Corp. El Cajon, CA	P	Fl	PO, C, MEG	1 C	—	
Princeton Univ. Princeton, NJ	P	O	MEG, C	2R	—	
Pyrox Japan	P, G, C	Fl	MEG	1C	—	Derived from Bailie process

Rockwell International Canoga Park, CA 91304	P	MS	MEG, C	1P	16	
Univ. of West Virginia at Wheelebrator Morgantown, WV	P, G, C	Fl	MEG	1P	—	Bailie fluidized bed system
9.2.8 Coal Gasification*						
Babcock & Wilcox Co. Barberton, OH	A/O	S	LEG/MEG	1P (1-20 atm pressure)	400M	Semicommercial unit of 15 ft ID (400 tons/day) operated for one year in 1955. Slurry feed is pumped to raise pressure and then spray dried by recycle gas. Still in development.
Battelle-Columbus Battelle Mem. Inst. 505 King Ave. Columbus, OH	PG	Dual Fl	MEG	1P (7 atm pressure)	25M	Agglomerating ash is heated in an air-blown combustor and recirculated to a steam-blown pyrolyzer.
BCR Bituminous Coal Research, Inc.	PG	3-Fl	LEG	1P (16 atm pressure)	1.2M	Three-stage process: Devolatilization/gasification/char combustion.
Bi-Gas Bituminous Coal Research, Inc. 350 Hockberg Rd. Monroeville, PA 15146	O-S	S	MEG	1P (34-100 atm pressure)	120M	

*There are dozens of systems being investigated for the gasification of various kinds of coal. We include here those that have long been commercialized or are presently being actively developed, for comparison with biomass gasifiers.

Organization	Gasifier Type		Fuel Products	Operating Units	Size (Btu/h)	Comments
	Input	Contact Mode				
CO_2 Acceptor Conoco Coal Dev. Co. Research Div. Library, PA	PG	2-Fl	MEG	1P (16 atm pressure)	30M	Char is burned to regenerate CaO/(CO_2 acceptor), which is recirculated to gasifier.
DOE-METC Morgantown Energy Technology Center Collins Ferry Rd. Box 880 Morgantown, WV	A-S	U(Stirred)	LEG	1P (20 atm pressure)	20M	
FW Stoic Stoic Combustion Pty. Ltd. Johannesburg, South Africa	A-S Two-Stage	U	LEG	4	22-90M	Diameter available: 6.5, 8.5, 10, 12.5 ft
Hydrane DOE-MERC Morgantown, WV	H_2	S	HEG	1P (200 atm pressure)	0.2M	Laboratory scale
Koppers-Totzek Koppers Co., Inc. Koppers Bldg. Pittsburgh, PA	O-S	S	MEG	39P (1-30 atm pressure)	450M-860M	

Lurgi American Lurgi Corp. 377 Rt. 17 Hasbrouck Heights, NJ	O-S	U	MEG	66P (30 atm pressure)	800M	
McDowell-Wellman Eng. Co.	A/O-S Single-Stage	U	LEG/MEG	15	3-100M	Standard sizes available: 3.5, 6.5, 8, 10 ft diam.
Riley-Stoker Corp. Riley Morgan Gasifer Riley Morgan Gasifier	A/O-S Single-Stage	U	LEG/MEG	10	100M	More than 9000 units sold through 1940s
SYNTHANE DOE-PETC 4800 Forbes Ave. Pittsburgh, PA	O-S	Fl	MEG	1P (70 atm pressure)	72M	
Wellman-Incandescent Applied Technology Corp. Houston, TX	A-S Two-Stage	U	LEG	30	14-100M	Mostly in South Africa Diameter available: 4.5, 5.5, 6.5, 8.5, 10, 10.75, 12 ft
Wilputte Corporation	A-S Single-Stage	U	LEG		67M	More than 250 units operated from 1913 to 1945
Winkler Davy Powergas, Inc. P.O. Drover 5000 Lakeland, FL	O/A-S	Fl	LEG/MEG	41P (1 atm pressure)	1100M	None in the United States
Woodall-Duckham	A/O-S Two-Stage	U	LEG/MEG	40	100M	

DIRECTORY OF GASIFIERS

BIOMASS AIR GASIFIER DIRECTORY

Organization
Alberta Industrial Developments Ltd.

Address
704 Cambridge Building
Edmonton, Alberta
Canada T5J 1R9

Personnel
Richard P. Assaly

Phone
(403) 429-4094

Type of Gasifier (up/down draft, size, fuel, application, etc.)

Thermex-Reactor- (Fluid Bed) 70 ton/day
30 million Btu/hr. Design and module size unlimited.

Status (research, pilot scale, commercial, etc.)

PROTOTYPE - Now ready for commercial use.

General Information (description, photo, sketch, etc.)

Gas Generator Process by Fluid Bed (Pyrolysis) includes flash drier/ feed bin/gasifier (Thermex-Reactor) operates on air, close couple gas connection for boilers, driers, etc.

Process can maximize gas or charcoal production. High efficiency process with low operating cost system can operate on very fine raw material higher heating values of gas than other systems.

Plans for Future

Short Term - 1979-80 Three to six reactor installations up to 10 tons/hr.

Long Term - High pressure (400-600 GPSI) system for SynGas.

Name Richard P. Assaly Date January 16, 1979

BIOMASS GASIFIER DIRECTORY

Organization	Address
Applied Engineering	1525 Charleston Hwy. Orangeburg, S. C. 29115
Personnel	**Phone**
J. F. Jackson	803-534-2424

Type of Gasifier (up/down draft, size, fuel, application, etc.)

Boiler retrofit of a continuous updraft unit sized to provide 25mm BTU/Hr. via the gasification of whole tree chips.

Status (research, pilot scale, commerial etc.)

Commercial application.

General Information (description, photo, sketch, etc.)

Proprietary grate and burner design gives the unit the capability of producing 25 MMBtu/h on a continuous basis. Commercial application comprises a turn-key installation consisting of wood chip storage and handling, gasification, boiler retrofit package, and control system.

Plans for Future

Commercial/Industrial Application - design, manufacture, and installation of biomass gasification equipment and related hardware.

Name James F. Jackson Date November 8 - 1978

BIOMASS AIR GASIFIER DIRECTORY

Organization	Address
Battelle-Northwest	P.O. Box 999, Richland, WA 99352
Personnel	**Phone**
L.K. Mudge	946-2268
P.C. Walkup	946-2432
~~D.G. Ham~~	~~946-2083~~

Type of Gasifier (up/down draft, size, fuel, application, etc.)

Updraft. Diameter: 1 ft; working bed height: 5 ft. Solids processed: corn stalks, grass straw, wood chips, wood pellets, industrial wastes, coke, ~~charcoal, coal~~

Status (research, pilot scale, commercial, etc.)

Operational at a small pilot scale.

General Information (description, photo, sketch, etc.)

The gasifier is refractory lined and is equipped with an eccentric, rotating grate and a mechanical feed distributor. Solid feed is introduced at the top of the reactor through a lock hopper and auger. A schematic of the gasifier is shown in Figure 1.

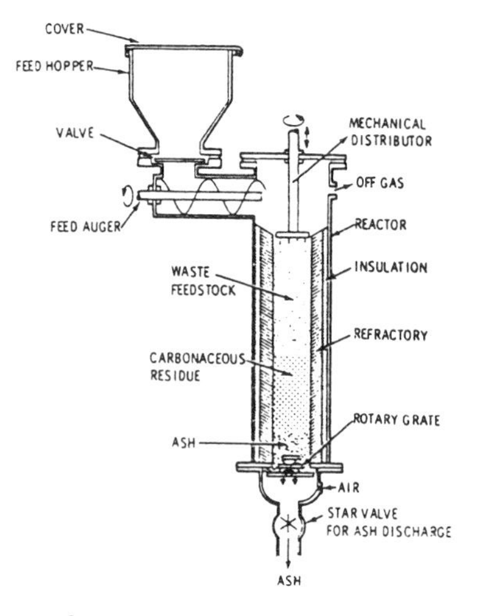

Figure 1 *Schematic of small gasifier*

Plans for Future

Continue operation of the gasifier to characterize gasification characteristics of different solids.

Name Lyle K. Mudge Date 9 January 1979

BIOMASS AIR GASIFIER DIRECTORY

Organization B.C. Research

Address 3650 Wesbrook Mall
Vancouver, B.C. V6S 2L2
Canada

Personnel Dr. Douglas W. Duncan

Phone (604) 224-4331

Type of Gasifier (up/down draft, size, fuel, application, etc.)

Fluidized bed wood waste gasifier using run-of-the-mill sawdust or hog fuel.

Status (research, pilot scale, commercial, etc.)

10^6 Btu/hr unit available at B.C. Research for research use.
$4x10^6$ Btu/hr unit at Saskatchewan Forest Products, Hudson Bay, Saskatchewan.

General Information (description, photo, sketch, etc.)

The B.C. Research unit has the dimensions shown in the attached sketch. Air is supplied below the pinhole grate by a 3 HP blower (150 CFM capacity). Run-of-mill hog fuel containing up to 50% moisture (total weight basis) is fed into the combustion zone just above the grate where the volatiles are driven off and consumed. The 5 ft bed consists of charcoal and ash. Surplus ash is withdrawn intermittently through the bottom of the unit. The raw gas (100-150 Btu/sdcf) exits via a port near the top of the reactor, passes through a dry cyclone to a furnace where it is burned.

The $4x10^6$ Btu/hr unit in Saskatchewan is similar except that the reactor has an expanded freeboard above the ash bed to aid in particulate removal and the raw gas exits from the top of the reactor where it passes through a cyclone and then through a gas cleaning system. The raw gas is intended to fire a diesel generator set.

The Btu gasifier is being commercialized by Lamb Cargate Industries Ltd., 1135 Queens Ave., New Westminster, B.C., V5L 4Y2.

Figure 2:
B.C. RESEARCH FLUIDIZED BED GASIFIER

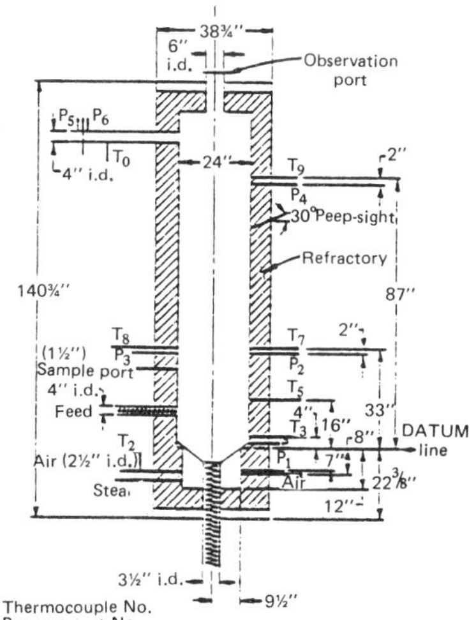

T = Thermocouple No.
P = Pressure port No.
═ = Orifice plate on pipe

Plans for Future

Continue research studies on research reactor. Generate financing to build $20x10^6$ Btu/hr prototype.

Name [illegible] Date January 24, 1979

BIOMASS AIR GASIFIER DIRECTORY

Organization

Biomass Corporation

Address

951 Live Oak
Blvd., Yuba City, Ca. 95991

Personnel

Theodore H. Crane, President
Robert O. Williams, Vice President Engineering

Phone

(916) 674-7230

Type of Gasifier (up/down draft, size, fuel, application, etc.)

Downdraft, fuel from prune pit size to 2x2x2 "hay-cubes"
5000 Btu per pound and up heating value, biomass or coal.

Status (research, pilot scale, commercial, etc.)

Commercial system. 1 to 15 million Btu per unit.
Manifold units to 70 million Btu.

General Information (description, photo, sketch, etc.)

The BIOMASS GASIFIER is a down draft, co-current flow, fixed bed reactor for conversion of solid carbonaceous fuel to low-Btu fuel gas. The fuel gas may be directly substituted for natural gas or fuel oil in existing or new boilers with only a change in the burner. Available standard low Btu gas burners are standard commercial products in sizes up to 100 million Btu.

The Biomass gasifier discharges no tar, oils or liquors which could require expensive or hazardous disposal by the operator. The char residue contains carbon and inorganic matter suitable for blending with conventionally produced charcoal for briquettes or as a low sulfur metallurgical carbon source. The residue is inert and may be land filled if there is no other use for it.

A large internal fuel hopper and a system of sealed external hoppers, augers and knife gate valves allow continuous operation with full automation of the fuel cycle and no possibility of gas leaks at any time.

The design analysis of the various sized Biomass gasifiers includes a detailed thermal stress study. The suspended design of the gasifier shall allow full expansion of the gasifier eliminating stress build-up, a subsequent shell cracking. Details of system designs, system sizing and economic analysis of the benefits of gasifier ownership available upon application.

Plans for Future

Detailed studies of the use of the biomass gasifier as a fuel source for internal combustion engines. These studies will include complete mass and energy balances and the wear factor upon the engines.

Name THEODORE H. CRANE Date January 16, 1979

BIOMASS AIR GASIFIER DIRECTORY

Organization
Bio-Solar Research & Development Corp.

Address
1500 Valley River Drive, Suite 220
Eugene, Oregon 97401

Personnel
35

Phone
(503) 686-0765

Type of Gasifier (up/down draft, size, fuel, application, etc.)

Updraft, tank size 12' x 25', burns WOODEX® solid fuel pellets to produce gas for any heat application.

Status (research, pilot scale, commercial, etc.)

Commercial and research

General Information (description, photo, sketch, etc.)

Bio-Solar Research & Development Corp. manufactures producer gas equipment burning WOODEX® pelletized solid fuel, and producing a gas of high heat value from a non-fossil derivative. The gas is called G-GAS, and a patent has been applied for. The gas can be used to produce heat for any purpose, and when cleaned by proprietary methods, can be used in glass smelting.

Plans for Future Bio-Solar Research & Development Corp. will continue to build WOODEX® plants with G-GAS producers providing heat for dehumidification of biomass in the manufacture of WOODEX® pellets. Gasifiers will also be utilized by joint-venture plants built with major companies and through license agreement.

Name Ted Carpentier Date 31 January, 1979

BIOMASS AIR GASIFIER DIRECTORY

Organization

Century Research, Inc.

Address

16935 S. Vermont Avenue, Gardena, Calif. 90247

Personnel

Dr. Steve S. Hu
Mr. Howard R. Amundsen

Phone

(213) 327-2405

Type of Gasifier (up/down draft, size, fuel, application, etc.)

Up-draft, Layer-zoned, Oxi-reduction Minimax Gas Producer, 10 ft diameter for standard model, Fuel: animal waste, agriculture waste, forest waste, paper waste, etc. Gas fuel for electricity, steam, cement/brick plant, chemical feedstock for ammonia/alcohol manufacturing.

Status (research, pilot scale, commercial, etc.)

Commercial

General Information (description, photo, sketch, etc.)

Overall dimension of standard 10 ft diameter unit: 35 ft tall represented by 15 ft of hopper and gravity feed system, 10 ft of combustion chamber, and 10 ft of residue cone and residue discharge system.

The unit can process approximately 100 tons of feed stock per day and produce 50 to 100 million btu equivalent of producer gas per hour.

The producer gas is composed of approx. 20-25% CO, 10-15% H2, 2%± CH4, and 5-10% CO2 and 50-60% N2 (by volume). It contains 125-165 btu per cu ft under std temp and pressure condition. It can reach 2700 deg F flame temperature.

A typical Century Research/Bainien gasification plant is composed of 5 component systems: Frontend feed stock processing system, Gasification system, Test and automatic control system, Environmental cleanup system, and End product synthetization or utilization/application system.

Marketable product on the basis of 1978 calculations is priced at $2. to $2.50 per million btu.

Plans for Future

Development of semi-portable or portable version of the standard model, so that the gasifier can process lower daily tonnage with high efficiency and on site to site basis.

Name Steve Hu / Howard R. Amundsen Date January 16th, 1979

BIOMASS AIR GASIFIER DIRECTORY

Organization

Davy Powergas Inc.

Personnel

1500 in USA
Worldwide

Address

P.O. Box 36444
Houston, Texas 77036

Phone

(713) 782-3440

Type of Gasifier (up/down draft, size, fuel, application, etc.)

Up draft fixed bed type, up to 13' 6" producing both gas engine fuel and ammonia synthesis gas.

Status (research, pilot scale, commercial, etc.)

Commercial - More than twenty gasifiers built & operated

General Information (description, photo, sketch, etc.)

This fixed bed "Waste Refuse Producer" is an offshoot of the Powergas Corp. Ltd. fixed bed producer of which more than one thousand gasifiers were built and operated. This biomass unit has operated on wood, wood bark, cotton seeds, bagasse, etc. Most of these units have been shut down due to the availability of natural gas and oil. We believe that one or two are still operating in Southern Africa.

Plans for Future

Davy is still promoting biomass gasification with air and now with oxygen. We are presently proceeding with the design of a 2000 TPD methanol plant based on wood gasification.

Name Edgar E. Bailey
Edgar E. Bailey
Product Manager

Date Jan 16 1979

BIOMASS AIR GASIFIER DIRECTORY

Organization

Deere & Company

Personnel

N. A. Sauter

Address

Technical Center
3300 River Drive
Moline, IL 61265

Phone

309/757-5275

Type of Gasifier (up/down draft, size, fuel, application, etc.)

Continuous, portable, downdraft unit for converting agricultural residues to gas and to electricity via 100 kW diesel generator set

Status (research, pilot scale, commercial, etc.)

Research Tool

General Information (description, photo, sketch, etc.)

Unit is generally described in Chapter 8, Solid Wastes and Residues - Conversion by Advanced Thermal Processes, American Chemical Society Symposium Series, Washington, D. C. 1978.

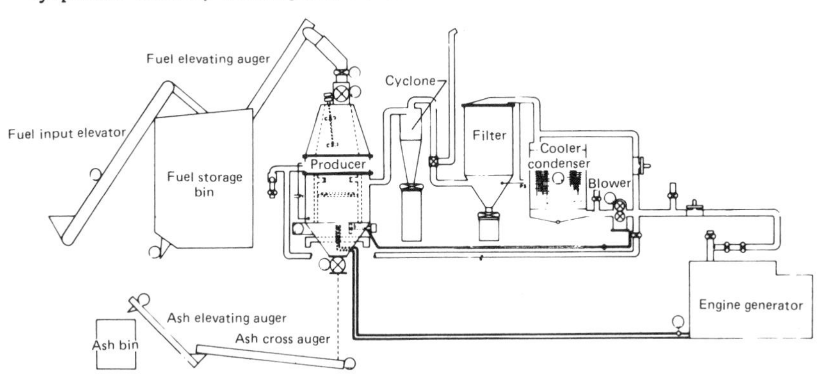

Schematic of portable 100 w farm power plant

Plans for Future

Not currently active

Name [signature] Date 11 January 1979

BIOMASS GASIFIER DIRECTORY

Organization DEKALB AgResearch, Inc. **Address** DeKalb, Illinois

Personnel Stan Bozdech
Harold Zink

Phone 815 758-3461

Type of Gasifier (up/down draft, size, fuel, application, etc.)

Up-draft with combustion system to dry seed
Fuel-dry corn cobs

Status (research, pilot scale, commerial etc.)

Pilot scale at 1.6 million BTU's/hour proven in actual drying tests. Scale-up to 6 million BTU's on line in fall of 1980.

General Information (description, photo, sketch, etc.)

Gasifier System was designed to overcome slagging at the grates and, through a close-coupled arrangement, with primary air mixed in a Commercial Burner Head, to complete combustion in a torroidal chamber. Clean combustion gases are tempered to 110°F for drying as they exit combustion chamber. Complete system operates as a vacuum.

Plans for Future

Name

Date November 7, 1979

BIOMASS AIR GASIFIER DIRECTORY

Organization	Address
Eco-Research Limited	P.O.Box 200, Station A Willowdale, Ontario. M2N 5S8
Personnel	**Phone**
John W. Black	416-226-7351

Type of Gasifier (up/down draft, size, fuel, application, etc.)

Fluidized Bed Gasifier
Application - wood, municipal refuse

Status (research, pilot scale, commercial, etc.)

25 TPD pilot plant - ready for commercialization Sept.'79

General Information (description, photo, sketch, etc.)

The pilot plant started up in May '76 and has been used both as a combustion unit with in-bed steam generation and a gasification system for the production of a low BTU fuel gas. Materials gasified have included tires, wood, wood wastes, agricultural biomass and municipal refuse.

Plans for Future

Plans for the near term include a continuous demonstration test of about 3 months and oxygen gasification

Name John W. Black Date January 16, 1979

BIOMASS GASIFIER DIRECTORY

Organization

Energy Resources Company Inc. (ERCO)

Address

185 Alewife Brook Parkway
Cambridge, MA 02138

Personnel

Herbert M. Kosstrin

Phone

(617) 661-3111

Type of Gasifier (up/down draft, size, fuel, application, etc.)

Continuous fluidized bed pyrolysis unit for conversion of agricultural and industrial wastes to produce low Btu gas, char and oil.

Status (research, pilot scale, commerial etc.)

Pilot scale unit available for client testing
Commercial units now under construction

General Information (description, photo, sketch, etc.)

Pilot unit described in paper given at Institute of Gas Technology Symposium: "New Fuels and Advanced Combustion Technologies," March, 1979.

Plans for Future

Continued commercialization for waste to energy units

Name [signature] Date November 1, 1979

Herbert M. Kosstrin

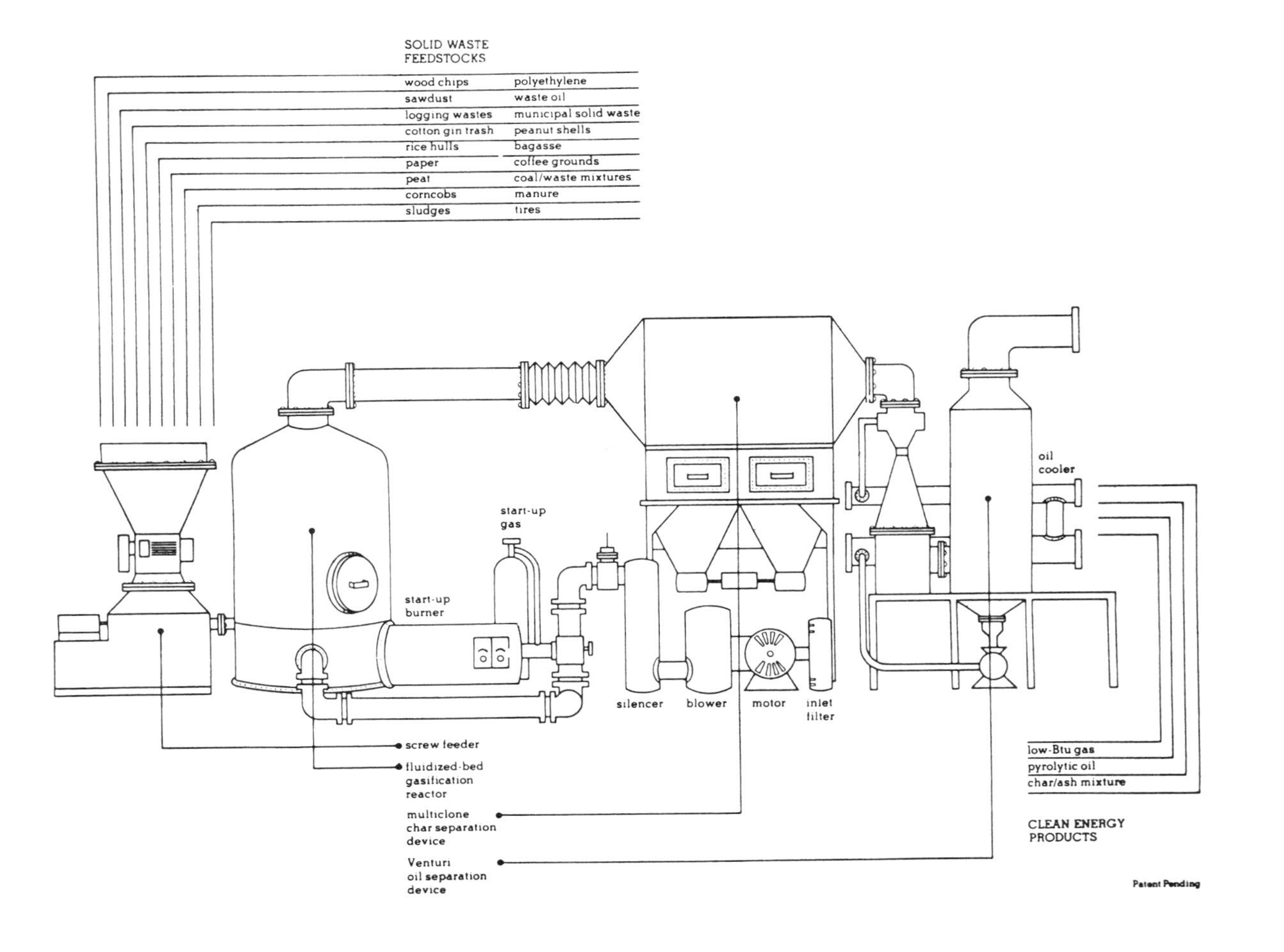
SOLID WASTE FEEDSTOCKS
wood chips
sawdust
logging wastes
cotton gin trash
rice hulls
paper
peat
corncobs
sludges
polyethylene
waste oil
municipal solid waste
peanut shells
bagasse
coffee grounds
coal/waste mixtures
manure
tires
start-up gas
start-up burner
silencer
blower
motor
inlet filter
oil cooler
screw feeder
fluidized-bed gasification reactor
multiclone char separation device
Venturi oil separation device
low-Btu gas
pyrolytic oil
char/ash mixture
CLEAN ENERGY PRODUCTS
Patent Pending

BIOMASS AIR GASIFIER DIRECTORY

Organization	Address
Environmental Energy Engineering Inc.	P.O. Box 4214, Morgantown, W.Va. 26505
Personnel	**Phone**
Dr. Richard C. Bailie	(304) 983-2196

Type of Gasifier (up/down draft, size, fuel, application, etc.)

Downdraft gasifier operating on char, wood blocks or pelletized wood. Operates commercial burner that can be used for crop drying, furnace industrial heat and internal combustion engine. Cap. 100,000/hr to 500,000 Btu/hr.

Status (research, pilot scale, commercial, etc.)

Batch system ready for commercial application but no manufacturer exists. Continuous system requires additional development.

General Information (description, photo, sketch, etc.)

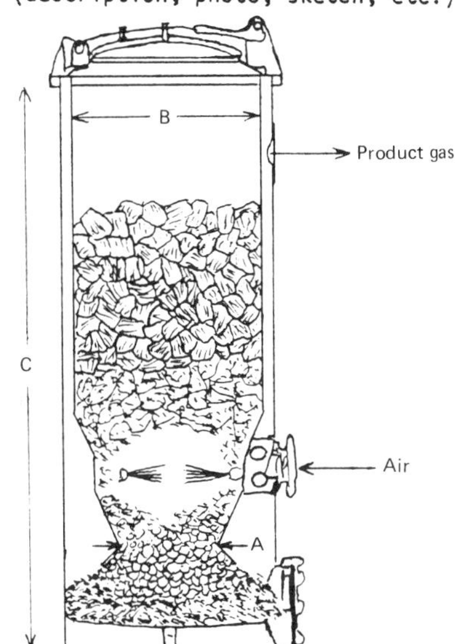

A = 5 inches
B = 18 inches
C = 50 inches

Plans for Future

Test in small commercial operations replacing natural gas. Test oxygen enriched air systems. Modify for continuous operation.

Name R. C. Bailie Date Jan 24, 1979

BIOMASS AIR GASIFIER DIRECTORY

Organization

Environmental Energy Engineering Inc.

Address

P.O. Box 4214, Morgantown, W.Va. 26505

Personnel

Dr. Richard C. Bailie

Phone

(304) 983-2196

Type of Gasifier (up/down draft, size, fuel, application, etc.)

Fluidized bed operating on wood blocks, sawdust or pellets. Operates commercial burner which can be used for crop drying, furnace industrial heat and internal combustion engine. Air blown, Cap. 3 x 10^6 Btu/hr.

Status (research, pilot scale, commercial, etc.)

Pilot plant test facility for different feed stocks.

General Information (description, photo, sketch, etc.)

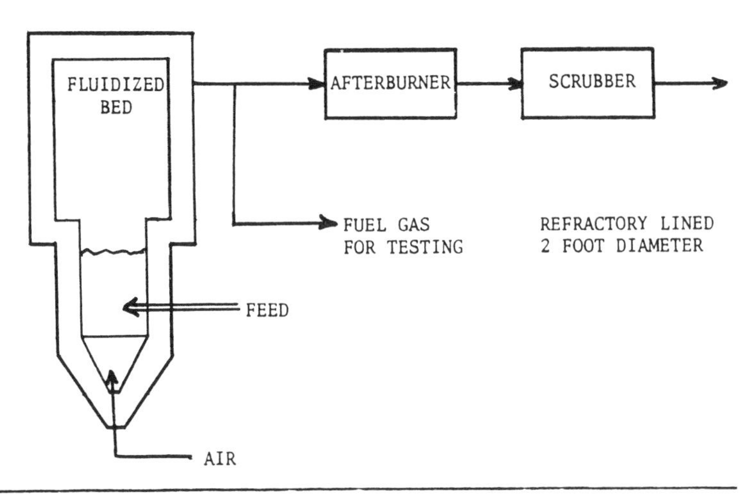

Plans for Future

Available for commercial development.

Name R. C. Bailie Date Jan. 24 1979

BIOMASS AIR GASIFIER DIRECTORY

Organization

Environmental Energy Engineering, Inc.

Address

P.O. Box 4214, Morgantown, W.Va 26505

Personnel

Dr. Richard C. Bailie

Phone

(304) 983-2196

Type of Gasifier (up/down draft, size, fuel, application, etc.)

Two fluidized beds which can produce 300 Btu/ft^3 gas not diluted with N_2 without need for oxygen plant. Operates on most any cellulosic feed.

Status (research, pilot scale, commercial, etc.)

Research - Pilot facility

General Information (description, photo, sketch, etc.)

Sketch of commercial system is shown below. Test facility adds heat electrically instead of circulating sand as shown.

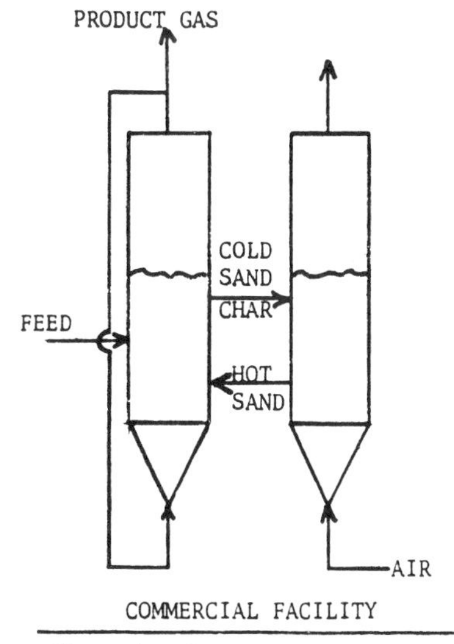

COMMERCIAL FACILITY

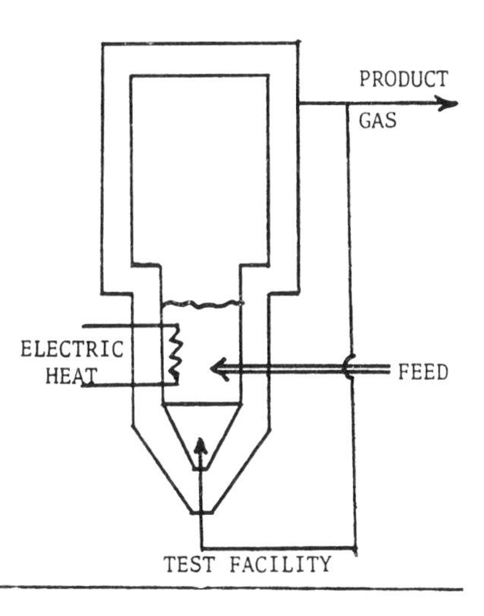

TEST FACILITY

Plans for Future

Demonstrate with sand circulation.

Name R. C. Bailie Date Jan. 24, 1979

BIOMASS GASIFIER DIRECTORY

Organization

Forest Fuels, Inc.

Address

Technical Center
Antrim, N. H. 03440

Personnel

M. H. Stevens
R. A. Caughey

Phone

603-588-2994

Type of Gasifier (up/down draft, size, fuel, application, etc.)

up draft, moving grate, close-coupled, using pulp chips, log saw dust, planer shavings, sized debarking waste - dried to 10-20% dry weight basis- to run package, sectional boilers, or direct fire to provide plant or process heat for kilns, factories, schools. 2mm BTU/hr. to 25mm BTU/hr.

Status (research, pilot scale, commerial etc.)

Pilot and limited commercial

General Information (description, photo, sketch, etc.)

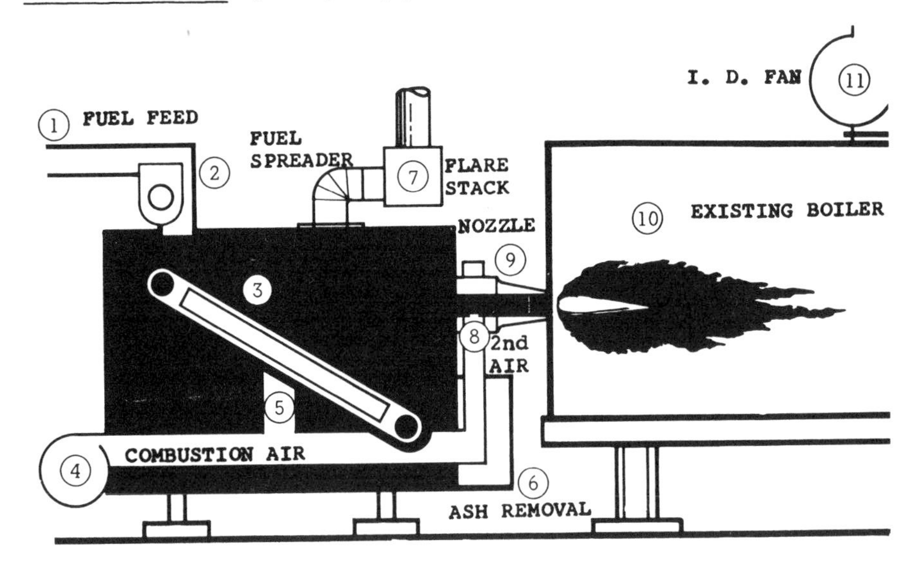

Plans for Future

Prove market readiness and increased sales in Northeast - and elsewhere on qualified basis.

Name [signature]

Date November 2, 1979

BIOMASS AIR GASIFIER DIRECTORY

Organization Foster Wheeler Energy Corp.

Address 110 South Orange Avenue Livingston, New Jersey 07039

Personnel

Roger J. Broeker

Phone

201-533-2667

Type of Gasifier (up/down draft, size, fuel, application, etc.)

updraft

Status (research, pilot scale, commercial, etc.)

Gasifier is commercial on coal. Have bench scale gasifier and 2-ft diameter test gasifier available for test work on wood.

General Information (description, photo, sketch, etc.)

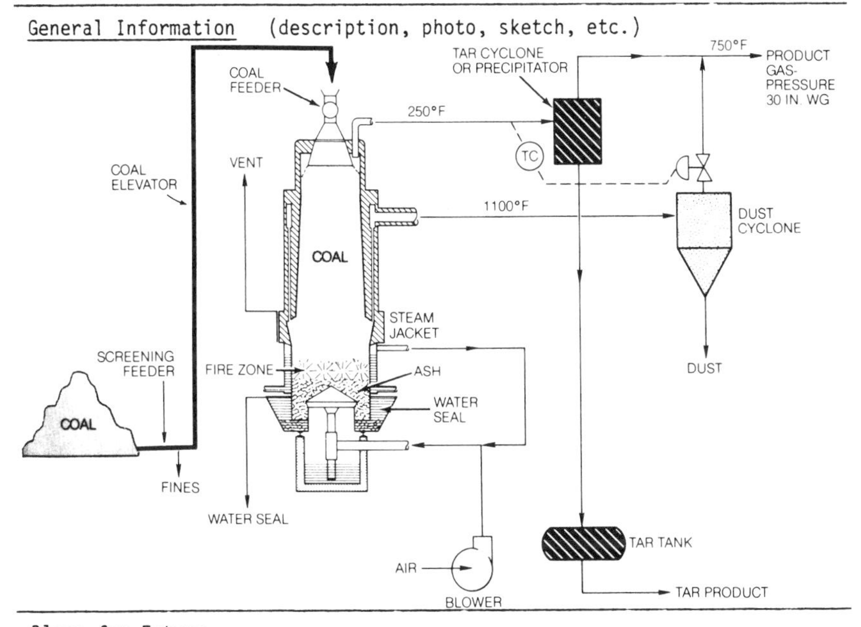

Plans for Future

Name R. J. Broeker Date 1/12/79

BIOMASS AIR GASIFIER DIRECTORY

Organization

Georgia Institute of Technology
Engineering Experiment Station

Address

Room 1512-A C&S Building
33 N Avenue - Atlanta, Ga. 30332

Personnel

Jerry L. Birchfield
Tomas F. McGowan

Phone

(404) 894-3448

Type of Gasifier (up/down draft, size, fuel, application, etc.)

Up draft, 1/2 million Btu/hr, textile drying

Status (research, pilot scale, commercial, etc.)

Research, under design and construction

General Information (description, photo, sketch, etc.)

Up draft gasifier operating under forced draft. Product gas will be burned in a closecoupled arrangement. Hot combusted gases will be mixed with air for textile drying and curing tests.

Plans for Future

Experiments with pellets, dry and wet chip wood fuels.

Name Thomas F. McGowan Date 2-22-79

BIOMASS AIR GASIFIER DIRECTORY

Organization	Address
Halcyon Associates, Inc.	Maple Street, East Andover, N.H. 03231
Personnel	**Phone**
William G. Finnie, President	(603) - 735 - 5356

Type of Gasifier (up/down draft, size, fuel, application, etc.)

Up draft - 6 MMBTUH through 50 MMBTUH - Green or dry wood waste or biomass fuel - For direct heating, boiler firing & direct power generation.

Status (research, pilot scale, commercial, etc.)

Commercial - 4 units sold, others being negotiated.

General Information (description, photo, sketch, etc.)

The Halcyon Gasifier produces cool clean gas using green or dry hogged size fuel or biomass. Calorific value is around 150 BTU per cubic foot. When burned, particulates are less than .02 pounds per million BTU with low Nox, well within E.P.A. requirements, without any cleaning of flue gases.

The gasifier operates below ash fusion temperatures and the grates are automatically self-cleaning. Ash removal is automatic.

Series of controls on the gasifier allows for automatic operation with little supervision.

A burner of up to 100 MMBTUH capacity, which can be adapted to fit most existing oil or natural gas fired boilers, can be supplied. The burner is capable of firing oil and/or natural gas as well as producer gas.

Output of the gasifier and burner(s) is controlled by regulating the gas flow actuated by boiler steam pressure or dryer/furnace temperature. Full modulation and flame failure safety features to meet insurance company requirements are included.

On power generation or direct drives, the gas is further cleaned to remove sub-micron size particles, and directly fuels internal combustion or compression ignition engines. This further cleaning may be used also when gas is burned where extremely low particulates are required.

Maintenance and power requirements are low.

Plans for Future

To engineer, manufacture, and apply units for commercial and industrial requirements.

Name William G. Finnie Date January 18, 1979

BIOMASS AIR GASIFIER DIRECTORY

Organization IMBERT AIR GASIFIER

Address Steinweg Nr. 11, 5760 Arnsberg 2, Germany

Personnel Walter Zerbin

Phone (0 19 31) 35 49
Telex 84 222 ins d

Type of Gasifier (up/down draft, size, fuel, application, etc.)
Downdraft air gasifier for diesel power generation

Status (research, pilot scale, commercial, etc.)
500,000 built and used over last 40 years

General Information (description, photo, sketch, etc.)
10 to 10,000 kw gasifier power plants. complete.

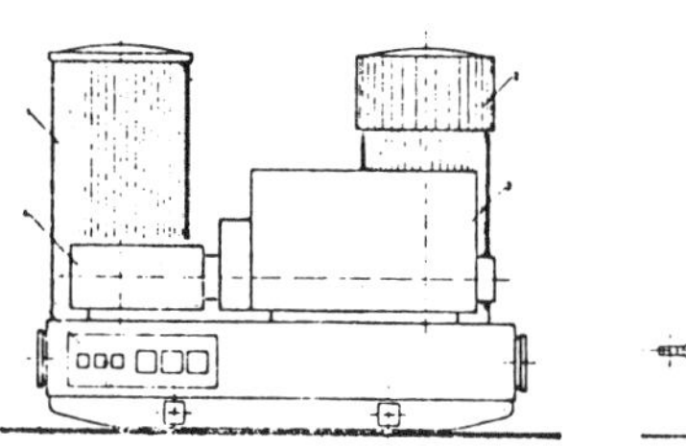

Power plant TSG 10 to 60 KVA

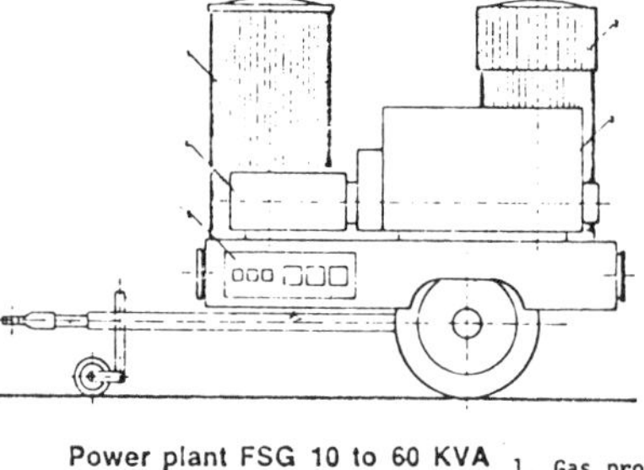

Power plant FSG 10 to 60 KVA

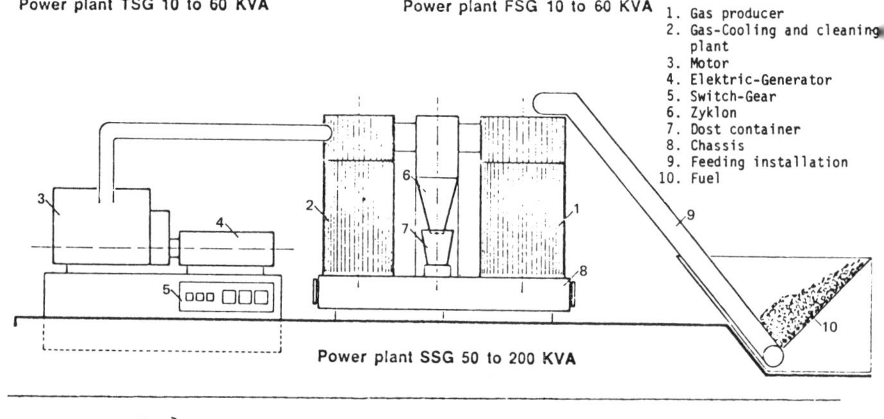

Power plant SSG 50 to 200 KVA

Name (TBR) Date 3/27/79

BIOMASS AIR GASIFIER DIRECTORY

Organization
INDUSTRIAL DEVELOPMENT
AND PROCUREMENT INC.

Address ONE OLD COUNTRY ROAD
CARLE PLACE, N.Y. 11514

Personnel Representing: Moteurs Duvant

Phone 516-248-0880

Jules A. LUSSIER, Vice-President

Type of Gasifier (up/down draft, size, fuel, application, etc.)
Down draft - 1 to 8 million BTU per unit. Fuel: Wood waste, chips, bark, corn cobs, rice husks, cotton gin residues, coffee shells, coconut shells and husks, sun flower seed residues, paper mill sludge, other miscellaneous organic waste.

Status (research, pilot scale, commercial, etc.) Commercial
Several Duvant Dual Fuel Engine systems have been delivered and installed in Europe, Africa, South Pacific, Asia, Central America.

General Information (description, photo, sketch, etc.)
Complete energy systems consisting of a low BTU gas production unit, a filtering and cooling unit and a dual fuel engine - generator set. Range 100 to 750 KW.
Possibility of Manifold Units.

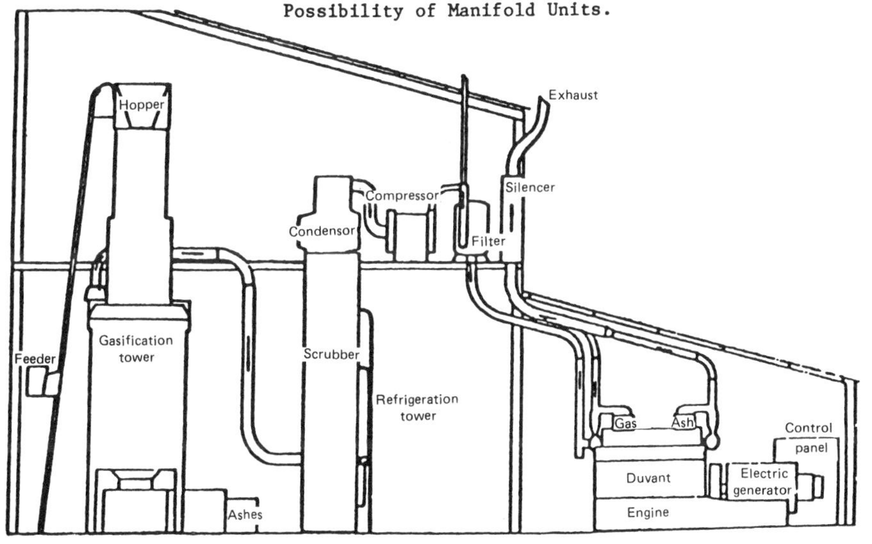

Plans for Future

Promote and develop sales in North America.

Name Philippe Santini **Date** March 27, 1979

BIOMASS AIR GASIFIER DIRECTORY

Organization	Address
Lamb-Cargate Industries Ltd.	1135 Queens Avenue NewWestminster, B.C.
Personnel	**Phone**
F.H. Lamb, President	604/521-8821

Type of Gasifier (up/down draft, size, fuel, application, etc.)

Up-Draft, 4 x 10^6 B.T.U./hour, clean hog fuel

Status (research, pilot scale, commercial, etc.)

Pilot Scale

General Information (description, photo, sketch, etc.)

Semi-fluid bed reactor, complete with fuel metering and continuous ash discharge. Fuel metering adjacent to the grate.

Equipped with gas cleaning station consisting of:

a) cyclone
b) wet centrifugal scrubber
c) gas dryers.

Hudson Bay, Saskatchewan, installation includes gas engine generation.

Plans for Future

Package generation unit for small isolated communities, dry kilns, dryers, etc.

Name F.H. Lamb   Date 1979 February 21

BIOMASS AIR GASIFIER DIRECTORY

Organization	Address
Lamb-Cargate Industries Ltd.	1135 Queens Avenue New Westminster, B.C.
Personnel	**Phone**
F.H. Lamb, President	604/521-8821

Type of Gasifier (up/down draft, size, fuel, application, etc.)

Up-draft, 25 x 10^6 Net BTU, Green Hog Fuel.

Status (research, pilot scale, commercial, etc.)

Commercial

General Information (description, photo, sketch, etc.)

The Lamb Wet-Cell Burner is a double chamber system. The fuel is fed in up through the bottom of the grates. The lower chamber gasifies the green hog fuel and the gases are burned in the second chamber with a close control of excess air. There are two 25 x 10^6 BTU/hour units in commercial services. One in British Columbia directly fires two lumber kilns and one in New Zealand fires a pulp flash dryer at a new TMP mill.

Plans for Future Going up to 150 x 10^6 BTU/hour and firing lime kilns, waste heat boilers, veneer dryers, rotary dryers, etc.

Name F.H. Lamb Date 1979 February 21

BIOMASS GASIFIER DIRECTORY

Organization
Morbark Industries, Inc.

Address
P.O. Box 1000, Winn, MI. 48896

Personnel
Ivor Bateman

Phone
517-866-2381

Type of Gasifier (up/down draft, size, fuel, application, etc.)

25 Million BTU/HR Cyclone Suspension gasifier, sawdust up to 25% moisture - 1/4" wood chips.

Status (research, pilot scale, commerial etc.)

Commercial Model Under Test

General Information (description, photo, sketch, etc.)

The gasifier produces low BTU gas at below ash fusion temperatures. It is ideally suited for direct coupling to a boiler, drier or any application where heat is required and also as a retro fit for gas or oil burners. Ash removal is continuous and automatic. Particulate emmision is in the order of 500 parts per million. Gasification is achieved with a partial burning process primary air required for gasification is 1 1/4 pounds air per pound fuel.

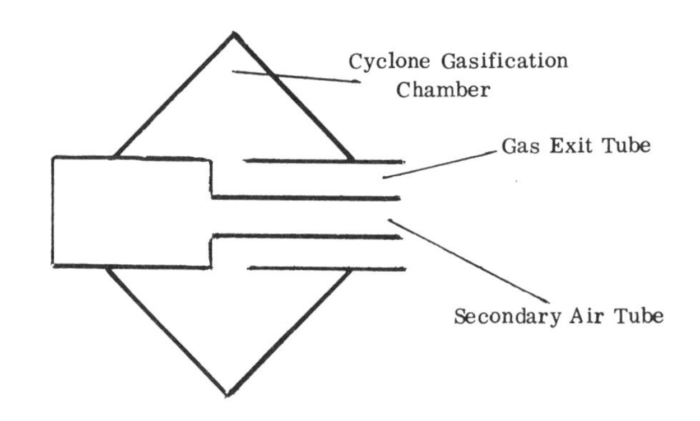

Plans for Future

To engineer and apply units for commercial and industrial requirements

Name Ivor Bateman

Date Nov. 5, 1979

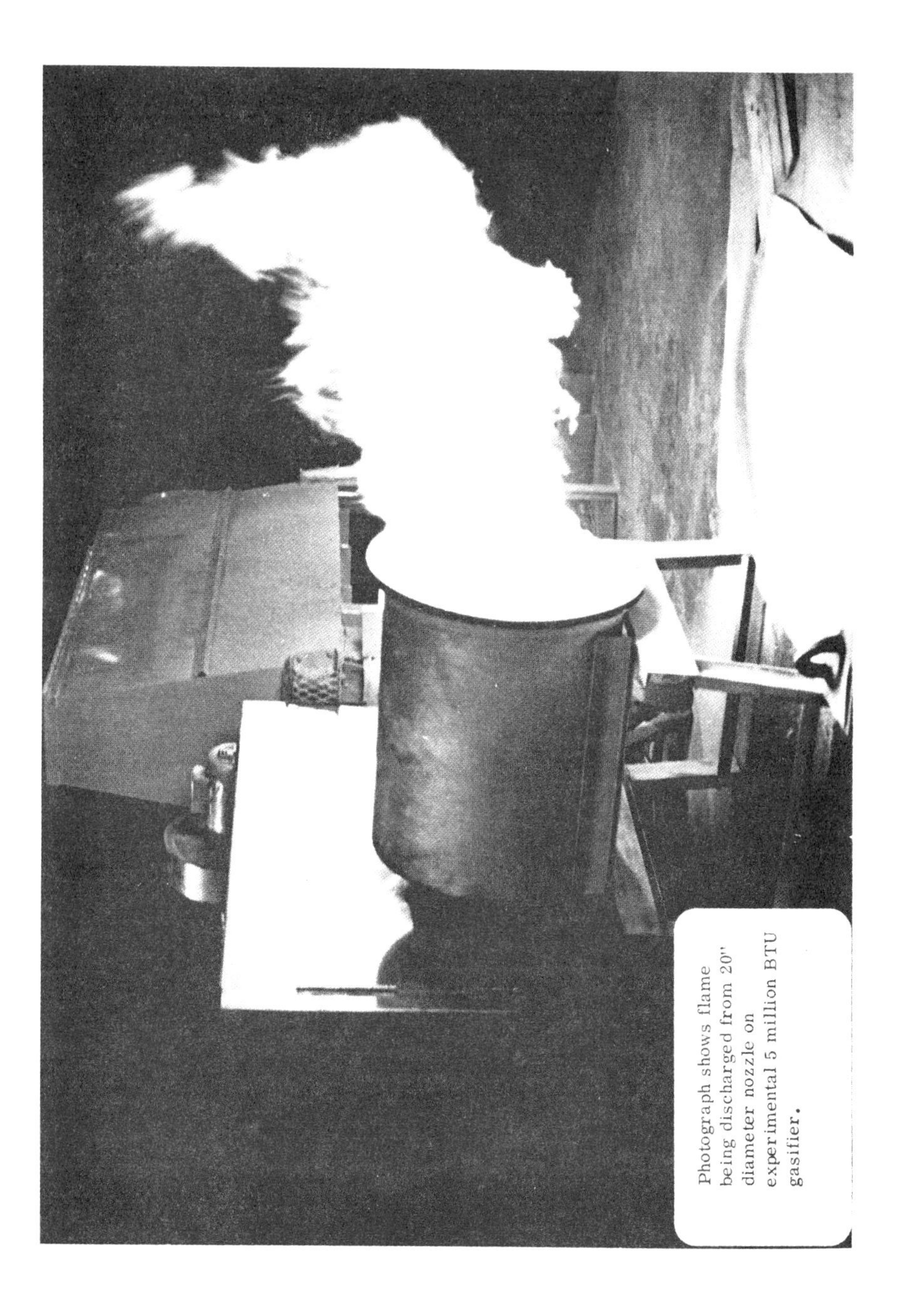

Photograph shows flame being discharged from 20" diameter nozzle on experimental 5 million BTU gasifier.

BIOMASS AIR GASIFIER DIRECTORY

Organization
PIONEER HI-BRED INTERNATIONAL, INC.

Address
5700 MERLE HAY ROAD, JOHNSTON, IA. 50131

Personnel
Walter Stohlgren

Phone
1-515-245-3721

Type of Gasifier (up/down draft, size, fuel, application, etc.)

Down Draft 9×10^6 Btu/Hr. Corn Cobs. Seed Dryer.

Status (research, pilot scale, commercial, etc.)

Research, Commercial

General Information (description,

Testing close coupled burner.

Looks good for eliminating the tar problem.

Plans for Future

Redesign grate to eliminate the ash caking problem.

Name [signature] Date February 6, 1979

BIOMASS AIR GASIFIER DIRECTORY

Organization
Pulp and Paper Research Institute of Canada

Address 570 St. John's Blvd., Pointe Claire, Quebec, Canada H9R 3J9

Phone (514) 697-4110

Personnel
S. Prahacs and M.K. Azarniouch

Type of Gasifier (up/down draft, size, fuel, application, etc.)
Down draft reactor, 316 SS, 12 in. diameter, 15 ft. high, suitable for spent pulping liquors and lignocellulosic material, pressure - 45 psig/atmospheric, temperature - 1450°F/1650°F.

Status (research, pilot scale, commercial, etc.)
Pilot scale (presently not operated).

General Information (description, photo, sketch, etc.)

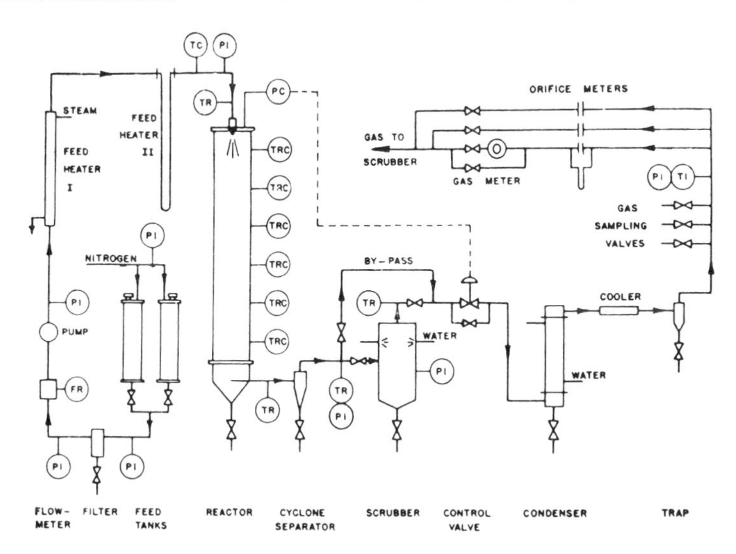

Plans for Future
To carry out gasification tests on lignocellulosic material.

Name S. Prahacs

Date January 15, 1979

BIOMASS AIR GASIFIER DIRECTORY

Organization

Purdue University

Address

Agricultural Engineering Department
W. Lafayette, IN 47907

Personnel

Robert M. Peart, Michael Ladisch

Phone

(317) 749-2971

Type of Gasifier (up/down draft, size, fuel, application, etc.)

Downdraft, corn cobs, for direct firing of corn dryer.

Status (research, pilot scale, commercial, etc.)

Research, crude operational model only

General Information (description, photo, sketch, etc.)

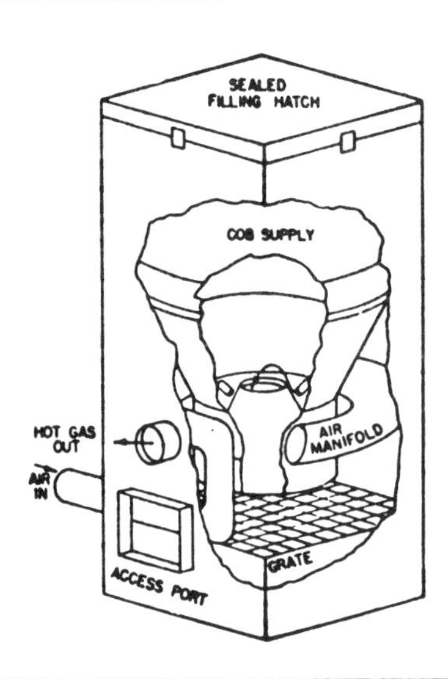

This batch unit holds about 300 pounds of cobs, is about 8 feet tall, 31.5 inches square, mild steel except for stainless steel support cone. Air flow 25 scfm, heat output estimated approximately 250,000 Btu/hr (50 pounds of cobs/hr). We have gasified cobs of from 15-25% moisture, wet basis.

Plans for Future

1) Build continuous flow unit for more accurate measurement of input/output.
2) Test turn-down ratios, cob moisture, air flow, insulation.
3) Build bench test unit for more accurate tests on composition as affected by operating variables.

Name Robert M. Peart Date January 16, 1979

BIOMASS GASIFIER DIRECTORY

Organization	Address
Saskatchewan Power Corporation	2025 Victoria Avenue, Regina, Sask. S4P 0S1
Personnel	**Phone**
G.A. Weisgerber	(306) 525-7611

Type of Gasifier (up/down draft, size, fuel, application, etc.)

Updraft unit for converting wood waste to gas and to electricity via 150 kW diesel generator set.

Status (research, pilot scale, commerial etc.)

The 1.2 MW unit has been operated with various wood feeds. An industrial burner and diesel generator set have been successfully run.

General Information (description, photo, sketch, etc.)

The wood gasification plant, located at the Saskatchewan Forest Products Corporation's plywood plant in Hudson Bay, Saskatchewan, is a joint venture of the Saskatchewan Power Corporation, Saskatchewan Forest Products Corporation, and the Federal Government of Canada.

The objectives of the current project are: i) to investigate the feasibility, economics, environmental acceptability and practicability of power generation via wood gasification in isolated northern communities, ii) to process wood waste from forest product industries to produce fuel gas, and iii) to develop a Canadian technology.

Plans for Future

Immediate plans are to operate on a continuous basis for an extended period to demonstrate commerciability.

Name G.A. Weisgerber (signature) Date 1979 December 10

BIOMASS AIR GASIFIER DIRECTORY

Organization
Texas Tech University

Address
Dept. of Chemical Engineering
Lubbock, TX 79409

Personnel
Harry W. Parker

Phone
(806) 742-3553

Type of Gasifier (up/down draft, size, fuel, application, etc.)

Prototype is up-draft batch, but subject to change. Objective is to utilize gin trash for fueling internal combustion engines on irrigation wells

Status (research, pilot scale, commercial, etc.)

pilot scale

General Information (description, photo, sketch, etc.)

The present gasifier is a simple up-draft batch gasifier 20 inches in diameter. This gasifier will have to have significant modifications to succeed in gasifying gin trash for operation of irrigation wells. Another type of gasifier may be selected.

Plans for Future

Determine feasibility of gasifying un-cubed gin trash for powering irrigation wells. If it is feasible a cost estimate will be made.

Name Harry W. Parker Date January 15, 1979

BIOMASS AIR GASIFIER DIRECTORY

Organization
University of California at Davis
Department of Agricultural Engineering

Address
University of California
Davis, CA 95616

Personnel
John R. Goss, Professor

Phone
(916) 752-1421/0102

Type of Gasifier (up/down draft, size, fuel, application, etc.)
Downdraft, 4-foot firebox, 54 ft^3 fuel capacity including active firebox volume 500 to 1100 lb/hr of hogged kiln dried lumber waste and other agricultural and forest residue.

Status (research, pilot scale, commercial, etc.)
Pilot scale for research and demonstration.

Pilot plant gas producer mounted on semi-trailer for transport to various test locations. Removal of upper cylinder and fuel feed assembly to meet 13 ft 6 inch transport height. Operation is monitored and fuel feed and ash removal automatically controlled from control and instrument panel mounted in cabin at front of trailer. Firebox volume - 38 ft^3. Ash grate basket - 143 ft^3. Ash pit - 69 ft^3. Gas producer weighs 3.9 tons. Firebox and lower outer cylinder constructed from A515 steel flat stock. Lower cylinder insulated with 2" thick J-M Thermo 12. Normal output 4 to 6 million Btu/hr on dry wood chips. Maximum output about 8 million Btu/hr (NTP) of combustible gases. To left of gas producer are the hot gas cyclone and three hot gas fiberglass bag filters. Combustion air blower and gasoline engine drive on ground at rear of trailer.

Plans for Future

Property of California Energy Commission awaiting further program development. Inquire Commission at 1111 Howe Avenue, Sacramento, CA 95825. (916) 920-6033.

Name John R. Goss   Date January, 1979

BIOMASS AIR GASIFIER DIRECTORY

Organization	Address
University of California at Davis Department of Agricultural Engineering	University of California Davis, CA 95616
Personnel	Phone

Type of Gasifier (up/down draft, size, fuel, application, etc.)

Downdraft, 12-inch firebox, 1.8 ft^3 fuel capacity, 30 to 80 pounds/hour fuel rate with agricultural and forest residues.

Status (research, pilot scale, commercial, etc.)

Laboratory scale gas producer to investigate gasification characteristics of fuels and test variations in design parameters.

General Information (description, photo, sketch, etc.)

The gas producer fuel is batch fed by opening the gasketed cover at the top. Fuel cylinders with different configurations can be inserted for particular physical characteristics of fuel. A fuel column 32 inches high is accommodated above the firebox. Tuyere nozzle sizes and lengths and elevation of choke plate and choke diameter can all be changed. Ash grates of various configurations can be interchanged with the one shown. Hand turning of the grate has been replaced with a small fractional horsepower motor, gear reducing box and roller chain drive.

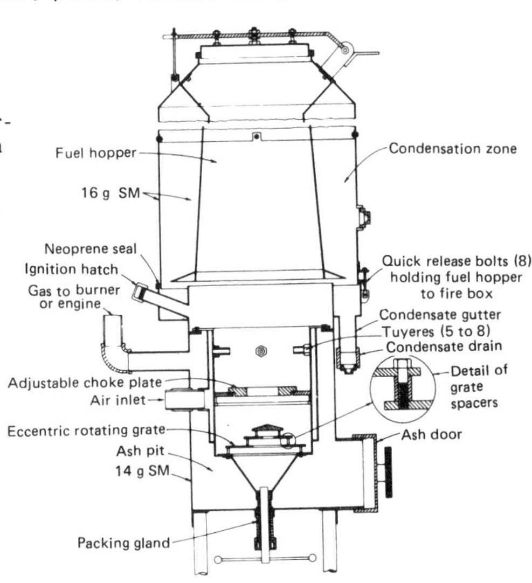

Plans for Future

Continue investigating gasification characteristics of agricultural and forest residues and low-Btu gas utilization before and after solid particulate filtration and then after cooling and condensing.

Name John R. Goss Date January, 1979

BIOMASS GASIFIER DIRECTORY

Organization

GROW Project: University of Mo. Rolla

Address

University of Missouri-Rolla
207 Harris Hall
Rolla, Mo. 65401

Personnel

Y. Omurtag,
Project Manager

Phone

Office: 314-341-4560
SITE: 314 341-4857

Type of Gasifier (up/down draft, size, fuel, application, etc.)

Fluid bed using sand and air as fluidizing medium, 40 in ID x 14ft. 2000lb/hr sawdust feed.

Status (research, pilot scale, commerial etc.)

Phase I: Low BTu gas pilot plant operation,data almost complete, Medium energy and other research is being planned.

General Information (description, photo, sketch, etc.)

The overall objective of the GROW program is to conduct a research and development program which will lead to the early commercialization of wood gasification technology to process wood residues typical of those found in the Missouri Ozark regions. Optimum commercialization parameters for low and medium BTU gas production as a substitute for natural gas will also be determined. The facility can be used in conducting research or providing training in the areas of fluidized bed reactor operations, feed stack handling, and marketing of various products resulting from operating such systems. First and foremost, however, it will allow for the determination of optimum design for energy conversion systems which use wood and other bio-energy sources. The equipment is suitable for gasification research of all types of biomass including, but not restricted to, wood chips, sawdust, animal manure, or corn cobs and other agricultural by-products. The project is expected to take from 18 months to two years after the start of testing and could prove to be invaluable in providing information about such energy conversion and its possible contribution to society.
The GROW project has the largest capacity reactor involved in the Bio-mass Thermochemical Conversion Program. As such, the GROW project has the potential to become the showcase project for the entire Thermochemical Program.

See the Experimental Facility Flow Diagram which follows:

Plans for Future

Phase II: Medium Btu Gas with re-cycle to be completed by August, 1980.

Name Yildirim Omurtag Date 11/5/1979

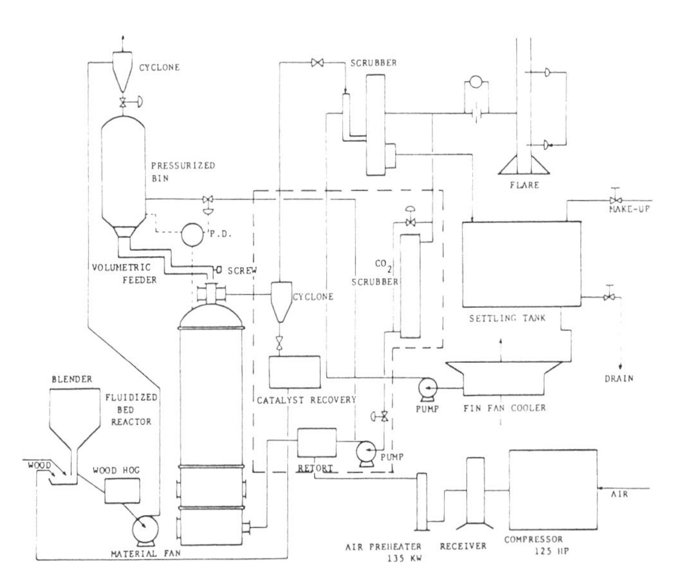

UMR-Coors Experimental Facility Flow Diagram

BIOMASS AIR GASIFIER DIRECTORY

Organization

The Vermont Wood Energy Corporation

Address

P.O. Box 280
Stowe, VT 05672

Personnel

J. Phillip Rich, President
Peter H. Bauer, Project Engineer
Cadwallader E. Brooks, Treasurer

Phone

802-253-7220

Type of Gasifier (up/down draft, size, fuel, application, etc.)

Close-coupled, down draft, semi-automatic (wood chips or pellets, manually loaded), thermostatic on/off operation, roughly 80,000 BTU/hr. output

Status (research, pilot scale, commercial, etc.)

One semi-automatic test model under development, about 2/3 of the way to successful operation.

General Information (description, photo, sketch, etc.)

The gasifier is intended for use with a home-size furnace, to convert a used or new furnace from oil flame to wood gas flame, or possibly as an adjunct installation with oil burner gun still in place.

The gasifier, about the size of a small suitcase, is surrounded by an insulating enclosure and has a chip hopper above it. Combustible gases are led through about 3 feet of pipe to the combustion chamber of a former oil burning furnace. The gasifier has been operating successfully using forced draft, and an induced draft system is under development.

When the thermostat signals for heat, the electrical/electronic control system begins a timed sequence of events, operating an electric fuel igniter, and then blowers, solenoid operated valves, tickler shaft motor, low fuel level detector motor, and the gas igniter electrodes. The controls shut off and turn on the system when signalled by the thermostat. Safe shutdown occurs upon electric supply failure or in case of various system failures or low fuel level.

Plans for Future Completion of development of semi-automatic test model... Development of automatic test model by replacing chip hopper with a surge bin, and adding a conveyor and storage bin for the fuel... Testing, prototype installations, modifications, marketing, production, and sales of one or both types of gasifiers

Name Peter H. Bauer **Date** January 12, 1979

BIOMASS O_2 GASIFICATION DIRECTORY

BIOMASS O_2 GASIFICATION DIRECTORY

Organization
Battelle-Northwest

Address
P.O. Box 999, Richland, WA 99352

Personnel
L.K. Mudge
P.C. Walkup
D.G. Ham

Phone
946-2268
946-2432
946-2083

Type of Gasifier (up/down draft, size, fuel, application, etc.)

Updraft. Diameter: 3 ft; working bed height 10 ft. Solids processed: wood municiple wastes, industrial wastes, coal, charcoal, coke.

Status (research, pilot scale, commercial, etc.)

Operational at pilot scale.

General Information (description, photo, sketch, etc.)

The gasifier is refractory lined. Solid feed is introduced at the top of the reactor through a lock hopper. A drag chain conveyor feeds the lock hopper arrangement. Steam and air, or oxygen, is introduced into the bottom of the reactor through a stationary grate. Continuous solids discharge is not provided with this gasifier. Ash is removed from the gasifier bottom after accumulation of an ash layer of about 3 ft. in depth.

Plans for Future

Continue operation of the gasifier to characterize gasification characteristics of different combustible solids.

Name D. G. Ham Date 3/5/79

BIOMASS O_2 GASIFICATION DIRECTORY

Organization
Energy Systems Group
Rockwell Molten Salt

Address
8900 De Soto Avenue
Canoga Park, California 91304

Personnel
C. Trilling, D. McKenzie
S. Yosim, J. Ashworth

Phone
C. R. Faulders, Marketing Rep.
(213) 341-1000, Extension 2045

Type of Gasifier (up/down draft, size, fuel, application, etc.)
Molten salt gasifier, currently being applied to coal gasification; can be operated air-blown or oxygen-blown. The salt used is sodium carbonate.

Status (research, pilot scale, commercial, etc.)
Molten Salt Test Facility (MSTF) is used to gasify ~500 lb/hr of coal or other carbonaceous fuels. Process Development Unit (PDU) for coal gasification, 1 ton per hr, now in operation under contract to DOE.

General Information (description, photo, sketch, etc.)

1) The MSTF gasification unit is 3 ft ID, 4 ft OD, stainless steel vessel lined with monofrax brick. This unit can be operated air-blown, up to a few atmospheres pressure, and includes facilities for continuous fuel preparation and feed of both fuel and carbonate. The melt can be continuously withdrawn through an overflow nozzle, but there is no melt regeneration system.

2) The molten salt coal gasification PDU is a completely integrated system including coal and carbonate feed, coal gasifier, melt overflow and quench, ash filtration, sulfur removal, and regeneration of sodium carbonate.

Plans for Future
The PDU will be operated on the current contract the remainder of this year. Follow-on effort to include oxygen gasification is expected.

Name [signature]
C. R. Faulders

Date
March 6, 1979

BIOMASS PYROLYSIS SYSTEMS DIRECTORY

PYROLYSIS SYSTEMS DIRECTORY

Organization Angelo Industries
A & P Coop Co.

Address PO Box 212,
Jonesboro, Ark, 72401

Personnel
J. F. Angelo Jr.

Phone 501 935 1234
932 7733

Type of Gasifier (up/down draft, size, fuel, application, etc.)
Rotary Pyrolyser for wet and dry biomass; produces char and process heat

Status (research, pilot scale, commercial, etc.)
Process operated since 1971 for commercial charcoal production, 40 tons/day. Joint project with U. of Arkansas to increase energy yields and determine energy balance (see U. of Arkansas).

General Information (description, photo, sketch, etc.)

Plans for Future

Name TBR

Date 3/26/79

PYROLYSIS SYSTEMS DIRECTORY

Organization	Address
Battelle-Northwest	P.O. Box 999, Richland, WA 99352

Personnel	Phone
L.K. Mudge	946-2268
P.C. Walkup	946-2432
D.H. Mitchell	946-3791
R.J. Robertus	946-3622

Type of Gasifier (up/down draft, size, fuel, application, etc.)

Agitated Fluid bed. Diameter: 11 in; working bed height 4.5 ft. Wood chips are processed in this gasifier.

Status (research, pilot scale, commercial, etc.)

Operational as a process development unit.

General Information (description, photo, sketch, etc

The gasifier is refractory lined and is equipped with a mechanical agitator. The wood chips are fluidized in the reaction zone. The agitator is provided to "stir" catalysts used in the production of methane, ammonia synthesis gas, hydrocarbon synthesis gas, hydrogen, or carbon monoxide. Wood feed is introduced into the bottom of the reaction zone with an auger. A schematic of the reactor is shown in Figure 1.

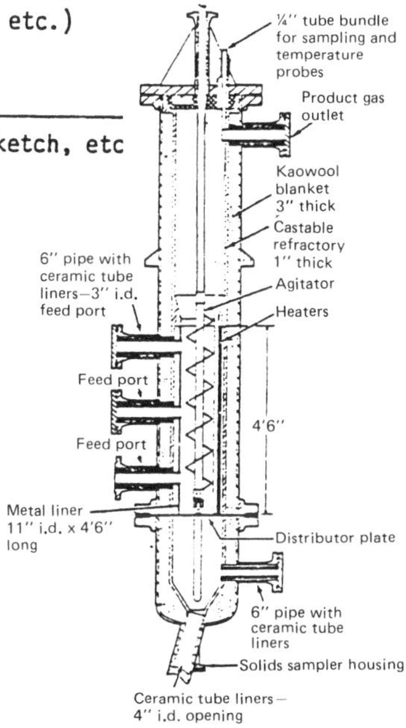

Figure 1. Biomass Gasification Reactor

Plans for Future

Unit will be used for the development of catalyzed biomass gasification processes.

Name L.K. Mudge Date 5 March 1979

PYROLYSIS SYSTEMS DIRECTORY

Organization

Enerco Incorporated

Address 139 A. Old Oxford Valley Road
Langhorne, PA 19047

Personnel
Miles J. Thomson
Eugene W. White

Phone 215/493-6565

Type of Gasifier (up/down draft, size, fuel, application, etc.)
Continuous, portable, cross-current pyrolytic converter for converting biomass into charcoal, pyrolysis oil, and medium BTU gas.

Status (research, pilot scale, commerial etc.)

Commercial

General Information (description, photo, sketch, etc.)

The unit is unique in its means of recirculating hot gases to accomplish pyrolysis without using air or oxygen in the reactor. A general description is available from a paper given as part of a symposium on Thermal Conversion of Solid Waste and Biomass, American Chemical Society Annual Meeting September 9-14, 1979, Washington, D.C.

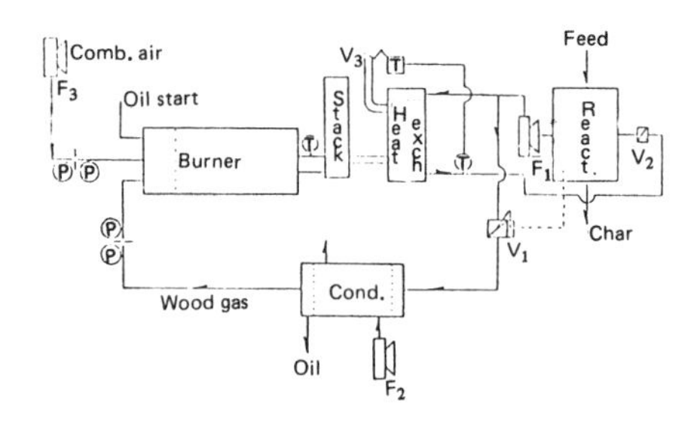

Schematic Diagram of Enerco Pyrolysis Unit

Plans for Future

Complete the demonstration of our commercial unit with the T.V.A. at Maryville College and install up to 45 commercial sites with the assistance of U.S.D.A. guaranteed loans.

Name Miles J. Thomson **Date** November 5, 1979

PYROLYSIS SYSTEMS DIRECTORY

Organization

Energy Resources Company, Inc.

Address 185 Alewife Brook Parkway
Cambridge, MA 02138

Personnel

Dr. Herb Kosstrin - Manager of Research & Engineering
Daniel R. Traxler - Marketing Manager

Phone (617) 661-3111

Type of Gasifier (up/down draft, size, fuel, application, etc.)
Fluidized Bed Gasification utilizing a wide variety of agricultural, forest products, industrial and municipal wastes.

Status (research, pilot scale, commercial, etc.) Pilot scale fluidized bed reactor in operation with 18 feedstocks utilized since 1976 (20" I.D. reactor, 16 MM Btu/hr). Second generation plant (20 MM Btu/hr) under construction and due for operation in second quarter of 1979.

General Information (description, photo, sketch, etc.)

In June of 1978 Energy Resources received a contract to design, build and operate a trailer mounted, transportable, fluidized bed gasification plant. The plant can convert agricultural wastes and forest residue into storable and transportable fuel products, pyrolytic oil and char. This competitive procurement was awarded jointly by EPA (Cincinnati) and the State of California's Solid Waste Management Board and Energy Commission. The plant is nominally rated at 90 tons per day of dry waste. The plant is scheduled for operation the third quarter of 1979 in California.

Commercial product offerings include Fluidized Bed Combustion Steam Boilers up to 100,000 pounds per hour and Fluidized Bed Gasification Systems. The FBG Systems are capable of handling a wide range of feedstocks including agricultural, wood, industry and municipal wastes with up to 60% moisture content. Modular, skid mounted systems are available in 50 and 100 MM Btu/hr output sizes. Custom applications are up to 250 MM Btu/hr. Complete materials handling equipment is available in addition to emission control equipment to meet all federal and state regulations.

Plans for Future: Further commercialization of Fluid Bed Gasification Systems to industries having a combustible waste product and an internal energy demand requiring oil and gas. In addition, various types of industrial and agricultural wastes are continually being tested and evaluated to become an economical feedstock for a Fluid Bed Gasification System.

Name Daniel R. Traxler Date March 2, 1979

PYROLYSIS SYSTEMS DIRECTORY

Organization / Address

Garrett Energy Research & Engineering, 911 Bryant Pl., Ojai, Ca. 93023

Personnel / Phone

Donald E. Garrett, President 805-646-0159
Ritchie D. Mikesell, Project Mgr.
Dinh Co. Hoang, Pilot Plant Supervisor

Type of Gasifier (up/down draft, size, fuel, application, etc.)

Multiple hearth. This not an air gasifier, as all heating is indirect. Agricultural wastes are processed to produce a medium - BTU gas.

Status (research, pilot scale, commercial, etc.)

Pilot scale. Shake-down stage.

General Information (description, photo, sketch, etc.)

Predrying, direct contact drying, pyrolysis, combustion, and water gas reaction are done sequentially in this device.

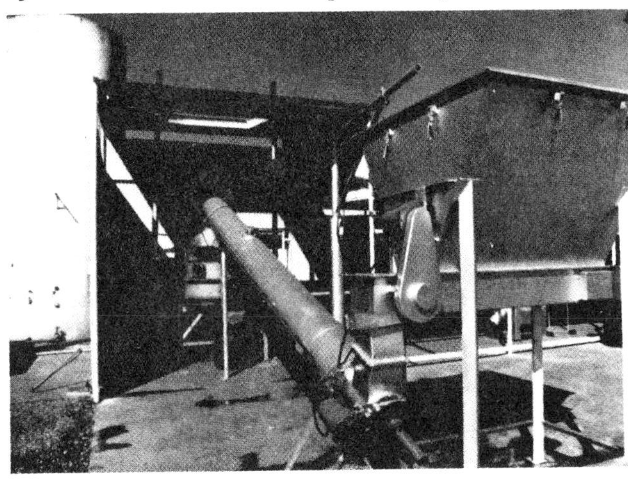

Plans for Future

Name Ritchie D. Mikesell Date 2/27/79

PYROLYSIS SYSTEMS DIRECTORY

Organization	Address
Prime Contractor - Gilbert Associates, Inc.	P.O. Box 1498 Reading, PA 19603
Major Subcontractors	**Phone**
West Virginia University and Environmental Energy Engineering, Inc.	(215) 775-2600

Type of Gasifier (up/down draft, size, fuel, application, etc.)

A two foot ID fluid bed gasifier operated with biomass and solid waste for research and development application.

Status (research, pilot scale, commercial, etc.)

Process development unit (PDU)

General Information (description, photo, sketch, etc.)

The 2'-0" ID fluidized bed gasifier can be operated with a biomass feed rate of up to 2 TPD biomass or solid waste. The hot gases leaving the top of the gasifier pass through a cyclone to remove particulates to a splitter where the stream is split into a product stream and a recycle stream. The gasifier can be modified so that it can operate as packed bed, entrained bed or free-fall bed. The hot product gas is scrubbed and is analyzed for the gas composition.

Plans for Future

Tests will be performed using 4 to 5 biomass feedstocks in combustion, pyrolysis and gasification modes of operation.

Name James T. Stewart
Manager, Fuels Conversion
Energy Research Division

Date February 26, 1979

PYROLYSIS SYSTEMS DIRECTORY

Organization

Princeton University

Address

D-215 Engineering Quadrangle
Princeton, New Jersey 08544

Personnel

M. J. Antal
F. E. Rogers
W. E. Edwards

Phone

(609) 452-5136

Type of Gasifier (up/down draft, size, fuel, application, etc.)

batch, electrically heated, zoned, tubular plug flow reactor

Status (research, pilot scale, commercial, etc.)

research, bench scale system

General Information (description, photo, sketch, etc.)

The one inch diameter, tubular quartz reactor has 3 zones of uniform temperature and is operated in a batch mode using 0.25 g samples of selected biomass material. It was designed to provide kinetic data on the gas phase reactions of pyrolytic volatile matter in steam. Rates of production as a function of temperature for CO_2, CO, H_2, CH_4, C_2H_4, C_2H_6, and C_3H_6 have been measured for cellulose and a selected wood species.

Plans for Future

Research on the effects of pressure on gasification rates and products. Research on the use of very high heating rates for biomass gasification.

Name Michael J. Antal, Jr. Date March 7, 1979

PYROLYSIS SYSTEMS DIRECTORY

Organization

Tennessee Valley Authority

Personnel

E. Lawrence Klein

Address

Division of Land and Forest Resources
Forestry Bldg., Norris, Tennessee 37828

Phone

(615) 494-9800

Type of Gasifier A continuous, portable, recirculating, pyrolysis unit capable of producing 1 ton of charcoal, 90 gallons of char oil and 8 million Btu's of medium Btu gas per hour from 3 tons of wood, designed to produce a fuel from wood to fire a natural gas/oil boiler.

Status

The unit is currently in the research/testing stage.

General Information

TVA purchased this unit from ENERCO, Inc., of Langhorne, Pennsylvania, original designer and manufacturer.

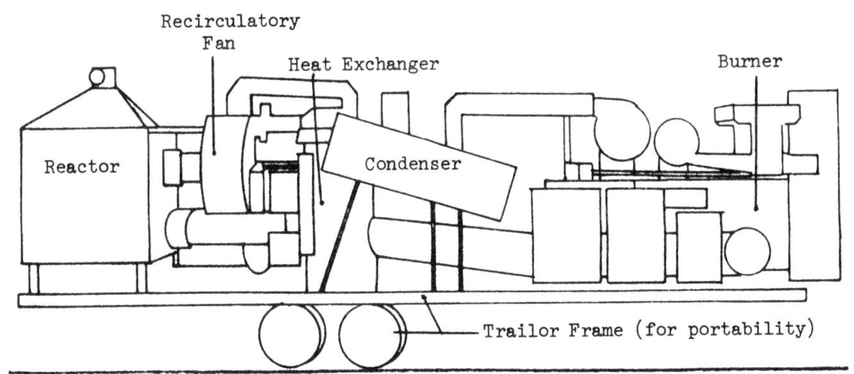

Plans for Future After extensive testing and any necessary modifications have been made, this unit will be taken to Maryville College to fuel the boiler.

Name E. Lawrence Klein Date 11/9/79

PYROLYSIS SYSTEMS DIRECTORY

Organization
Texas Tech University
Department of Chemical Engineering

Address
Lubbock, Texas 79409

Personnel
Steven R. Beck
Uzi Mann

Phone
(806) 742-3553

Type of Gasifier (up/down draft, size, fuel, application, etc.)

Fluidized Bed, 50 lb/hr, any biomass for conversion to medium-GTU gas

Status (research, pilot scale, commercial, etc.)

Pilot scale testing has been in progress for 2 years.

General Information (description, photo, sketch, etc.)

A counter current pyrolysis reactor for cattle wastes has been invented which allows volatile organic compounds to escape from the heating zone very rapidly. This results in a different product mix than has been observed in other pyrolysis research, containing unusually high concentrations of ethylene. Fuel values of gases plus the sparing of petroleum needs by ethylene, if economically feasible, would supplement petroleum supplies. The work includes studies in an existing 1/2 ton/day test reactor to determine the influence of temperature, residence time, pressure, and feedstock materials on the yield and quality of the products of reaction. The scope of work includes economic assessments of the process, utilizing animal manures and other biomass materials as feedstocks. Studies include the effects of reactor geometry and solid/gas contact in cold models. Relationships for the design of a staged reactor will be developed. This work may benefit programs on coal hydrogasification and coal gasification.

Plans for Future

Evaluate other feedstocks. Develop kinetic model of reactor.

Name Steven R. Beck Date 1/15/79

PYROLYSIS SYSTEMS DIRECTORY

Organization
University of Arkansas
Pyrolysis Project

Address
Fayetteville, Ark 72701

Personnel
Prof. Henry Hicks, ME Principle Investigator
Jas. Kimzey, James Turpin, Robt. Maccalum

Phone 501 575 3153

Type of Gasifier (up/down draft, size, fuel, application, etc.)

1 Ton/Day Rotary Kiln Pyrolysis Unit

Status (research, pilot scale, commercial, etc.)

Research being conducted on wood pyrolysis

General Information (description, photo, sketch, etc.)

1) Evaluation of commercial (A & P Coop) rotary kiln (Hicks)
2) Construction and operation of pilot scale rotary kiln (1 ton/day) to determine scale factors (Turpin)
3) Wood pyrolysis basic studies and service to above (Mccalum)

Program funded by DoE

Plans for Future

Name (TBR) Date 3/27/79

BIOMASS HYDROGASIFICATION DIRECTORY

BIOMASS HYDROGASIFICATION DIRECTORY

Organization Battelle Columbus Laboratories

Address 505 King Avenue Columbus, OH 43201

Personnel H. F. Feldmann

Phone (614) 424-4732

Type of Gasifier (up/down draft, size, fuel, application, etc.)

3-in. I.D. externally heated rated at 2000 F at 1000 psig with provision for continuous operation as fluid bed, free fall or moving bed. Can be fed H_2, syngas, or steam to simulate various gasification atmospheres.

Status (research, pilot scale, commercial, etc.)

Research reactor

General Information (description, photo, sketch, etc.)

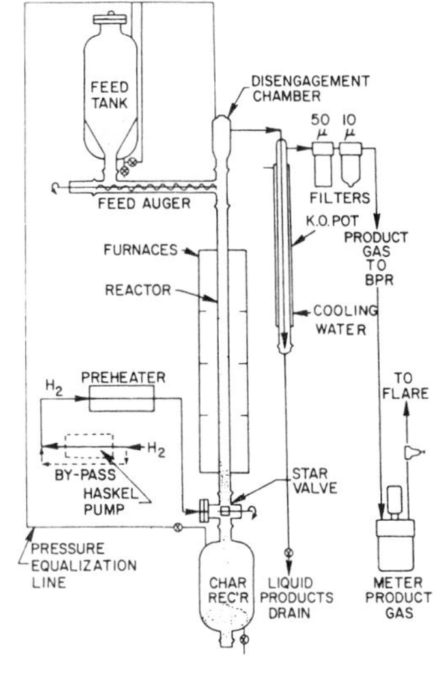

Pressurized Continuous Gasification System

Plans for Future

Coal and biomass gasification

Name Herman Feldmann Date February 22, 1979

OXYGEN GASIFICATION OF SMW

SMW OXYGEN GASIFICATION DIRECTORY

Organization
Union Carbide Corporation
Linde Division

Address
Post Office Box 44
Tonawanda, New York 14150

Personnel
G. F. Hagenbach
Product Manager - Purox

Phone
716/877-1600

Type of Gasifier (up/down draft, size, fuel, application, etc.)

Oxygen-blown slagging pyrolysis in a moving-burden shaft furnace

Status (research, pilot scale, commercial, etc.)

Commercial (for municipal solid waste)

General Information (description, photo, sketch, etc.)

Materials are fed near the top of the furnace and descend as a moving burden, in countercurrent contact with generated gases, through subsequent drying, pyrolysis and partial oxidation-melting zones. Pyrolysis of organic materials yields reducing gases and char. The char is subsequently burned in the hearth area, where nearly-pure oxygen is introduced. Non-volatile inorganics are slagged within the hearth, and tapped continuously.

Heat recovered from the rising hearth gases drives the endothermic pyrolysis and drying steps. Gas withdrawn from the top of the furnace - consisting primarily of carbon monoxide, hydrogen, carbon dioxide, light hydrocarbons and moisture - is further processed according to its intended use as a fuel or synthesis gas.

Commercial scale experience to date has been limited to processing of municipal solid waste and codisposal of sewage sludge with refuse. Laboratory scale tests have been carried out on additional materials.

Plans for Future

Commercially market Purox Systems for processing municipal wastes. Expand the technology for processing wood wastes and other biomass materials when warranted by market conditions.

Name G. F. Hagenbach (signature)
G. F. Hagenbach

Date 3/2/79
March 2, 1979

Chapter 10

Survey of Current Gasification Research

T. B. Reed, D. Jantzen,
R. Desrosiers, T. Milne
SERI

10.1 INTRODUCTION

The art of gasification is two centuries old, yet research in gasification has hardly begun. This paradoxical situation has arisen from the relative ease with which operating gasifiers can be built and run, so that research may at first appear to be redundant and unnecessary. The argument is fallacious, and both fundamental and process research are needed.

10.1.1 Fundamental Research

The most significant research in biomass gasification was done in Sweden during and after World War II (Generator Gas 1979). A small group at the Swedish Agricultural Machinery Institute has continued this work, but primary emphasis has been on air gasification and minor improvements in small air gasifiers.

Modern techniques of thermogravimetric analysis, calorimetry, and gas analysis make possible a better understanding of the pyrolysis process itself and of post-pyrolysis reactions. Modern understanding of the thermodynamics and kinetics of gasification reactions can enhance the degree of control and the yield of char reactions.

10.1.2 Process Research

Modern methods for achieving high-intensity heating will permit more rapid pyrolysis than could be attained earlier, resulting in very different products. Modern fluidized and suspended bed operation promises to greatly enhance unit yield and to decrease tars and char. Current catalytic techniques can give higher yields of valuable products at lower temperatures, and molten salt approaches can produce specific compounds in high yield.

New materials of insulation and fabrication will permit construction of more reliable gasification units with longer lifetimes. Modern gas separation techniques will make possible more efficient gas separation and reduced emissions. Microprocessors and new methods of measuring temperature and pressure will permit close control of gasification processes for higher efficiency and lower emissions. New methods of oxygen production will permit simple production of medium energy gas for pipeline or synthesis use. New biomass preprocessing technologies, such as densification, will permit gasification of previously unuseable materials. The development of the gas turbine will make possible generation of electric power in small units with high efficiency. New catalytic processes will permit the production of methanol, ammonia, gasoline, methane, glycol, and other chemicals from biomass.

10.2 CURRENT BIOMASS GASIFICATION RESEARCH PROCESSES

The following pages summarize the experimental approach and results for a number of current biomass gasification processes. Representative processes presently in an active research phase were chosen for each of the major types of biomass gasification presented in the Ch. 9 survey (air gasification, oxygen gasification, etc.). The R&D survey presented in this chapter is not intended to be comprehensive, and the inclusion or exclusion of a process does not reflect the merit of that process in comparison to other processes. Process descriptions, product distributions, and product compositions were obtained from the open literature; references are given for those wishing to study these processes in greater detail.

10.2.1 Air Gasification

GASIFICATION CASE SUMMARY

PROCESS: Molten Salt Air Gasification (Rockwell International Corp.).

FEEDSTOCK: Sawdust, rubber, nitropropane, sucrose, coal, X-ray film.

HEAT SOURCE: Air combustion of portion of feedstock.

GAS/FUEL CONTACT: (Figure 10-1) Feed and makeup Na_2CO_3 are transported pneumatically by air to molten salt combustion furnace, where the air and feed are injected into the molten salt bath. A portion of the feed is combusted with the transport air. Gas generated in the process leaves through the furnace head for downstream processing.

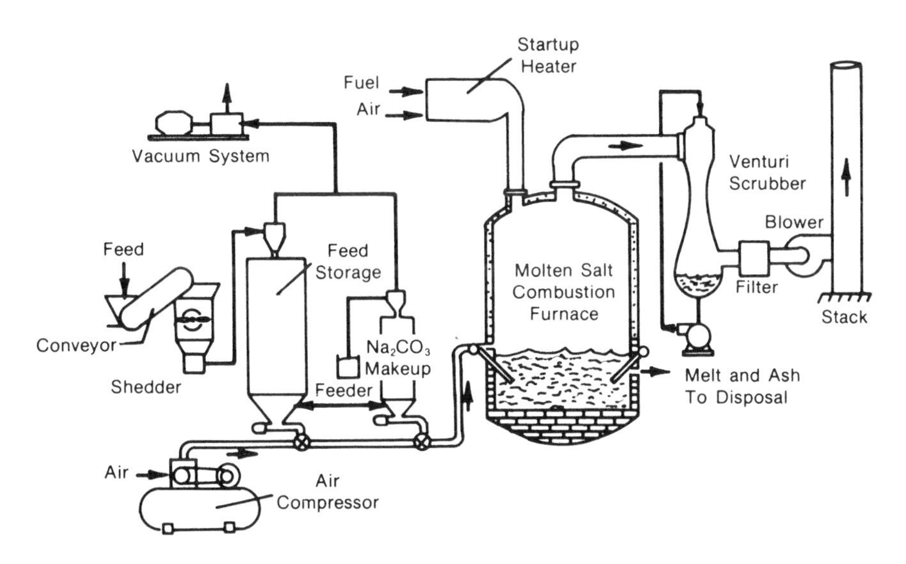

Figure 10-1. Schematic of Molten Salt Pilot Plant, Rockwell International Corporation.

ASH/CHAR: No char is produced, and the ash is removed with molten salt.

PRODUCTS: Low-Energy Gas - Compositions of product gases for various feedstocks and operating conditions are given below:

Table 10-1. GASIFICATION OF WASTES, ROCKWELL INTERNATIONAL CORP., MOLTEN SALT PROCESS

Waste	Temperature °C	Air feed rate (SCF/min)	Fuel feed rate (lb/h)	Percent theoretical air[a]	Composition of off-gas (vol. %)					Higher heating value[b] (Btu/SCF)
					CO_2	CO	H_2	CH_4	C_2	
Rubber	920	1.63	1.81	33	4.0	18.4	16.0	2.4	1.1	155
Wood	951	1.00	2.08	30	14.5	20.3	21.1	3.0	0.9	181
Nitropropane	1000	2.50	2.58	75	11.0	8.0	9.0	NM[c]	NM[c]	55
Film	1015	4.50	5.34	51	16.5	12.0	11.7	2.6	0.2	107
Film	958	2.50	6.58	22	16.0	18.3	14.1	5.2	1.2	179

[a]Percentage of air required to oxidize material completely to CO_2 and H_2O.

[b]Calculated from composition of off-gas.

[c]Not measured.

OPERATING CONDITIONS:	Temperature	= 920-1015 C
	Pressure	= atmospheric
	Salt	= Na_2CO_3
	Air, superficial velocity	= 0.5-2.0 fps
	Air, required for complete combustion	= 18-75%

SIZE:	I D	= 2 ft
	Length	= 10 ft
	Salt charge	= 1 ton

FUNDING, LOCATION, PERSONNEL: The process was developed by the Atomics International Division, Rockwell International Corporation at Canoga Park, Calif., under an Energy Research and Development Administration (ERDA) contract.

REFERENCE: Yosim, S. J; Barklay, K. M. 1977. "Production of Low-Btu Gas from Wastes, Using Molten Salts." Ch. 3 in Fuels From Wastes. Anderson, L. L.; Tillman, D. A., eds. New York: Academic Press.

COMMENTS: The process eliminates char disposal by consuming char in the combustion furnace. This is advantageous in gasifying feedstocks where any char produced would have high ash content with minimal or no market potential. The molten salt is reported to act as a sulfur or chlorine scavenger, which should help to alleviate pollution problems in gasifying a high sulfur feedstock such as coal or municipal solid waste containing high levels of plastics (e.g., PVC).

The gasification process has been shown to be technically feasible, but process economics have not been presented.

GASIFICATION CASE SUMMARY

PROCESS: SERI Air Gasification Test Facility.

FEEDSTOCK: Wood pellets.

HEAT SOURCE: Partial oxidation.

GAS/FUEL CONTACT: Cocurrent, Countercurrent, and Fluidized bed.

ASH/CHAR: Dry ash.

PRODUCTS: Low-energy gas.

OPERATING CONDITIONS: Atmospheric pressure.

SIZE: 0.5 MBtu/h

FUNDING, LOCATION, PERSONNEL: SERI Task No. 3356.20
1617 Cole Blvd.
Golden, Colo. 80401

PERSONNEL: R. Desrosiers, T. Reed, F. Posey (SERI)
M. Graboski (Colo. School of Mines - Consultant)

COMMENTS: The product of the gasification reactor studies will be process information for several reactor types, all based on a common set of fuels. The reactor types being considered are updraft and downdraft fixed bed, entrained flow, and fluidized bed reactors. In addition to mass and energy balances, temperature and gas composition profiles will be obtained as well as residence time distribution data. The plan is to design a system with flexible peripheral components to accommodate the entire spectrum of reactor types. The emphasis in this phase of the program is not on optimized reactor design but on precise analytical and kinetic data. Each reactor will be simply constructed to provide the desired gas-solid contacting method, and after preliminary runs to define a set of stable operating conditions, a comprehensive set of physical, chemical, and rate data will be collected. As the data is gathered, reactor models will be continuously tested and updated.

GASIFICATION CASE SUMMARY

PROCESS: Texas Tech University - Syngas from Manure (SGFM)

FEEDSTOCK: Feedlot cattle manure.

HEAT SOURCE: Partial oxidation of feedstock.

GAS/FUEL CONTACT: Steam and air are fed to the bottom of the fluidized bed through a distribution plate, and the feed manure is fed from the top of the reactor. The reactor is termed a falling bed reactor; there is no circulating refractory material.

ASH/CHAR: Dry char is removed from the bottom of the reactor and can be used to satisfy heat requirements for the process.

PRODUCTS: Ammonia syngas to yield about 0.5 kg ammonia per kg of dry, ash-free manure; ethylene with a yield of 21-70 g per kg of dry, ash-free manure.

OPERATING CONDITIONS: Atmospheric pressure and 600-700 C.

SIZE: Reactor is 2.5-m long, with a main body 1.5-m long and 15 cm in diameter, and a top section 20 cm in diameter by 60-cm long for separation of the solids and gas. A schematic of the system is shown in Fig. 10-2 .

Table 10-2. SUMMARY OF OPERATING CONDITIONS AND PRODUCT GAS DATA FOR SGFM PROCESS

Operating Conditions	Run Number						
	1	6a	6b	7	8	9	10
Manure feed rate (kg dry, ash-free/h)	5.22	7.21	16.15	12.97	12.34	12.70	8.26
Manure feed rate (kg as received/h)	7.76	10.60	23.61	18.95	18.05	18.01	11.75
Air feed rate (kl/h)	1.149	1.700	1.487	1.904	1.402	4.249	0.765
Steam feed rate (kg/h)	5.44	4.54	4.54	3.63	3.08	2.72	3.72
Particle size (in)	>0.95	>0.95	>0.95	>0.95	>0.95	>0.32	>0.32
Average temperature (°C)	711	695	641	617	629	668	628
Product gas data[a]							
Total dry gas (l/g dry, ash-free)[b]	1.19	(0.667)	0.580	0.406	0.455	(0.718)	0.318
Heat value (HHV) (cal/l)	2855	2918	3790	3380	3523	2624	3345
Gas composition (vol %)							
H_2	25.2	22.2	20.0	28.2	17.4	15.1	20.9
N_2	14.6	27.8	15.1	23.2	26.7	36.8	24.2
CH_4	12.8	7.7	12.6	9.2	14.1	8.9	11.7
CO	11.6	15.3	21.3	16.4	21.2	20.3	22.4
CO_2	30.8	20.7	22.1	15.4	14.1	14.2	14.8
C_2H_4	4.7	6.4	8.5	4.9	5.8	4.2	5.5
C_2H_6	0.3	0.5	0.4	2.7	0.7	0.5	0.5

[a]All data are average values from at least two samples. Individual gas samples were analyzed on the gas chromatograph using at least two injections.

[b]Values in parentheses are back-calculated values using a nitrogen balance.

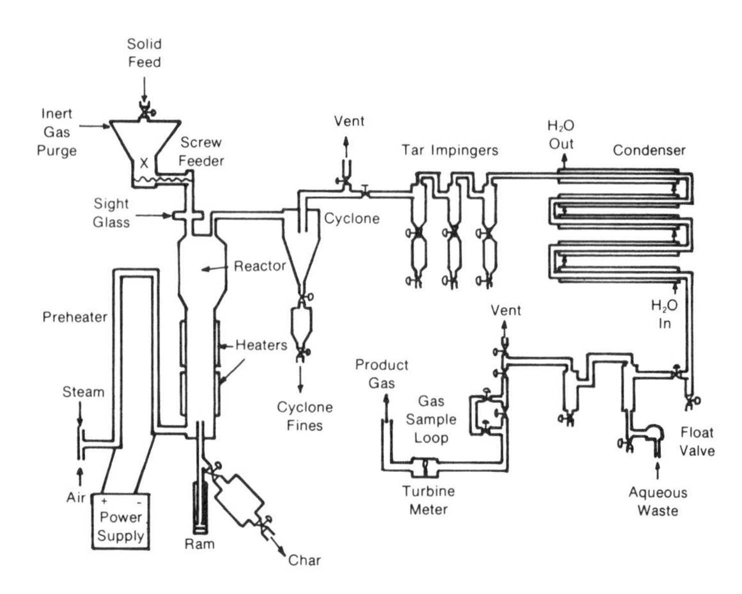

Figure 10-2. SGFM Pilot Plant, Texas Tech University

FUNDING, LOCATION, PERSONNEL:

The reactor construction and testing was done by Texas Tech University in Lubbock, Tex., from January 1974 to June 1977 under EPA grant No. S 802934. Additional data to better define heat and mass balances were obtained with support from ERDA contract E29-2-3779. Bechtel National, Inc. developed two conceptual plant designs, to produce ammonia syngas and ammonia syngas plus ethylene, from 1000 dry tons per day of manure, on subcontract from Texas Tech. Phase II of the ERDA-DOE contract is now in progress, seeking to develop data for partial oxidation and pyrolysis of wood, wood residues, and agricultural residues.

REFERENCES:

1. Huffman, W. J. et al. 1978. Conversion of Cattle Feedlot Manure to Ethylene and Ammonia Synthesis Gas. EPA-600/Z-78-026. Feb.

2. Hipkin, H. G.; Basuino, D. J. 1978. Syngas From Manure - A Conceptual Plant Design. Prepared for Texas Tech University by Bechtel National, Inc.; Final Report; July.

3. Huffman, W. J. et al. 1977. "Ammonia Synthesis Gas and Petrochemicals from Cattle Feedlot Manure." Presented at Symposium on Clean Fuels from Biomass. Orlando, FL: Jan. 27.

4. Huffman, W. J. et al. 1978. "A Review of Heat/Mass Balances and Product Data for Partial Oxidation of Cattle Feedlot Manure." Presented at AIChE National Meeting. Atlanta, GA; Feb. 26.

5. Beck, S. R. 1979. "Application of SGFM Technology to Other Feedstocks." 3rd Annual Biomass Energy Systems Conference Proceedings: The National Biomass Program. Colorado School of Mines, Golden, CO; June 1979. Golden, CO: Solar Energy Research Institute; p. 339.

COMMENTS: The Bechtel study (a conceptual plant design) concluded that the process is not competitive with natural gas re-former plants at the present but will become economical as the price of natural gas increases. The process would be competitive with syngas from coal.

Removal of ethylene is not justified under present economic conditions, but as the cost of syngas decreases, recovery does become economical.

There are a number of changes in design which can reduce the cost substantially.

10.2.2 Oxygen Gasification

GASIFICATION CASE SUMMARY

PROCESS: Battelle - Pacific Northwest Laboratory
Gasification of Biomass in the Presence of Catalyst
(lab-scale and pilot demonstration unit).

FEEDSTOCK: Wood, bark.

HEAT SOURCE: Electric radiation heaters, hot feed gas.

GAS/FUEL CONTACT: Lab-scale: Steam and feed cocurrent flow through reactor, pilot demonstration unit—stirred fluidized bed.

ASH/CHAR: Dry ash.

PRODUCTS: Variable, depending on catalyst and operating conditions. Conditions for optimizing CH_4, H_2, CO, hydrocarbon synthesis gas, and ammonia synthesis gas will be investigated.

OPERATING CONDITIONS: Up to 800 C at 1 atm.

SIZE: Pilot demonstration unit—20 kg/h dry wood.

FUNDING, LOCATION, PERSONNEL: Pacific Northwest Laboratory (Richland, Wash.) laboratory studies—L. J. Sealock. Pilot demonstration unit design, procurement, installation—R. J. Robertus. Technical and economic feasibility studies—L. K. Mudge

Funded by DOE, Nov. 1977 to Sept. 1979. Contract EY-76-C-06-1830. Continuing.

REFERENCES

1. Sealock, L. J., Jr., et al. 1978. "Catalyzed Gasification of Biomass." Presented at 1st World Conference on Future Sources of Organic Raw Materials. Toronto, Canada; June 16.

2. Mudge, L. K. et al. 1979. "Catalytic Gasification of Biomass." 3rd Annual Biomass Energy Systems Conference Proceedings: The National Biomass Program. Colorado School of Mines, Golden, CO; June 1979. Golden, CO: Solar Energy Research Institute; p. 351.

COMMENTS: The work at PNL is aimed at determining the ability of selected catalysts to alter the kinetics of biomass gasification; to produce methane, hydrogen, carbon monoxide, or synthesis gas for generation of ammonia, methanol, or hydrocarbons; and at determining the technical and economic feasibility of catalyzed biomass gasification. The work will culminate with the operation of a pilot demonstration unit to demonstrate the selected reaction systems and an economic analysis of these systems.

GASIFICATION CASE SUMMARY

PROCESS: Downdraft Gasifier (Swedish Hessleman Model 50/13 generator) operated with air or oxygen-enriched air by Environmental Energy Engineering.

FEEDSTOCK: Charcoal, hardwood blocks, pine blocks, wood pellets.

HEAT SOURCE: Combustion of char and tars.

GAS/FUEL CONTACT: Air is injected into the middle of the gasifier where combustion occurs. A constriction in this zone results in higher temperatures and greater decomposition of tars. Pyrolysis occurs in the top zone of the gasifier, and chars and pyrolytic tars pass downward through the combustion and reduction zones. Product gases recirculate through the pyrolysis zone, providing heat for pyrolysis, and are removed for use in an industrial burner or internal combustion engine. A schematic of the gasifier is given in Fig. 10-3.

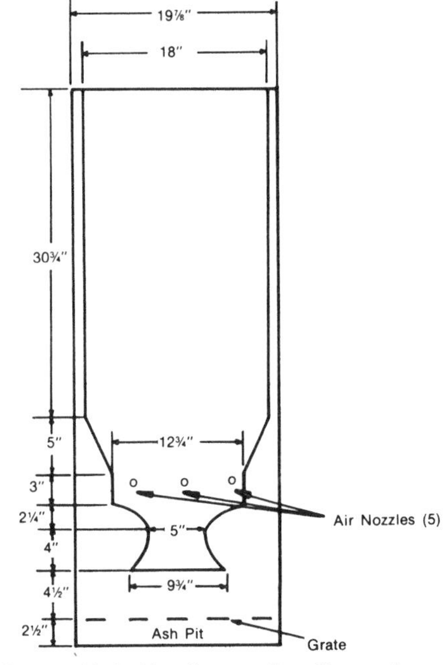

Figure 10-3. Hessleman Gas Generator, Environmental Energy Engineering

ASH/CHAR: Ash goes through a grate at the bottom of the gasifier and is collected in an ash pit.

PRODUCTS: Low-Btu Gas (heating value 110-295 Btu/SCF). The gas composition and heating value are functions of the oxygen concentration of the combustion gas used. Figure 10-4 shows the effect of oxygen concentration upon product gas composition, and Fig. 10-5 shows the effect upon gas heating value.

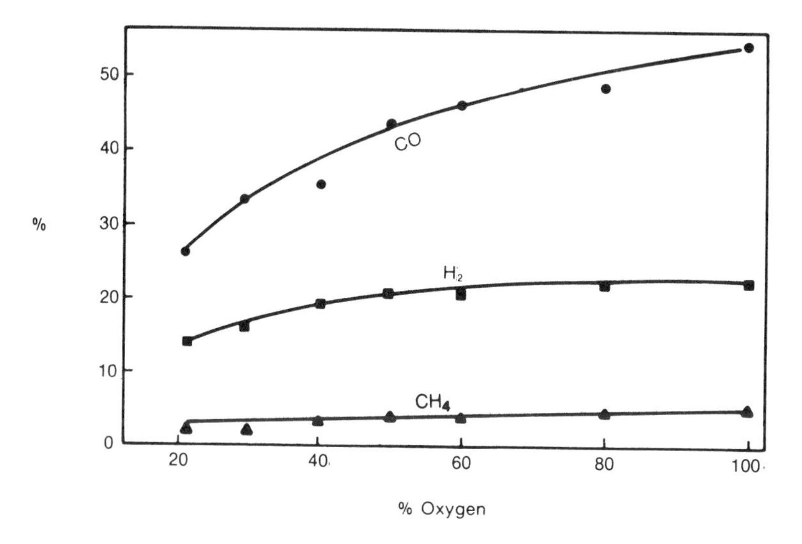

Figure 10-4. Effect of Oxygen Concentration on Gas Composition, Hessleman Gas Generator

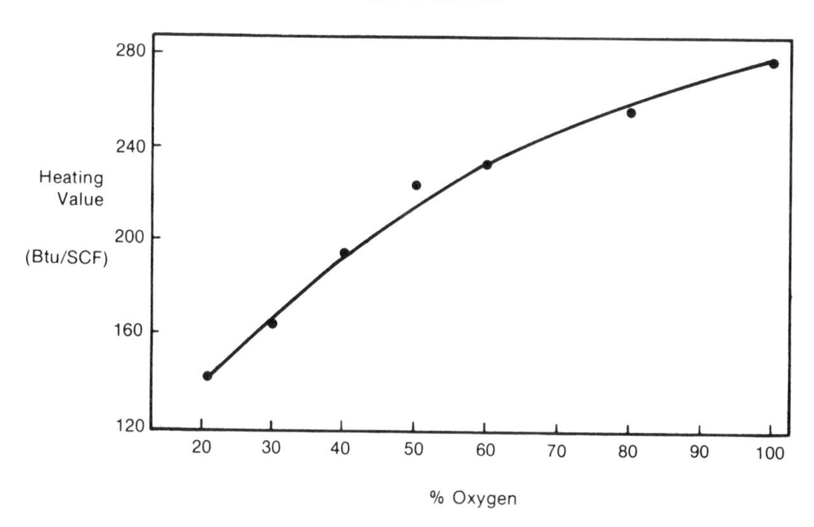

Figure 10-5. Effect of Oxygen Concentration on Gas Heating Value, Hessleman Gas Generator

OPERATING CONDITIONS:	Combustion Zone Temperature = 2000-2300 F Combustion Gas Oxygen = 21-100 vol %
SIZE:	Hessleman Vedgasierk, Type T-500, NR 110964/10 with a throat diameter of 5 in.
FUNDING, LOCATION, PERSONNEL:	The gasifier was operated by personnel of Environmental Energy Engineering, Inc., Morgantown, W. Va., under the supervision of Dr. R. C. Bailie, under a grant from the Solar Energy Research Institute (Contract No. AH-8-1077-1).
REFERENCE:	Environmental Energy Engineering, Inc. 1979. "Hessleman Gas Generator Testing for Solar Energy Research Institute." P. O. No. AH-8-1077-1.

GASIFICATION CASE SUMMARY

PROCESS:	SERI Oxygen Biomass gasifier.
FEEDSTOCK:	Initially pellets, other coarse forms in final process.
HEAT SOURCE:	Oxygen (or air) combustion.
GAS/FUEL CONTACT:	Downdraft gasifier, 10 atmosphere pressure.
ASH/CHAR:	Dry ash.
PRODUCTS:	Medium Btu, clean syngas (CO, H_2) for oxygen operation, low-Btu gas for air operation.
SIZE:	Prototype, 1-5 MBtu/h (100-500 lb biomass/h) 100-300 ton/day in final process.
FUNDING, LOCATION, PERSONNEL:	SERI Task no. 3356.20 Solar Energy Research Institute, 1617 Cole Blvd. Golden, CO 80401 T. Reed and M. Graboski (consultant) Colo. School of Mines.
COMMENTS:	Oxygen pressurized gasification can provide a medium Btu gas from farm or forest residues for synthesis of methanol or ammonia to give fuel or fertilizer. Small gasification systems recover in mass production, and lower transport and handling the higher investment and labor required for smaller plants.

GASIFICATION CASE SUMMARY

PROCESS: Purox Process (Oxygen-fed Slagging Pyrolysis), Union Carbide Corporation.

FEEDSTOCK: Municipal solid waste.

HEAT SOURCE: Combustion of pyrolytic char, tars, and liquids.

GAS/FUEL CONTACT: In the Purox process, municipal solid waste (shredded and magnetically sorted) is fed into the top of a shaft furnace and oxygen is fed at the bottom. Pyrolytic char is combusted with the oxygen at the bottom of the gasification furnace, providing enough thermal energy to produce temperatures in the range from 2900 to 3100 F and to produce a molten slag from all noncombustible materials. This molten slag is romoved for quenching and disposal.

Combustion gases rise counter currently through the MSW producing gas, liquids, and char. The liquids and char are burned in the combustion zone. The pyrolytic gas rises through the furnace, drying and preheating the feed. A diagram of the process is given in Fig. 10-6. Gases leave the furnace for further processing to produce a medium energy fuel gas.

ASH/CHAR: The char is consumed during the combustion step to provide process heat. The ash is removed in a molten form from the reactor and quenched to form a granular frit.

PRODUCTS: Medium Energy Gas: A comparison of this product gas with methane is given in Table 10-3.

OPERATING CONDITIONS:

Temperature (maximum) = 3100 F
Pressure = atmospheric.

SIZE: 200 ton/day Raw Refuse Conversion Facility.

FUNDING, LOCATION, PERSONNEL: The process was developed by Union Carbide Corporation in Tarrytown, N.Y., at a 5-ton/day scale. A 200-ton/day facility is located in South Charlestown, W. Va.

REFERENCES:

Shulz, H.M. (Principal Investigator) et al. 1976. Resource Recovery Technology for Urban Decisionmakers. New York: Urban Technology Center, Columbia University.

Tillman, D. A. 1976. "Mixing Urban Waste and Wood Waste for Gasification in a Purox Reactor." Thermal Uses and Properties of Carbohydrates and Lignius. Schafizadeh, F.; Sarkanen, K. V.; and Tillman, D. A., eds. New York: Academic Press.

Desrosiers, R. E. 1979. Process Designs and Cost Estimates for a Medium Btu Gasification Plant Using a Wood Feedstock. SERI/TR-33-151. Golden, CO: Solar Energy Research Institute.

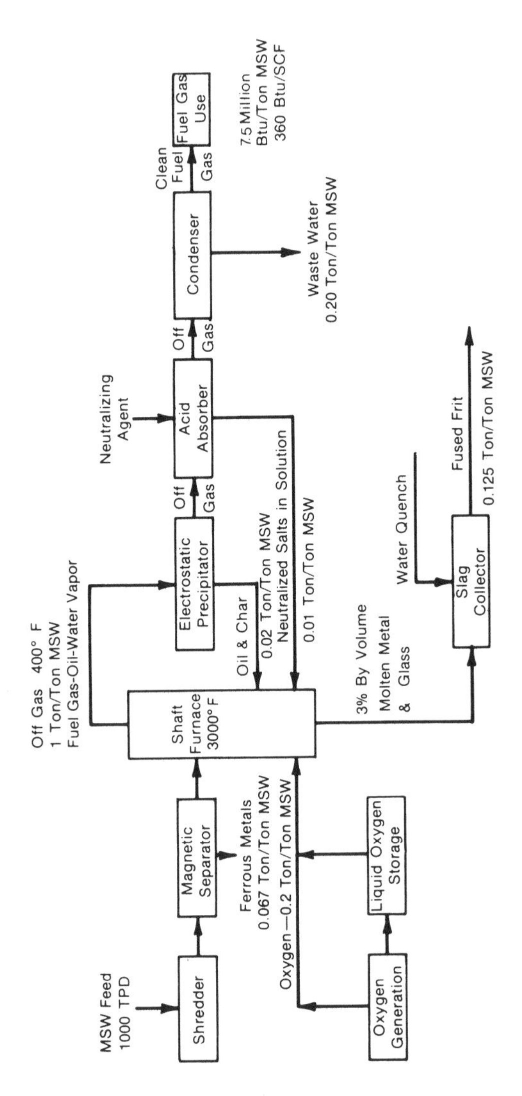

Figure 10-6. Union Carbide Corporation Purox System Oxygen-Fed Slagging Pyrolysis

Table 10-3. COMPARISON OF COMBUSTION CHARACTERISTICS OF PUROX GAS AND METHANE

Component	Volume %	Heat of Combustion (Btu/SCF)	Air Required for Combustion (SCF/SCF)	Volume of Flue Products (SCF/SCF)
Purox Gas				
CO	44	322	2.38	2.88
H_2	31	275	2.38	2.88
CO_2	13	0	0	1
CH_4	4	913	9.53	10.53
C_2H_4	1	1,513	14.29	15.29
N_2	1	0	0	1
H_2O	6	0	0	1
	100	280	2.43	2.97

Methane

Heat of Combustion	913 Btu/SCF
Air Required for Combustion	9.53 SCF/SCF
Volume of Flue Products	10.53 SCF/SCF

	Purox Gas	Methane
Feed (SCF/MBtu)	3,600	1,095
Air Required for Combustion (SCF/MBtu)[a]	8,700	10,440
Volume of Flue Products (SCF/MBtu)	10,500	11,530
Heat Release (Btu/SCF)	95	87
Compression Power (kWh/MBtu)[b]	5.7	1.8

[a]Based on a minimal amount of air needed to convert gas to CO_2 and H_2O.

[b]Gas compressed to 35 psig from 1 atm, 100 F, with 75% efficiency.

10.2.3 Pyrolysis Gasification

GASIFICATION CASE SUMMARY

PROCESS:	Arizona State University: Dual Fluidized-Bed Flash Pyrolysis System.
FEEDSTOCK:	Organic fraction of MSW, kelp residue, synthetic polymers, agricultural biomass sources.
HEAT SOURCE:	Recirculated inert and catalytic solids.
GAS/FUEL CONTACT:	Fluidized bed.
ASH/CHAR:	Char circulated to combustor for process heat. Dry ash separated from combustor.

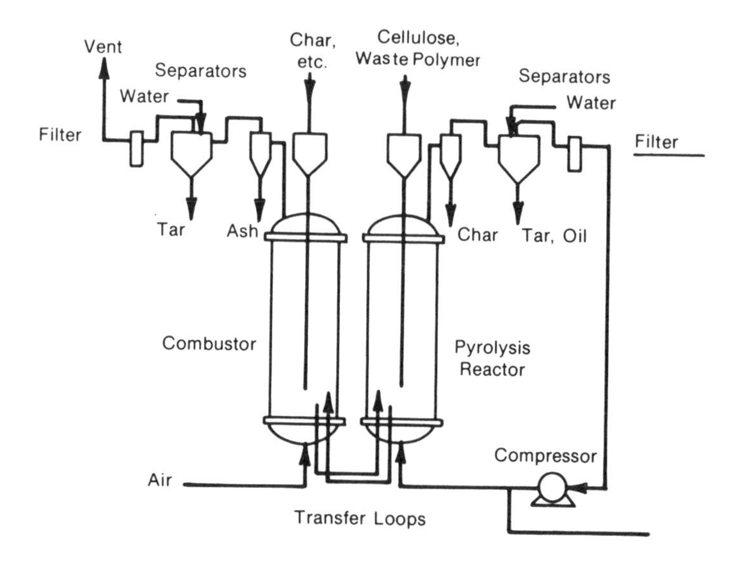

Figure 10-7. Thermal Gasification, Arizona State University

PRODUCTS: Typical gas phase yields of 75-85%. Typical pyrolysis gas composition (cellulose source) is:

	mole %
C_2H_4	5-15
CO	35-45
H_2	10-20
CH_4	10-15
C_2H_6	1-5
CO_2	15-30

OPERATING CONDITIONS: Temperatures of 500-1000 C. Pressures 0-5 psig. Inert and catalytic fluidizing solids.

SIZE: 25 lb/h.

FUNDING, LOCATION, PERSONNEL:

Prof. James L. Kuester
College of Engineering & Applied Sciences
Arizona State University
Tempe, AR
Supported for last three years by the EPA.

REFERENCE: Kuester, J. L. 1979. "Liquid Hydrocarbon Fuels From Biomass." Presented at Honolulu meeting of ACS, April 1-6.

GASIFICATION CASE SUMMARY

PROCESS:	Battelle-Columbus, Multi-Solid Fluid Bed Reactor, Batch-Solids Fluid-Bed Gasifier, Multiple Catalysts, Hydrogasification.
FEEDSTOCKS:	Forest residues, hard and soft woods.
HEAT SOURCE:	Circulatory bed material or external furnace.
GAS/FUEL CONTACT:	Fluid bed.
ASH/CHAR:	Dry ash.
PRODUCTS:	Wood ash and CaO shown to be effective gasification and shift catalysts. Hydrogasification has given up to 18% CH_4 (uncatalyzed.) Detailed studies in progress.
OPERATING CONDITIONS:	Temperatures of 625-825 C. Steam, H_2, recycle gas environment. Variety of catalysts and incorporation methods. Fluid and entrained bed operation.
SIZE:	10 lb/h.
FUNDING, LOCATION, PERSONNEL:	H. F. Feldman. Battelle Columbus Laboratories. Fuels from Biomass Systems Branch Contractor.

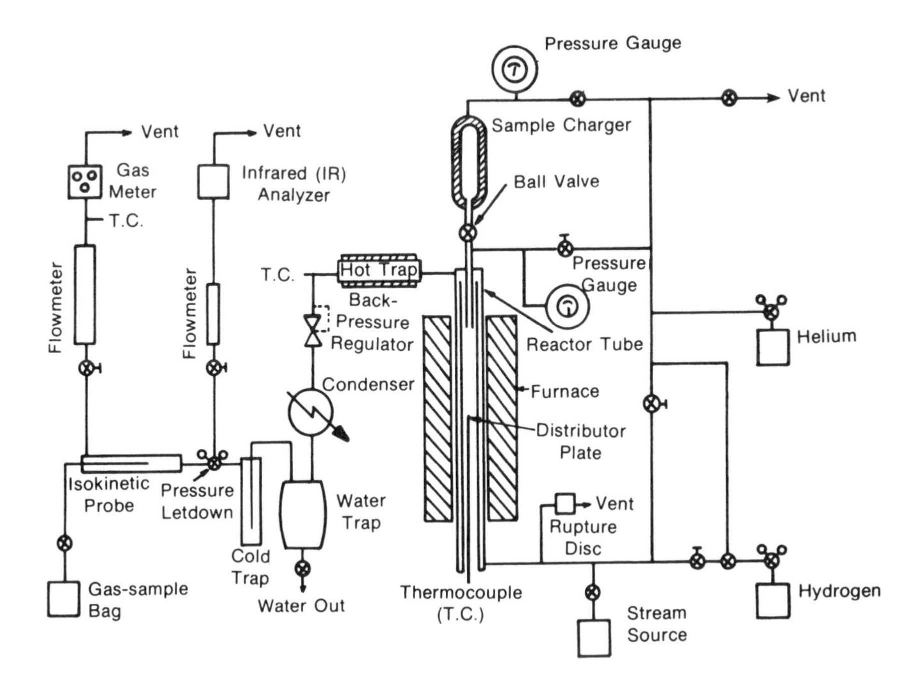

Figure 10-8. Bench-Scale Batch Reactor, Battelle-Columbus Laboratory

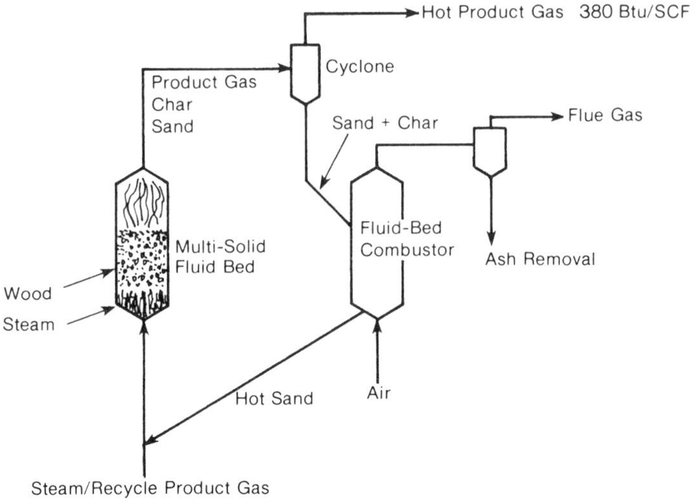

Figure 10-9. Schematic of MSFBG Process

REFERENCES:

1. Feldman, H. F. 1978. "Conversion of Forest Residues to a Methane-Rich Gas." Presented at IGT Symposium, Washington, D.C., Aug. 14-18.

2. Feldman, H. F.; Choi, P. S.; Liu, K. T. 1978. "Conversion of Forest Residue to a Methane-Rich Gas." Presented at Sixth Biomass Thermoconversion Contractors Meeting, Biomass Energy Systems. Univ. of Arizona, Jan. 16-17.

3. Feldmann, H. F., et al. 1979. "Conversion of Forest Residue to a Methane-Rich Gas." 3rd Annual Biomass Energy Systems Conference Proceedings: The National Biomass Program. Colorado School of Mines, Golden, CO; June 1979. Golden, CO: Solar Energy Research Institute; p. 439.

GASIFICATION CASE SUMMARY

PROCESS: Garrett Multiple Hearth Biomass Gasifier.

FEEDSTOCK: Any form of biomass that can be fed through a 14-in. diameter screw, including materials of high moisture content.

HEAT SOURCE: Recirculated hot char and heat transfer through metal wall.

GAS/FUEL CONTACT: Five hearths are used to accomplish drying, pyrolysis, two stages of char combustion, and ash cooling. Each chamber is isolated from the others. Drying is accomplished by countercurrent contact of feed with fuel gas from the combustion hearth. The dried feed is pyrolyzed by hot char delivered from the combustion chamber by a steam lift. Positive solids transport is achieved by internal hollow rakes. The char residue from the second hearth is dropped to the combustion hearths, which produce hot char and steam for pyrolysis. Ash from the combustion hearths drops to the ash cooler where combustion air is preheated.

ASH/CHAR: Dry ash exits from the bottom hearth. All char is used in satisfying heat requirements for the process.

PRODUCTS: Medium Btu Gas.

Feed Material	Manure		Sawdust	
Solids temperature (°C)	635	657	631	653
H_2O in feed (wt. fraction)	0.4316	0.052	0.0995	0.2995
H_2O in pyrolysis gas (vol. fraction)	0.6480	0.3976	0.1916	0.4090
g/g dry, ash-free feed (mol. fraction)				
CO_2	0.509	0.323	0.341	0.379
CO	0.118	0.118	0.312	0.307
H_2	0.035	0.017	0.016	0.017
CH_4	0.054	0.048	0.086	0.089
C_2H_4	0.016	0.014	0.016	0.020
C_2H_6	0.006	0.011	0.012	0.011
Total (g/g dry, ash-free feed)	0.738	0.531	0.783	0.823
LHV (Btu/SCF)	294	343	387	385

OPERATING CONDITIONS:

	Gas Temperature (°C)
Drying hearth	100-300
Pyrolysis	600-750
Combustion	1100-2000

Gas velocity 0.1 ft/s.

SIZE: (Pilot Demonstration Unit) Each hearth is 4 ft in diameter, 1 ft in height.
(Projected) Capital investment for 100-ton/day plant would be $1.9 million (1977).

FUNDING, LOCATION, PERSONNEL: An exploratory, bench-scale, pilot unit and laboratory study was completed by the Garrett Energy Research and Engineering (GERE) Co. from May 25, 1976 to June 24, 1977 under ERDA Contract No. E (04-3) -1241. This work included an evaluation of each of the processing steps required in the multiple hearth equipment. First, the jacketed, vacuum, screw-flight conveyor was tested. Then, a single hearth was used to study the design variables involved in direct contact drying, steam-char pyrolysis, and combustion.

Testing of the entire process is currently being performed under DOE contract EY-76-C-03-1241. The pilot plant is located in Hanford, Calif.

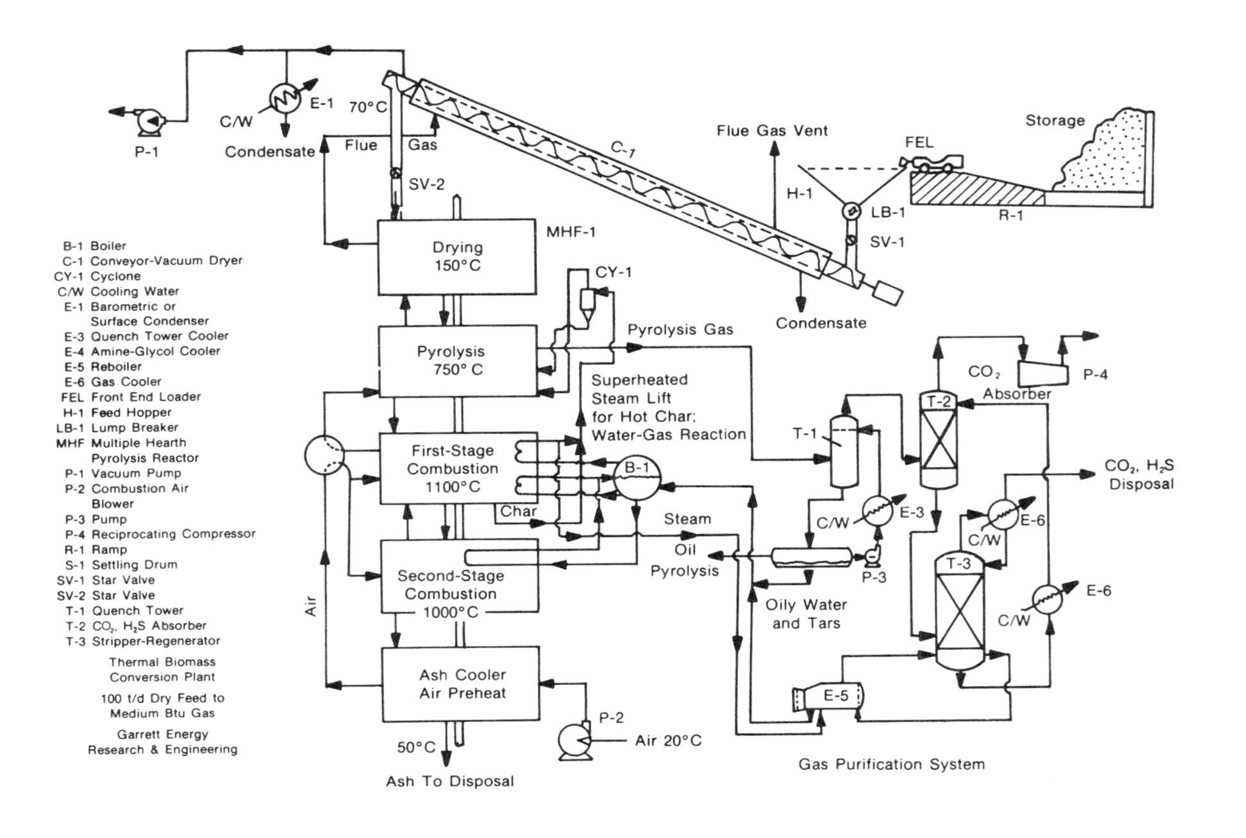

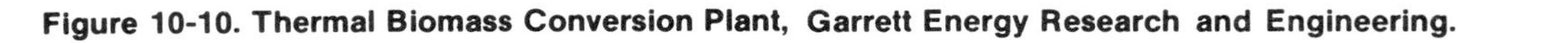
Figure 10-10. Thermal Biomass Conversion Plant, Garrett Energy Research and Engineering.

REFERENCES:

1. Garrett, D. E. 1977. Conversion of Biomass Materials into Gaseous Products, Final Technical Report. Work performed by Garrett Energy Research and Engineering for ERDA; Contract No. E(04-3) -1241, Oct.

2. Garrett, D. E. 1977. Thermochemical Conversion: Biomass Gasification. Presented at the Second Annual FFB Symposium, Troy, N.Y. June 20-22.

3. Garrett, D. E. 1979. "Conversion of Biomass Materials to Gaseous Products." 3rd Annual Biomass Energy Systems Conference Proceedings: The National Biomass Program. Colorado School of Mines, Golden, CO: June 1979. Golden, CO: Solar Energy Research Institute; p. 445.

COMMENTS: The incorporation of two stages of drying which use waste heat from flue gas makes the GERE process suitable for very moist feeds. It appears that the process could be economical even at a plant size of 50 ton/day.

GASIFICATION CASE SUMMARY

PROCESS: Naval Weapons Center, China Lake - Flash Pyrolysis Process

FEEDSTOCK: MSW (Ecofuel II, 200 μm minimum dimension;—probably any small-particle biomass form).

HEAT SOURCE: Kiln, heated with char, byproducts, etc.

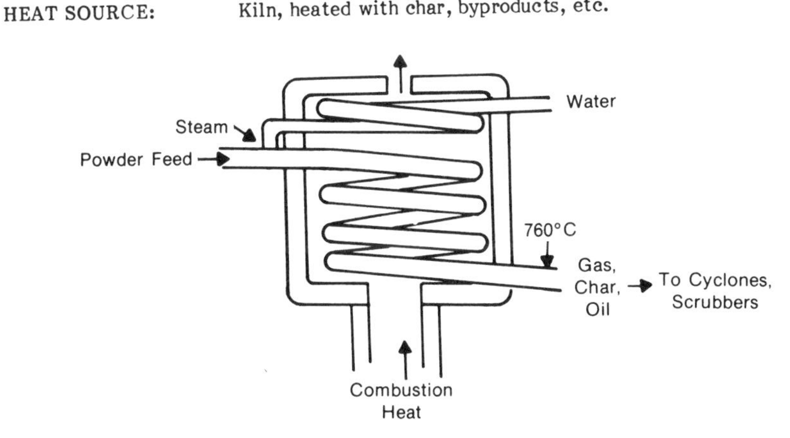

Figure 10-11. Flash Pyrolysis Process, Naval Weapons Center.

GAS/FUEL CONTACT: Suspended flow, typically 50 milliseconds.

ASH/CHAR: Dry ash.

PRODUCTS:

(Dry)	Mass (%)	Energy (%)
Gasoline precursors (C_2+)	24	53
CO	36	16
CH_4	4	11
H_2	1	6
CO_2	16	-
Char	19	14

NOTES: The MSW char energy content is 7000 Btu/lb (versus 14,000 Btu/lb for carbon) and is high in ash. The byproduct gases contain 415 Btu/SCF.

The process has been developed primarily for the production of gasoline. Pure ethylene was converted to a 90 motor octane number (MON) gasoline by thermal polymerization. The gasoline precursors were converted to a gasoline having virtually the same physical appearance and distillation characteristics.

SIZE: Bench scale, 10 lb/h maximum.

PROJECTIONS: From one ton (metric) of waste, the process would produce:

226 lb of gasoline (41 gal); 25 lb of light oil (5 gal); 228 lb of char and ash; 501 lb of by-product gases, some of which would be burned for process heat; 192 lb of CO_2; and 28 lb of tar.

The authors have used a preliminary evaluation made by Dow Chemical under contract to EPA and scaling techniques commonly used in the oil industry to produce economic projections of cost of gasoline from MSW. A few representative figures are:

Plant size (ton/day)	100	100	500	1000
Tipping fee ($/ton)	8	8	8	8
Rate of return (%)	—	15	15	15
Municipal Amortization (%)	8	—	—	—
Gasoline cost ($/gal)	0.80	1.35	0.55	0.38

Credits of $4.85/ton are taken for inorganics in waste.

PROCESS ADVANTAGES: Process can convert a wide variety of biomass feedstocks at 0-$2/MBtu to gasoline worth $5/MBtu with immediate product acceptance. Process steps are relatively simple and similar to present refinery practice. All medium Btu by-product gas, char, and tars would be consumed for process energy, so that only premium quality hydrocarbon fuels would be the final products.

PROCESS DISADVANTAGES: Process has only been demonstrated with finely divided feedstock. It is capital intensive and will require technical personnel for operation.

FUNDING, LOCATION, PERSONNEL: Process developed starting May 1975, under EPA contracts EPA-IAG-D5-0781 at the Naval Weapons Center, China Lake, Calif. 93555, under James P. Diebold, Charles Benham, and Garyl D. Smith. EPA Funding now withdrawn; process being discontinued at China Lake during 1979. Work is resuming at SERI under the direction of James Diebold and Tom Reed.

REFERENCES:

1. Diebold, J. P.; Benham, C. B.; Smith, G. D. Wastes to Unleaded, High-Octane Gasoline. EPA-IAG-D6-0781.

2. Diebold, J. P. 1980. Research into the Pyrolysis of Pure Cellulose, Lignin, and Birch Wood Flour in the China Lake Entrained Flow Pyrolysis Reactor. SERI/TR-332-586. Golden, CO: Solar Energy Research Institute.

3. Diebold, J. P.; Smith, G. D. 1979. "Noncatalytic Conversion of Biomass to Gasoline." ASME Solar Energy Conference. ASME 79-Sol-29. March.

4. Diebold, J. P. 1979. "Gasoline from Solid Wastes by a Noncatalytic Thermal Process." ACS Symposium on Thermal Conversion of Solid Wastes and Biomass. September.

5. Diebold, J. P.; Smith, G. D. 1979. "Thermochemical Conversion of Biomass to Gasoline." 3rd Annual Biomass Energy Systems Conference Proceedings: The National Biomass Program. Colorado School of Mines, Golden, CO: June 1979. Golden, CO: Solar Energy Research Institute; p. 139.

GASIFICATION CASE SUMMARY

PROCESS: Steam Gasification of Biomass, Princeton University.

FEEDSTOCK: Cellulose.

HEAT SOURCE: Electrical Resistance Heaters.

GAS/FUEL CONTACT: The pyrolysis unit (see Fig. 10-12) is operated in a semi-batch mode by passing steam over a small batch sample of biomass material at pyrolysis temperatures, then using gas-phase pyrolysis reactions to convert pyrolytic gases to synthesis gases.

ASH/CHAR: Char is collected and weighed at the end of the experiment.

PRODUCTS: Synthesis Gas - representative compositions are shown below for cellulose pyrolysis.

Table 10-4. STEAM PYROLYSIS OF CELLULOSE, PRINCETON
(Experimental Conditions and Results)

Pyrolysis Temp. (°C)	500	500	500	500	500
Gas Reactor Temp. (°C)	600	500	600	700	600
Gas Reactor Res. Time (s)	10	9	6	6	2
Gas Analysis (Vol. %)					
CO	55	40	52	53	55
H_2	10	11	10	13	10
CO_2	16	42	20	13	20
CH_4	8	2	8	12	6
C_2H_4	4	1	4	5	3
C_3H_6	1	1	2	1	1
C_2H_6	2	1	1	1	2
Other	4	2	3	2	3
Cal. Value (MBtu/ton)	6.2	0.98	5.4	9.7	3.6

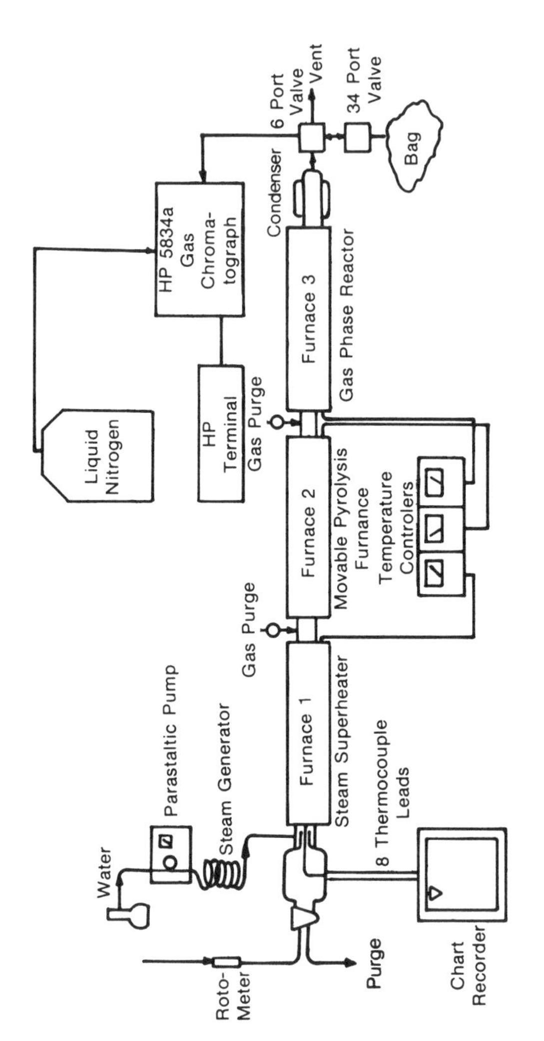

Figure 10-12. Schematic of the Tubular Quartz Reactor Experiment, Princeton University

SIZE: Bench scale.

FUNDING, LOCATION, PERSONNEL: Supported the last several years by the U.S. EPA. Michael J. Antal, Jr., Princeton University Department of Mechanical Aerospace Engineering.

REFERENCE: IGT Conference on Energy from Biomass and Waste. Aug. 1978. Wash., D.C.

GASIFICATION CASE SURVEY

PROCESS: Flash Pyrolysis and TGA Studies
Royal Institute of Technology, Dept. of Chemical Technology, Stockholm, Sweden.

FEEDSTOCK: Wood, straw, municipal solid waste, peat, coal, graphite.

HEAT SOURCE: TGA Studies
- electrical heating of biomass sample
- superheating of steam and other gases

Flash pyrolysis reactor - electrical heating.

GAS/FUEL CONTACT: Solids are fed to the pyrolysis reactor by means of a screw feeder and mixed with steam or other gas at the inlet of an electrically heated, down-flow pyrolysis reactor. Steam or inert gas can be added at any level in the reactor.

ASH/CHAR: Char and ash are removed by a cyclone at the exit of the pyrolysis reactor.

PRODUCTS: The major products are a medium energy gas and tar. Figure 10-13 (a) shows the amount of gas produced during flash pyrolysis of peat and solid waste. Figure 10-13 (b) shows the product distribution during pyrolysis of solid waste. Figure 10-13 (c) gives the composition of product gas during solid waste pyrolysis. Figure 10-13 (d) shows gas composition for various biomass materials.

Figures 10-14 (a, b) present TG-curves and DTG-curves for TGA pyrolysis of various biomass materials.

OPERATING CONDITIONS:

Temperatures	-	to 1000 C
Heating Rate	-	to 100 C/min in TGA
	-	to 1000 C/s in flash pyrolysis reactor
Pressure	-	atmospheric

SIZE: Pilot demonstration unit: 0.1 - 1.0 kg/h.

FUNDING, LOCATION, PERSONNEL: The pyrolysis studies are being performed by personnel at the Royal Institute of Technology, Department of Chemical Technology, Stockholm, Sweden, under the direction of E. Rensfelt.

Grant support is provided by the Swedish National Board for Energy Source Development and the Swedish Board for Technical Development.

REFERENCE: Rensfelt, E. et al. 1978. "Basic Gasification Studies for Development of Biomass Medium - Btu Gasification Process." Energy from Biomass Wastes. Chicago, IL: Institute of Gas Technology.

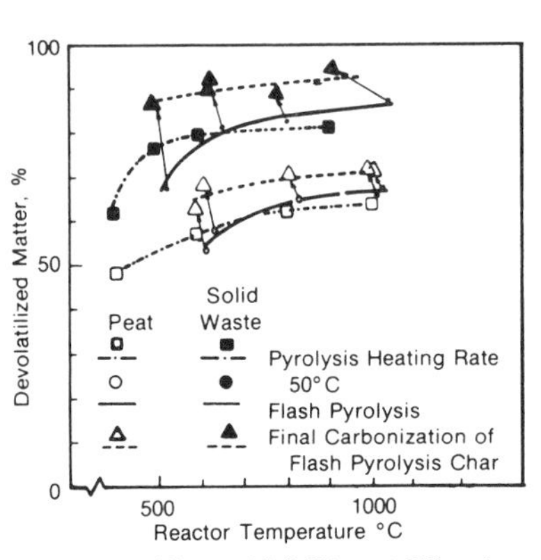

(a) Pyrolysis of Peat and Solid Waste at Different Heating Rates and Residence Times. Percentage Devolatilized of m.f. Peat (Low Moor Peat II) resp. m.a.f. Solid Waste Versus Pyrolysis Temperature

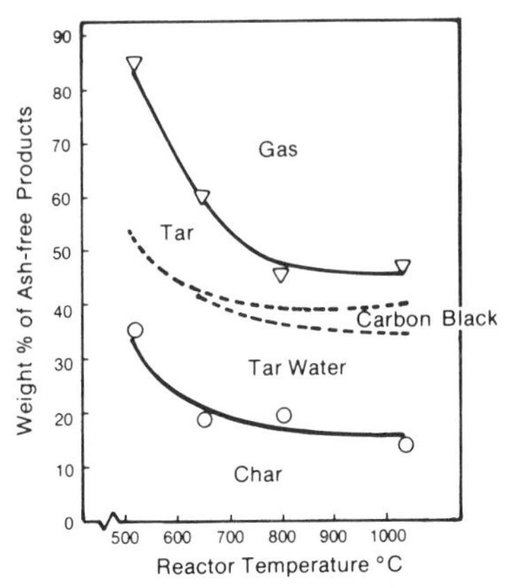

(b) Flash Pyrolysis of Solid Waste. Product Distribution at Different Reactor Temperatures

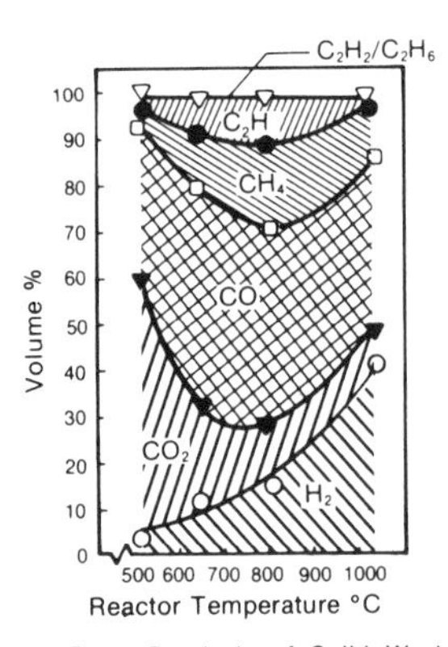

(c) Flash Pyrolysis of Solid Waste. Composition of Product Gas at Different Reactor Temperatures

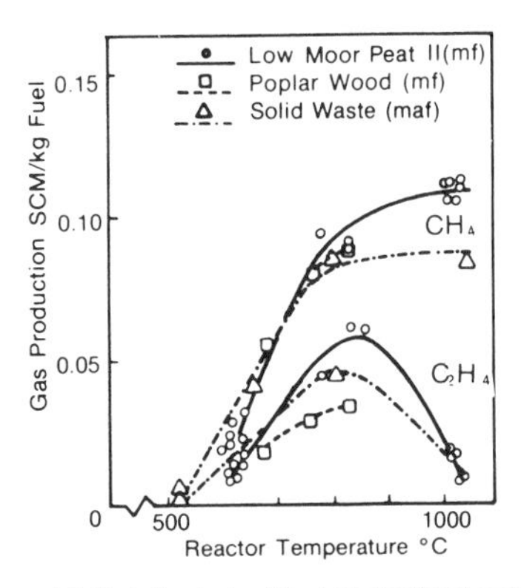

(d) Flash Pyrolysis of Peat, Poplar Wood, and Solid Waste. Production of Methane and Ethylene Versus Reactor Temperature

Figure 10-13. Flash Pyrolysis Yields, Royal Institute of Technology, Stockholm, Sweden

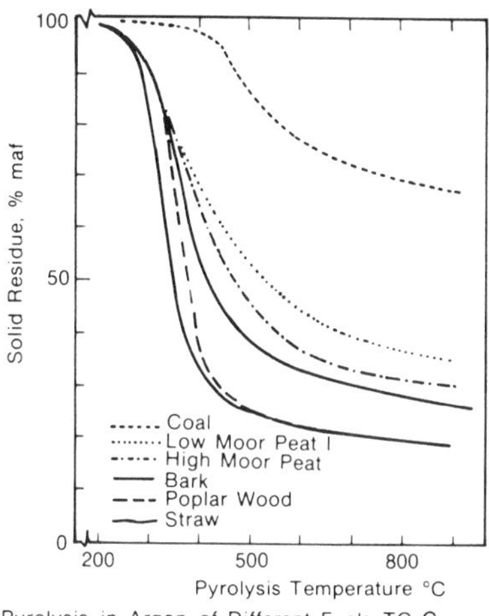

(a) Pyrolysis in Argon of Different Fuels, TG-Curves. Solid Residue (maf) as Percentage by Weight of Fresh Fuel (maf). Heating Rate 20°C/min for Bark, Straw, Coal, and Low Moor Peat and 50°C/min for High Moor Peat and Poplar Wood

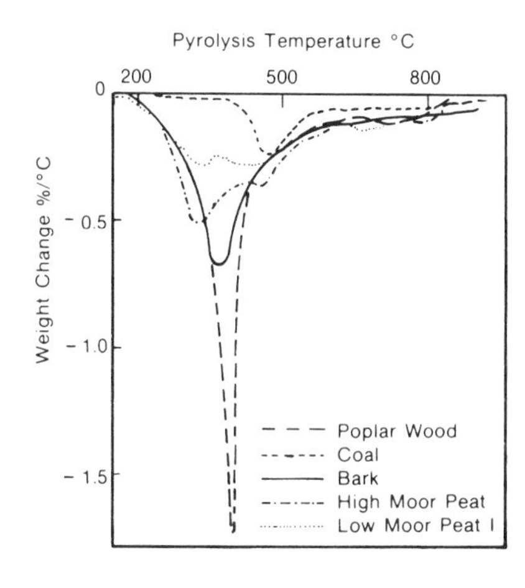

(b) Pyrolysis in Argon of Different Fuels, DTG-Curves as Percentage by Weight (maf) per Centigrade Heating Rates, See (a)

Figure 10-14. TGA Results for Pyrolysis of Biomass, Royal Institute of Technology, Sweden

---

GASIFICATION CASE SUMMARY

PROCESS:	Solar Energy Research Institute, fundamental studies of flash pyrolysis kinetics and mechanisms.
FEEDSTOCK:	Finely divided (10-1000 μm) powders of wood, cellulose, lignin.
HEAT SOURCE:	Variety of experimental approaches ranging from contact heating, through transport-line reactors to radiant heating.
GAS/FUEL CONTACT:	Short residence time reactors (1 to 10^{-3} s).
ASH/CHAR:	May reach fusion temperatures.
PRODUCTS:	Olefins, other unsaturates.
OPERATING CONDITIONS:	500-2000 C; 1 to 10^{-3} s; 1 atm; inert, steam, and H_2 environments; fast quenching and collection of gaseous, liquid, and solid products.
SIZE:	Laboratory scale.
FUNDING, LOCATION, PERSONNEL:	SERI Project 3356.10, Fundamental Studies in Thermal Conversion. T. Milne, M. Soltys.
COMMENTS:	Experimental work initiated in October 1979.

GASIFICATION CASE SUMMARY

PROCESS: SERI/Naval Weapons Center flash pyrolysis to olefins.

FEEDSTOCK: Ligno-cellulose materials.

HEAT SOURCE: Externally heated tube reactor.

GAS/FUEL CONTACT: Feed is entrained in a steam carrier and passed through a hot tube at such a rate as to achieve rapid heatup at millisecond residence times.

ASH/CHAR: Dry ash, char.

PRODUCTS: Char (1-20%) and olefin-rich gas (unsaturates about 25% wt).

SIZE: 20-30 lb/h.

FUNDING, LOCATION, PERSONNEL: SERI task no. 3356.30
1617 Cole Blvd.
Golden, Colo. 80401

J. Diebold, T. Reed

COMMENTS: In addition to optimizing the yield of olefins from the pyrolysis process, development work will be performed on olefin separation, thermal polymerization to gasoline, and hydration to mixed alcohols. Pyrolysis efforts will be directed toward the use of scalable reactor designs.

GASIFICATION CASE SUMMARY

PROCESS: Batch, quasi-steady-state, and pneumatically stirred reactors. University of California, Berkeley.

FEEDSTOCK: Wood, kraft black liquor, MSW.

HEAT SOURCE: External from laboratory furnaces.

GAS/FUEL CONTACT: Entrained flow, fixed bed.

ASH/CHAR:

PRODUCTS: The approximate weight percentage of organics is char 2.5%; tar 7%; and gases 90.5%: H_2 1%, CH_4 10%, C_2H_4 5%, CO 62%, and CO_2 13%.

OPERATING CONDITIONS: White fir particles, 20-40 mesh. Rate of heating, about 1000° C/s. Maximum temperature, 843 C. Residence time, 3 seconds.

SIZE: Various.

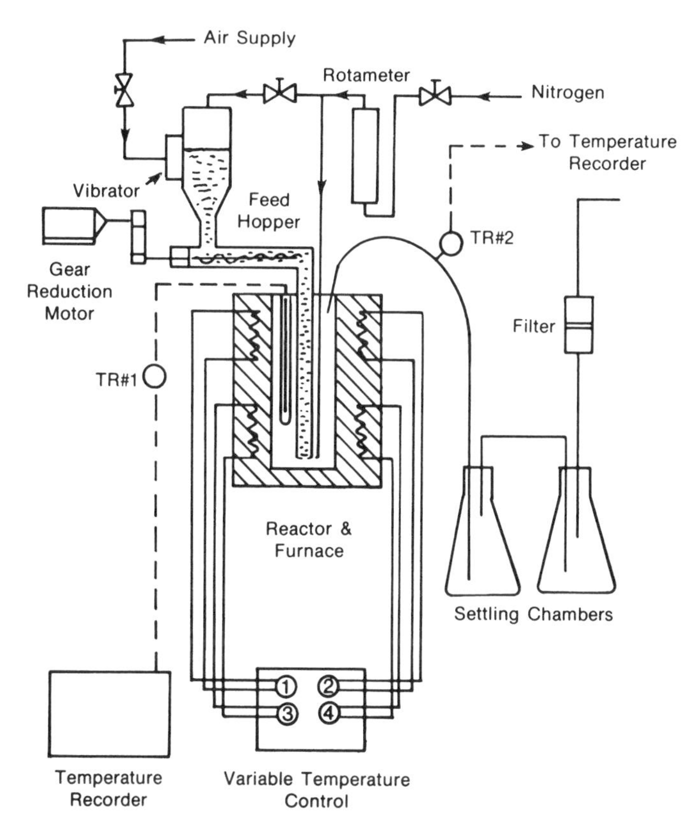

Figure 10-15. Bench-Scale Reactor, University of California, Berkeley

FUNDING, LOCATION, PERSONNEL: Prof. D. C. Brink
College of Natural Resources
University of California
Berkeley, CA

REFERENCES:

1. Brink, D. L.; Massoudi, M. S. 1978. J. Fire & Flammability. Vol. 9: p. 176.

2. Brink, D. L. 1976. Applied Polymer Symposium No. 28. New York: John Wiley & Sons; p. 1377.

GASIFICATION CASE SUMMARY

PROCESS: West Virginia University Fluid Bed Pyrolysis Process

FEEDSTOCK: MSW, sewage sludge, sawdust, manure, plastic, coal (partial drying, separation of noncombustibles, and size reduction to less than 1 in. are required).

Table 10-5. SOLID WASTE FEED ANALYSIS FOR WEST VIRGINIA UNIVERSITY STUDIES

Waste Material	Carbon[a] (Wt. %)	Hydrogen[a] (Wt. %)	Ash[a] (Wt. %)	Moisture (Wt. %)	Heating Value (Btu/lb Dry)
MSW[b]	30.25	4.03	40.17	5.49	5,500
Sawdust	47.20	6.49	0.97	2.62	8,114
Chicken manure	28.25	4.65	24.70	4.91	5,789
Cow manure	37.45	3.99	16.12	7.82	7,396
Animal fat	77.77	11.79	0.34	4.62	16,368
Tire rubber	76.11	7.15	4.40	1.91	15,401
PVC plastic	41.18	5.25	0.15	0.47	9,129
Nylon	84.18	10.07	0.08	1.48	13,481
Bituminous coal	73.36	5.34	7.57	3.42	13,097
Sewage sludge	18.43	2.21	62.95	42.16	3,900

[a]Dry basis; moisture is found by difference.

[b]Average of five tests.

HEAT SOURCE: Natural gas burner/sand bed (Pilot demonstration unit)
Char combustion in dual bed/recirculating sand (projected)

GAS/FUEL CONTACT: Fully fluidized, well-mixed sand bed.

ASH/CHAR: Dry ash and char elutriated from bed and separated from off-gas in a cyclone.

PRODUCTS:

Table 10-6. WEST VIRGINIA UNIVERSITY: PYROLYSIS OPERATING CONDITIONS AND RESULTS

Waste Material	Temperature (°F)	Dry Feed Rate (lb/min)	Gas Production (SCF/lb) Dry)	Gas Phase Thermal Efficiency
MSW[a]	1,420	0.40	9.34	0.72
Sawdust[b]	1,520	0.35	18.29	0.90
Chicken manure	1,280	0.39	9.53	0.51
Cow manure[c]	1,400	0.39	9.86	0.44
Animal fat	1,400	0.36	16.53	0.67
Tire rubber	1,370	0.36	5.36	0.22
PVC plastic	1,485	0.41	6.39	0.29
Nylon	1,530	0.31	8.59	0.26
Bituminous coal	1,440	0.34	10.92	0.36
Sewage sludge	1,420	0.22	9.48	0.88

[a]Average of five tests.

[b]Average of three tests.

[c]Average of two tests.

Table 10-7. **WEST VIRGINIA UNIVERSITY PYROLYSIS GAS ANALYSIS**

Waste Material	Gas Analysis (Dry Basis, Vol. %)								Low Heating Value (Btu/SCF)
	H_2	CO_2	CH_4	CO	C_2H_2	C_2H_4	C_2H_6	C_3H_8	
MSW[a]	44.47	15.78	6.96	24.76	4.97	1.49	0.66	0.91	421
Sawdust[b]	29.32	12.13	11.04	43.79	3.12	0.36	0.36	NM	398
Chicken manure	35.91	29.50	8.31	21.37	2.22	NM	0.61	NM	308
Cow manure[c]	31.07	20.60	7.70	38.06	1.86	NM	0.31	NM	328
Animal fat	11.57	27.63	18.12	14.72	25.05	NM	2.91	NM	683
Tire rubber	33.81	15.33	29.09	5.67	12.94	NM	3.17	NM	661
PVC plastic	41.02	19.06	14.51	20.76	4.02	0.21	0.43	NM	412
Nylon	45.38	6.03	15.47	34.64	0.0	NM	0.0	NM	403
Bituminous coal	46.88	11.68	16.63	21.72	2.08	NM	1.01	NM	435
Sewage sludge	47.01	22.88	11.22	15.57	3.12	NM	0.21	NM	360

[a]Average of five tests.

[b]Average of three tests.

[c]Average of two tests.

NM = not measured.

OPERATING CONDITIONS: T = 1400-1500 F, P = 0-10 psig
Superficial gas velocity: 1.5 ft/s
Feed rate: 40-80 lb/h-ft^2.

SIZE: (Pilot demonstration unit)
Bed ID: 15 in., 15-16 lb/h (0.7 tons/day)
Capital Investment: $150,000.

(Projected) Bed ID: 12 ft, 170 tons/day
Capital investment for a plant to process 1,000 tons/day of dried refuse: $19.6 million (1978).

FUNDING, LOCATION, PERSONNEL: Principal Investigator: Dr. Richard C. Bailie
Department of Chemical Engineering, West Virginia University
Morgantown, W. Va.

Funding initiated with HEW grant for waste disposal studies in 1966. Work completed under EPA Contract No. R01 EC 00399-03 EUH. Final report submitted August 1, 1972. Nonexclusive license granted to Wheelabrator Incineration.

REFERENCES:

1. Bailie, R. C. U.S. Patent 3,853,498. "Production of High Energy Fuel Gas From Municipal Wastes."

2. Bailie, R. C., Burton, R. S. 1979. "Fluid Bed Pyrolysis of Solid Waste Materials." Combustion. p. 13; Feb.

3. Alpert et al. 1972. "Pyrolysis of Solid Waste: A Technical and Economic Assessment." Prepared for WVU by SRI, Sept. NTIS PB 218-231.

COMMENTS: The work at West Virginia University was aimed at characterizing the pyrolysis behavior of solid waste components. Mass balances based on carbon were seldom closed to better than 90%. Gas analyses were precise, but the char and oil were not characterized. Gas yields were strongly dependent on bed temperature, increasing rapidly to 1400 F (760 C) and then leveling off at higher temperatures.

The dual fluidized bed system envisioned for the commercial scale plant was described by Bailie in his patent. The projected economics were reported by SRI. No prototype was built in this country; however, a plant using the same concept is now operating in Japan.

GASIFICATION CASE SUMMARY

PROCESS: Wright-Malta Steam Gasification Process

FEEDSTOCK: Any form of biomass that can be screw-fed, including very wet materials.

HEAT SOURCE: Condensing high-pressure steam.

GAS/FUEL CONTACT: Solids are transported by a slowly rotating screw and are maintained in close contact with a gas stream consisting mostly of steam (Fig. 10-16).

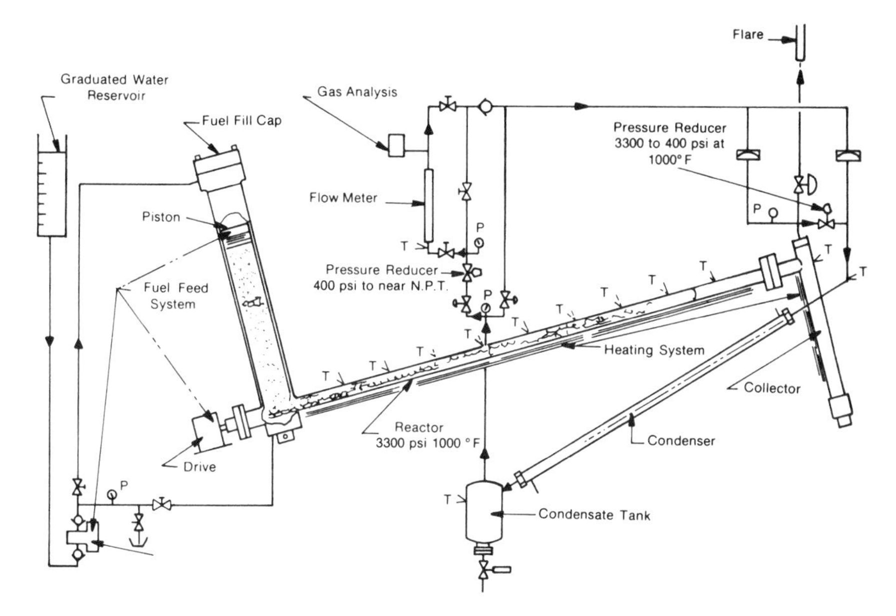

Figure 10-16. Biomass Gasifier Schematic, Wright-Malta Steam Gasification

ASH/CHAR: Residues are dropped from the end of the screw flight into a lock hopper.

PRODUCTS: Medium Btu gas consisting chiefly of H_2 and CO_2. In Fig. 10-17, the dependence of composition on temperature and pressure is illustrated (solid lines) and compared with calculated equilibrium compositions (dashed lines).

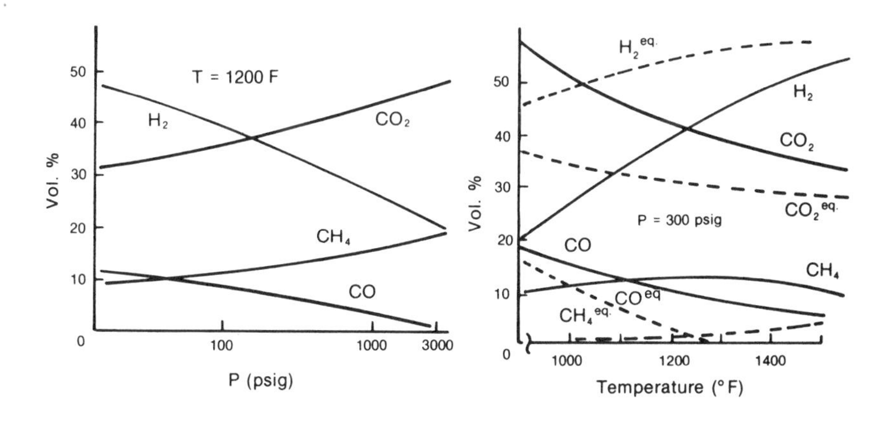

Figure 10-17. Medium-Btu Gas

OPERATING CONDITIONS: T = 400-1500 F
P = 0-3000 psig
Catalyst: Na_2CO_3.

SIZE: ID = 2.5 in. Length = 10 ft

FUNDING, LOCATION, PERSONNEL: The Wright-Malta Corp. is located in Ballston Spa, N.Y. Preliminary investigation of design variables, performed on a batch-fed minikiln gasifier (Fig. 10-18), was funded by the Empire State Electric Energy Research Corp. Product studies with MSW were funded by the U.S. EPA. Work on coal was sponsored by the N.Y. State Energy Research and Development Authority. DOE is currently funding further development work.

REFERENCES:

1. Hooverman, R. H.; Coffman, J. A. 1977. "Rotary Kiln Gasification of Biomass and Municipal Wastes." IGT Symposium on Clean Fuels from Biomass and Wastes. Orlando, FL; Jan. 25-28.

2. Wright-Malta Corp. 1979. Steam Gasification of Biomass. Progress Report No. C00/4124-4, for Fuels from Biomass Program. U.S. Dept. of Energy. Nov. 1.

3. Coffman, John A. 1979. "Steam Gasification of Biomass." 3rd Annual Biomass Energy Systems Conference Proceedings: The National Biomass Program. Colorado School of Mines, Golden, CO: June 1979. Golden, CO: Solar Energy Research Institute; p. 349.

COMMENTS: The behavior of the steam gasification system has been explored over a wide range of operating conditions both with and without an added catalyst.

Pressure - Solid to gas conversion was greatest in the pressure range from 400-500 psig. More residue was obtained at lower and higher pressures. The catalyst was effective in reducing char production below 600 psig; it was ineffective above this pressure. Also, the form of the residue changed from loose and granular below 600 psig to compact, 1-2 cm lumps at higher pressures.

Temperature - Above 1400 F (760 C) the gas composition is very near the equilibrium composition. Below this temperature, the steam re-forming reactions are not fast enough to convert CO and CH_4 to H_2. (Note: WM reports only metal wall temperatures and exit gas temperatures). The time-temperature history of the feed as it passes through the continuous reactor is a slow heating in the presence of steam. An interesting feature of the minikiln batch procedure is that the isolated events of pyrolysis and steam gasification can be followed, as illustrated in the plot of gas evolution and temperature vs. time in Fig. 10-19. Pyrolysis begins at 150 C and is complete at 400 C. Steam gasification of char begins at about 500 C. Note that the steam shifts all the CO to CO_2; in fact, one of the chief characteristics of the WM gas product is an extremely high H_2/CO ratio.

Particle Size - Tests in the minikiln indicated that the process was insensitive to the form of the biomass charge. However, only finely divided materials have been used in the continuous reactor for mechanical reasons. Related to this feeding problem is the ratio of water to solid: most of the data have been obtained by feeding a sawdust slurry. Work is under way to alter the feed system to permit lower water/charge ratios.

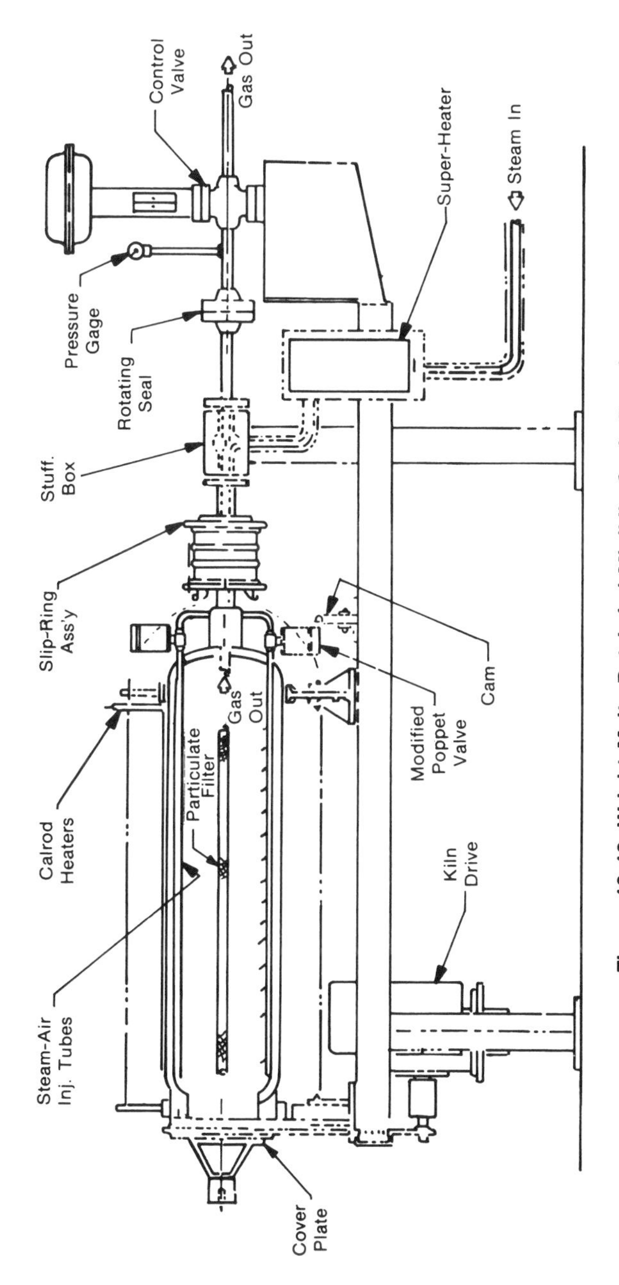

Figure 10-18. Wright-Malta Batch-fed Minikiln Scale Drawing

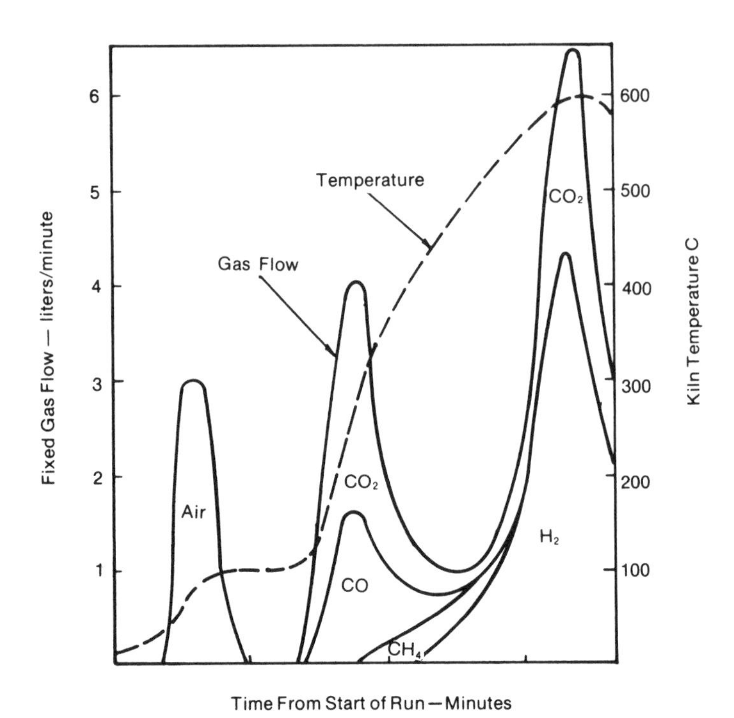

Figure 10-19. Fixed Gases Evolved, Typical Minikiln Run, Wright Malta

10.2.4 Hydrogengasification and Bromine Conversion

GASIFICATION CASE SUMMARY

PROCESS: Institute of Gas Technology Hydrogasification Process.

FEEDSTOCK: Peat, various coals.

HEAT SOURCE: Electrical resistance heaters.

GAS/FUEL CONTACT: Hydrogen (and steam if desired) is preheated and mixed with feed at the entrance of a helical coil reactor. The reactor is operated as an entrained flow reactor in an isothermal or a constant heat-up mode. A diagram of the PDU reactor system is given in Fig. 10-20.

ASH/CHAR: Char is removed by a cyclone and solids filter.

PRODUCTS: Products are hydrocarbon gases, heavy hydrocarbons and carbon oxides. Figures 10-21(a) and 10-21(b) show typical yields during peat hydrogasification in the bench-scale reactor.

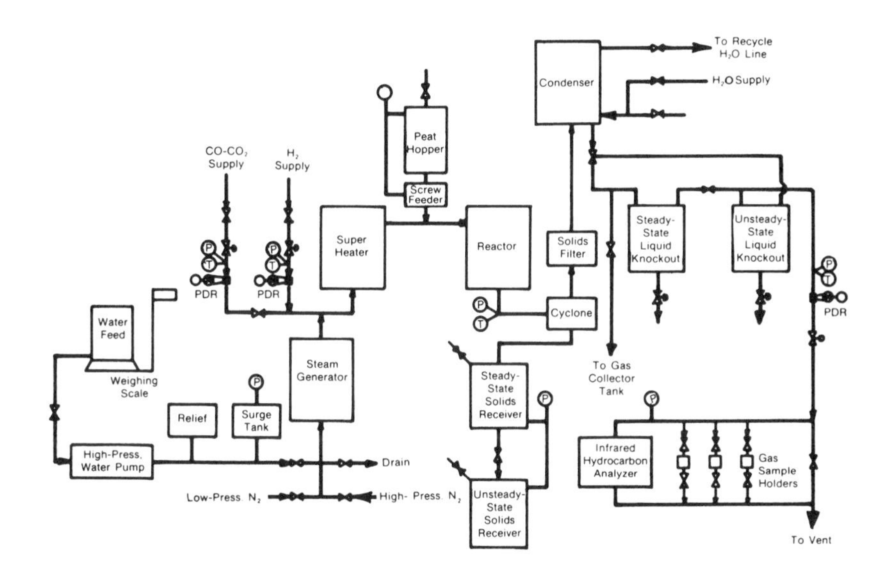

Figure 10-20. IGT Hydrogasification Process, Block Flow Diagram for the Process Development Unit (PDU)

OPERATING CONDITIONS:

	Laboratory-Scale Reactor	Process Development Unit
Peat feed rate (lb/h)	0.022-0.048	5-12.5
Feed gases	H_2, He	H_2, H_2-H_2O, Synthesis Gas
Hydrogen partial pressure (atm)	4-71	4.3-36
Maximum temperature (°F)	855-1500	1000-1500
Gas flow rate (SCF/h)	24-48	400-1030
Residence time (s)	4-7.7	8-14
Average feed peat Particle size, (in.)	0.005	0.011

SIZE: Laboratory scale (0.05 lb/h)
Process development unit (12.5 lb/h).

FUNDING, LOCATION, PERSONNEL: The IGT hydrogasification process has been developed by personnel at the Institute of Gas Technology, Chicago, Ill., under joint sponsorship of DOE and IGT.

REFERENCES:

Punwani, D. V.; Nandi, S. P; Gavin, L. W.; Johnson, J. L. 1978. "Peat Gasification - An Experimental Study." Presented at 85th National Meeting of the AIChE, Philadelphia, PA.

Weil, S. A.; Nandi, S. P.; Punwani, D. V.; Kopstein, M. J. 1978. "Peat Hydrogasification." Presented at 176th National Meeting of ACS, Miami, Fl.

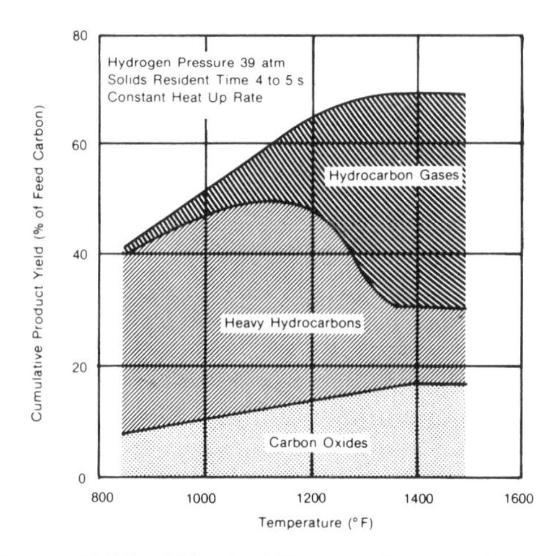

Figure 10-21a. Effect of Temperature on Product Yields Obtained During Peat Gasification, IGT Hydrogasification

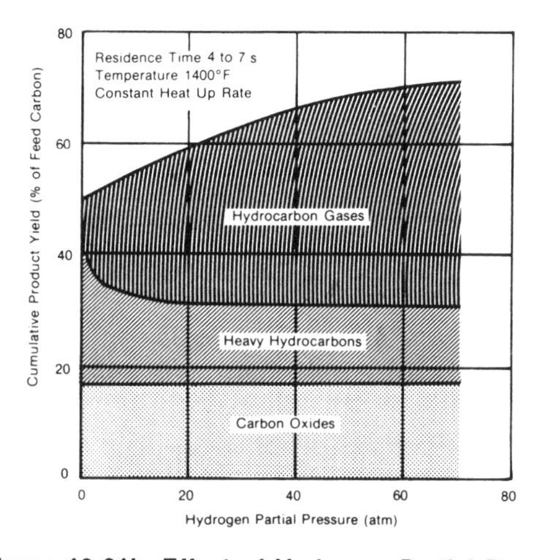

Figure 10-21b. Effect of Hydrogen Partial Pressure on Product Yield Obtained During Peat Gasification, IGT Hydrogasification

COMMENTS: The IGT hydrogasification has been shown to be technically feasible for gasification of peat and coal. Experimental data on biomass and process economics are needed before the usefulness of this process can be compared to the other major types of biomass gasification processes.

GASIFICATION CASE SUMMARY

PROCESS:	Bromine Conversion of Biomass to HBr Followed by Electrolysis to H_2. Rockwell Energy Systems Group.
FEEDSTOCK:	Wood, sugarcane, water hyacinth, kelp, lignin, Eco Fuel II.
HEAT SOURCE:	Small glass ampules in electric furnace.
GAS/FUEL CONTACT:	Aqueous bromination under pressure.
ASH/CHAR:	Filtered from product solution.
PRODUCTS:	Almost entirely CO_2 and HBr in bromination step.
OPERATING CONDITIONS:	At 250 C and 30 min. there is 95-96% conversion to HBr.
SIZE:	Laboratory tests on gram samples.
FUNDING, LOCATION, PERSONNEL:	SERI H_2 - Production Program. Canoga Park, CA. A. J. Darnell, principal investigator.
REFERENCES:	Paper to be presented at 1979 World H_2 Energy Conference.

10.2.5 Solar-Thermal Gasification

GASIFICATION CASE SUMMARY

PROCESS:	Gasification of biomass using an integral pyrolysis entrained flow reactor/solar receiver.
FEEDSTOCKS:	Agricultural wastes and products—straw, cornstalks, Sudan grass, sunflowers, etc.
HEAT SOURCE:	Lab studies: Electric tube furnace Field tests: 6-meter diameter parabolic dish solar concentrator.
GAS/FUEL CONTACT:	The biomass is entrained and transported through the stainless steel heat transport coil by either steam or pyrolysis gas.
ASH/CHAR:	The ash and char are collected from the quench water.
PRODUCTS:	Hydrogen, propylene, acetylene, methane, carbon monoxide, carbon dioxide, and ethylene; traces of butenes and saturated hydrocarbons.
OPERATING CONDITIONS:	700° C to 1500° C at 1 atm pressure.
SIZE:	10-20 lb/h
FUNDING, LOCATION, PERSONNEL:	SERI Task No. 3457.13 1617 Cole Blvd. Golden, Colo. 80401 C. Benham, G. Bessler, P. Bergeron, M. Bohn, R. Kemna, and R. Passamaneck.

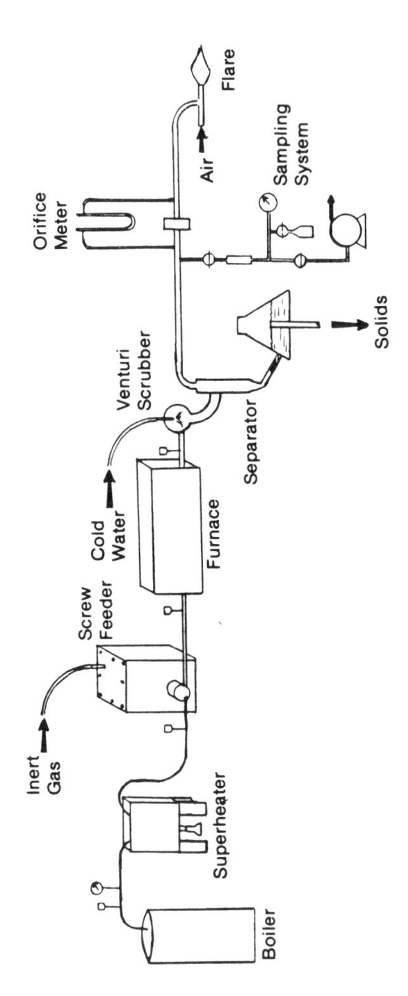

Figure 10-22. Laboratory Apparatus for Biomass Pyrolysis (SERI)

Chapter 11

Economics of Air Gasification for Retrofitting Oil/Gas Boilers

T. B. Reed, D. Jantzen,
W. P. Corcoran, R. Witholder
SERI

11.1 INTRODUCTION

Many industrial concerns converted from coal to natural gas or oil during the last decade to meet more stringent emission requirements. Now they are faced with much higher fuel prices and the possible curtailment or total interruption of supply. Their most obvious course is to convert those boilers that originally used coal back to coal or to wood or to replace new oil/gas package boilers with new coal/wood installations. Both options are relatively expensive and also will require less stringent emission controls.

A less obvious option is the use of a biomass (or coal) gasifier to retrofit the existing gas/oil boiler to an intermediate-energy gas generated in situ, using the "close-coupled gasifier" (described in Chapter 8). In this chapter we examine the technology and economics of biomass gasifiers and compare the economics of retrofit to the economics of complete combustion installations for biomass.

11.2 GASIFIER OPERATION

A partial list of manufacturers of gasifiers suitable for converting gas/oil boilers is given in Table 11-1, including the type of gasifier, the size, and status of development. A more complete list is given in Section 9.2.

Table 11-1. PARTIAL LIST OF BIOMASS GASIFIER MANUFACTURERS IN THE UNITED STATES

Name	Type	Status[a]	Size (MBtu/h)
Applied Engineering, Orangeburg, SC	Updraft	D	8
Biomass Fuel Conversion, Yuba City, CA	Downdraft	D	14
Century Research, Gardena, CA	Updraft	C	85
Davis Gasifier, U. of Calif., CA	Downdraft	D	14
DeKalb Agricultural Research, DeKalb, IL	Updraft	D	1.7
Forest Fuels, Keene, NH	Updraft	C	1-12
Foster-Wheeler, Livingston, NJ	Updraft	D	50
Halcyon, E. Andover, NH	Updraft	C	8
Pioneer Hi-Bred Inst., Johnston, IA	Updraft	D	7
Woodex Corp., Eugene, OR	Updraft	C	10

[a]Status of project: C-Commercial (at least one unit in field); D-Demonstration and testing.

The gases produced by these gasifiers contain CO, H_2, and hydrocarbon gases as their principal fuel ingredients and N_2, CO_2, and H_2O as diluents. If the gases are cooled and conditioned for use in engines or a pipeline, they have a typical energy content of 90 Btu/SCF and are called low energy gas (LEG), producer gas, and gen-gas or generator gas. A typical analysis shows: CO = 20.5%; H_2 = 15.3%; CO_2 = 7.4%; O_2 = 1.4%; hydrocarbons = 8.1%; N_2 = 47.4% (Williams and Gross 1977).

If these gases are to be used for heating, it is not desirable to remove the pyrolysis oil vapors and the sensible heat; these same gases then have an effective heat content of 140 to 200 Btu/SCF, depending on temperature, feedstock, type of gasifier, etc.

11.2.1 Efficiency of Combustion of Medium Energy Gas (MEG)

The energy content of a gas is very important if the gas is to be shipped by pipeline. However, the flame temperature and flue gas mass produced varies with energy content by only a small amount because large quantities of air must be added for combustion. The relative efficiency of boilers using gases of various energy contents are shown in Fig. 11-1 as a function of energy content of the gas (Bechtel Corporation 1975). Here it can be seen that efficiency is actually higher for the medium energy gases (MEG) (with energy content around 350 Btu/SCF) than it is for high energy gas (HEG) with energy content about 1,000 Btu/SCF. The efficiency falls rapidly below about 200 Btu/SCF. It can be seen that there is little loss for MEG, but considerably more for low energy gas (LEG).

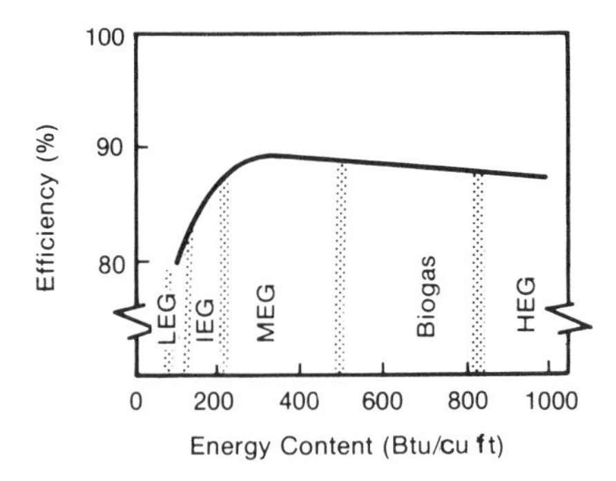

Figure 11-1. Boiler Efficiency Vs. Gas Energy Content

11.2.2 Scale of Close-Coupled Gasifiers

Table 11-1 shows that there are a number of close-coupled gasifiers being developed in the capacity range from 1 to 100 MBtu/h. There also may be some need for even smaller gasifiers, for example, for heating apartments and shopping centers. At present, there are no proven biomass gasifiers with operating capacities greater than 100 Btu/h, and there would seem to be a need for this size for large process steam installations, especially in the paper industry. However, coal gasifiers have been built at this larger scale and there seems to be no technical barrier to scaling gasifiers to larger or smaller sizes.

11.2.3 Efficiency of Close-Coupled Gasifiers

Since all the gas generated is burned and the sensible heat of the gas stream is also conserved in close-coupled gasifiers, these units can have very high efficiencies. Essentially complete combustion of the resulting gas is easily achieved as a result of the two-stage combustion in the gasifier and boiler. The only losses in the gasifier are the heat losses from the outer surfaces and heat to the ash, which is negligible. The Century gasifier is reported to have a thermal efficiency of 90% (Amundsen 1976), while the Davis gasifier operates at a typical efficiency of 85% (Gross 1978). The early gasifiers used for transportation in Europe had thermal efficiencies of 80% even after the tars had been cooled and scrubbed (Reed and Jantzen 1979).

11.2.4 Retrofitting Close-Coupled Gasifiers to Existing Boilers

The gases produced in the gasifiers listed in Table 11-1 can be burned in existing oil/gas installations, and a number of commercial installations have been made. The gas is somewhat more difficult to burn than natural gas and requires insulated piping to prevent condensation of pyrolysis oils and tars. A gas pilot flame or a flame holder is used to ensure combustion. However, the conversion problems are minimal.

In general, the modifications needed for retrofitting existing boilers are not documented, but a recent feasibility study at the California State Central Heating and Cooling Plant in Sacramento has used the Davis gasifier to power one of their boilers (Fuels Office 1978) for 158 h. The gasifier is 8 ft in diameter and 15 ft tall and produced 16 MBtu/h. Tests were run using two fuels: kiln dried wood chips purchased for $9/ton or $12.50/ton delivered; and pelleted white fir sawdust purchased for $25.50/ton or $35/ton delivered. The heating value of the gas varied from 182 to 206 Btu/SCF. Emissions were: 0% SO_2 observed (0.2% allowable); 130 ppm NO_x (200 ppm federal standard); and 0.703 lb/h particulates (4.09 lb/h allowable). Some condensate, tar, and charcoal were collected; however, the California Division of Water Quality concluded that they would not present a serious disposal problem.

Minor problems encountered during the test runs included burning out of an auger motor and some tar buildup in the delivery line. Most of the problems were associated with the temporary nature of the hookup for testing and should be no obstacle to commercialization. There was no noticeable deterioration of the metal parts. (Gasifiers that were built 60 years ago are still in operation.) During the tests, the gasifier production rate was controlled manually by controlling the intake air. Moreover, since gasifiers have been used to operate trucks, cars, and tractors, it has been proven that they can respond quickly to changes in load.

11.3 ECONOMICS OF RETROFITTING GASIFIERS TO EXISTING BOILERS

Two manufacturers with commercial experience have projected costs for commercial-sized gasifiers and their assumptions and costs are given in Table 11-2 (Gross 1978; Amundsen 1976). The gas costs derived ($0.73 and $1.06 per MBtu) are attractive relative to natural gas costs. However, the two biomass-derived gas costs cannot be compared to each other directly because of different assumptions used and the different sizes of the units.

In order to make these costs more directly comparable with each other and with other energy costs, we have used the cost analysis method developed at the Electric Power Research Institute (EPRI) for the Energy Research and Development Administration (ERDA) (Jet Propulsion Laboratory 1976). This method, developed initially to compare steam and power costs of fossil and nuclear fuels, has been used recently at SERI to develop a computer program for comparing various solar energy costs as well (Witholder 1978). The program uses certain assumptions (see Table 11-3) to determine anticipated capital flows and operating costs over the lifetime of the facility. These costs are then used to derive a fuel cost for the first year of the application and also a levelized cost over the assumed lifetime of the facility.*

We have used the EPRI/ERDA/SERI program to determine the cost of gas produced in the gasifiers described in Table 11-2 as a function of input fuel cost. These first-year fuel costs are shown in Table 11-3, derived from the assumptions listed. The levelized

---

*The levelized cost is the constant price at which the gas must be sold over the life of the project to produce the required rated return.

Table 11-2. OPERATING COSTS OF GASIFICATION

	Gasifier A[a]	Gasifier B[b]
Fuel	Walnut hulls	Chaparral
Rated gas production (MBtu/h)	14.1	85
Rated feed rate (ton/h)	1.19	7.87
Capital cost ($)	125,000	350,000
Efficiency (%)	85	90
Annual Operating Costs ($)		
Depreciation (10%)	12,580	35,000
Repairs and maintenance (3%)	3,774	10,500
Utilities (water, power)	---	38,795
Operating labor	6,000 (250 days)	14,600 (365 days)
Taxes and insurance (2%)	2,516	---
Interest (7%)	8,806	---
Profit	---	---
Gasification cost ($)	33,676	98,895
Fuel cost ($)	28,571 ($4/ton)	689,450($10/ton)
Total operating cost ($)	62,247	788,345
Annual gas production (MBtu)	85,000	744,600
Gasification cost ($/MBtu)	0.40	0.13
Gas cost ($/MBtu)	0.73	1.06

[a]Data from Gross (1978).
[b]Data from Amundsen (1976).

Table 11-3. FIRST-YEAR GAS COST AS A FUNCTION OF INPUT FUEL COST

	Biomass Cost		
	$10/ton	$20/ton	$30/ton
Gasifier A	1978 Cost $1.41	$2.58	$3.74
(14 MBtu/h)	Levelized Cost (2.08)	(3.78)	(5.49)
Gasifier B	1978 Cost $1.44	$2.72	$3.99
(85 MBtu/h)	Levelized Cost (2.12)	(3.99)	(5.86)

Assumptions:

20-year life of project
Capital and operating costs are given in Table 11-2
Plant capacity factor = 0.92
Tax and insurance rates:

Effective federal income tax rates	0.48	
Other taxes	0.82	(fraction of present value of capital investment)
Insurance premiums	0.0025	(fraction of present value of capital investment)

Capitalization Ratios:		Rate of Return
Debt	0.50	0.08
Common stock	0.40	0.12
Preferred stock	0.10	0.08

Rates of Inflation (%)	
General economy	5
Capital costs	5
Operating costs	6
Maintenance	6
Fuel costs	7

fuel costs are given in parentheses. In order to show the sensitivity of gas cost to the fuel, operating, and capital costs, these factors are listed separately in Table 11-4 for a fuel cost of $20/ton. Since the gas cost depends linearly on fuel costs, the gas cost can be computed for any other input fuel cost by multiplying the fuel contributions from Table 11-4 by the fuel cost and dividing by 20; gas costs for other capital or operating costs can be determined in the same manner.

Tables 11-3 and 11-4 demonstrate that the principal factor determining gas cost is the cost of the biomass fuel used, with operating costs and capital costs affecting gas cost to a much lesser extent; thus gasification of low-cost forest and agricultural wastes (costing $0 to $15/ton) is very attractive in these days of rising fuel costs. Other biomass feedstocks, such as cull trees and straw (costing $15 to $40/ton), are less attractive in comparison with today's natural gas costs but may soon be competitive. Other advantages for the use of gasifiers are that they can be operated intermittently when gas or oil is unavailable or too costly (depending on spot prices for both gas/oil and biomass), and that they dispose of unwanted biomass (which of itself would have a negative fuel value).

Table 11-4. DETAILED COST BREAKDOWN FOR $20/TON FUEL

	Gasifier A (15 MBtu/h)		Gasifier B (85 MBtu/h)	
	1978 Cost	Levelized Cost	1978 Cost	Levelized Cost
Operating costs	$0.11	$0.15	$0.13	$0.19
Capital costs	0.06	0.09	0.13	0.19
Fuel costs	2.55	3.75	2.32	3.40
Total costs	$2.72	$3.99	$2.58	$3.78

Assumptions:

20-year life of project
Capital and operating costs are given in Table 11-2
Plant capacity factor = 0.92
Tax and insurance rates:

Effective federal income tax rates	0.48	
Other taxes	0.82	(fraction of present value of capital investment)
Insurance premiums	0.0025	(fraction of present value of capital investment)

Capitalization Ratios:		Rate of Return
Debt	0.50	0.08
Common stock	0.40	0.12
Preferred stock	0.10	0.08

Rates of Inflation (%)	
General economy	5
Capital costs	5
Operating costs	6
Maintenance	6
Fuel costs	7

11.4 COMPARISON OF ALTERNATE FUEL CONVERSION OPTIONS

If it is difficult to establish cost guidelines for retrofitting gas/oil boilers with close-coupled gasifiers, it is even more difficult to compare these costs with those of other conversion options in a time of rapidly changing costs and varying availability of fossil fuels. In a recent study on wood combustion economics made by the Forest Products Laboratory (FPL), the authors explained that "the procurement cost of combustion equipment options is a dominant factor in their selection. In a combustion equipment survey, cost data were found to be very difficult to obtain without establishing point designs. Repetitive contact with manufacturers and review of published data ultimately resulted in a set of cost curves" (FPL 1976). We have used similar methods here to evaluate the use of gasifiers to retrofit existing gas/oil installations and to compare these costs to those of other options.

The options available today for converting from gas/oil are:

1. Reconversion to solid fuel of an originally solid-fueled installation (which had been converted from gas/oil). Where possible, this is probably the most economical conversion, yet often the solid fuel handling equipment will have been scrapped, new emission control equipment will have to be added, and the existing boiler is likely to be old and inefficient.
2. Replacement of the existing gas/oil boiler (often relatively new) and installation of a new solid fuel system burning coal or wood or other biomass. This will cost on the order of \$8 to \$30/lb steam/h and will require installation of new emission control equipment.
3. Installation of a close-coupled gasifier to operate the existing gas/oil equipment. This will cost on the order of \$4 to \$9/lb steam/h (see Tables 11-3 and 11-4) and will make use of much of the existing installation. It also permits using gas/oil where and when they are available and economical and permits use of biomass wastes that otherwise would not have value as fuels.

Figure 11-2 compares the costs of these options. It appears that the cost of adding a gasifier to an existing package boiler (Option 3) is about two-thirds the cost of installing a new wood-fired boiler (Option 2).

In general, the cost of package wood-fired boilers (\$8 to \$18/lb steam/h) is considerably less than that for field-erected boilers (\$15 to \$25/lb steam/h), which are required for generating steam in excess of about 10^5/lb steam/h as shown by the FPL (1976) results in Fig. 11-2. An early study for several paper industries in Maine indicated the advantages of close-coupled gasifiers for retrofitting very large existing boilers (typically 2-10 X 10^5/lb steam/h) with gasifiers (Reed and Stevenson 1975). At present, this attractive option for larger boilers is not available because there are no gasifiers with capacities greater than 10^5/lb steam/h. Development of such a gasifier would allow the paper industry to convert from gas/oil at a minimum cost.

11.5 COMPARISON OF NEW CONSTRUCTION ECONOMICS

If gasifiers are more economical for retrofit, it may be asked whether their combination with an inexpensive gas/oil boiler (two-stage combustion) may also be preferable to conventional package wood-fired boilers for new installations. Adding the lower two curves of Fig. 11-2 gives prices for a complete new gasifier-boiler system of \$6.90-\$19.00/lb steam/h as compared to \$6.20-\$18.00/lb steam/h for conventional package wood-fired boilers. The closeness of these numbers is probably fortuitous, and it would be premature to conclude that the two-stage combustion option using a gasifier is superior to the conventional package wood-fired boiler, yet this possibility cannot be ruled out and should be investigated further. The economics which could favor the gasifier-boiler combination are the very low price of conventional gas/oil boilers as compared to wood boilers and the relative simplicity and low cost of gasifiers as compared to wood furnaces. In addition, the emissions from gasifiers may be lower than for conventional wood firing, and the turndown ratio of gasifiers may be superior to that for wood firing. Use of gasifiers would permit return to fossil fuel (dual fuel capability) should that be desirable.

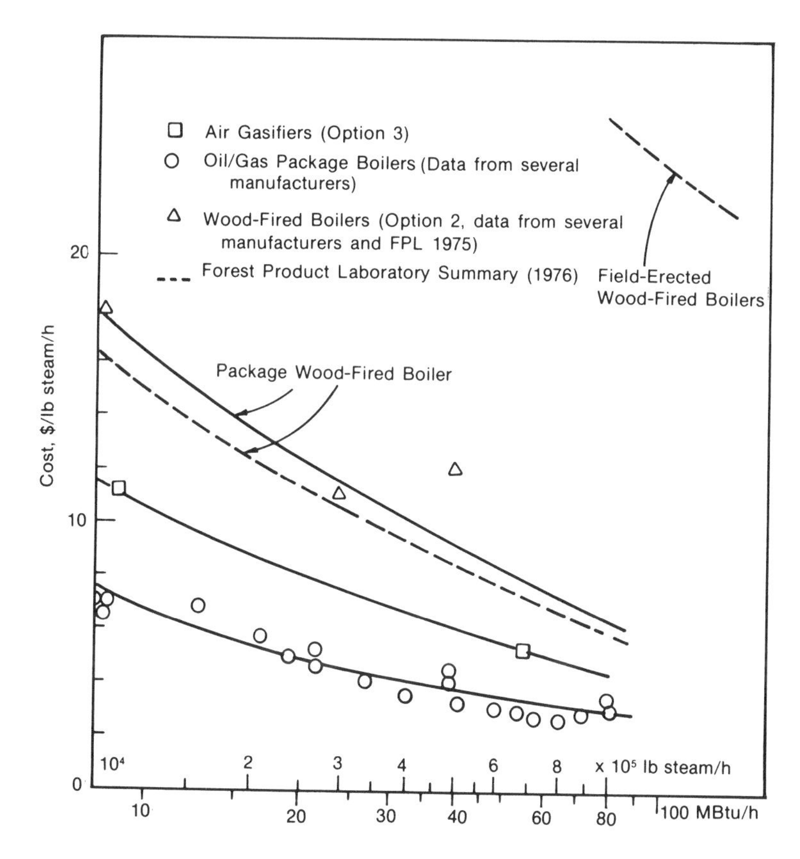

Figure 11-2. Cost Comparisons Between Retrofitting Existing Equipment and New Installations

A recent study on a fluidized-bed, medium energy gasifier now under development suggests that the combination of this more expensive technology with package boilers is at least comparable in cost to installation of solid fuel combustion equipment (Bailie and Richmond 1978).

11.6 CONCLUSIONS

- Gasifiers are now being developed for retrofitting existing boilers in the 10^4-10^5/lb steam/h (10-100 MBtu/h) range to use wood and biomass residues.
- The cost of gas from these gasifiers is estimated to be $1.40-$2.70/MBtu for biomass feedstock costing $10 to $20/ton.
- The addition of a close-coupled gasifier to an existing gas/oil boiler will cost on the order of two-thirds the cost of installing a new package wood-fired boiler.
- Although gasifiers larger than 100 MBtu/h (10^5/lb steam/h) are not presently available, they could probably be used to convert existing field-erected gas/oil boilers to biomass more economically than construction of new wood-fired boilers.
- The use of a gasifier plus a low cost gas/oil boiler for new construction is comparable in cost to wood package boilers and should be investigated for future installations, particularly where dual fuel operation is desired.

11.7 REFERENCES

Amundsen, H. R. 1976. The Economics of Wood Gasification. Chaparral for Energy Information Exchange Conference; Pasadena, CA; June 22, 1976. Sponsored by PSW Experiment Station, Angeles National Forest. p. 118.

Bailie, R.; Richmond, C. A. 1978. Economics Associated with Waste or Biomass Pyrolysis Systems. Presented at ACS Symposium; Anaheim, CA; Mar. 10-12, 1978.

Bechtel Corporation. 1975. Fuels from Municipal Refuse for Utilities: A Technical Assessment. Palo Alto, CA: Electric Power Research Institute; EPRI Report 261-1.

Forest Products Laboratory. 1976. The Feasibility of Utilizing Forest Residues for Energy and Chemicals. Washington, D.C.: U.S. Forest Service; Available NTIS as PB 258 630.

Fuels Office Alternatives Division. 1978. Commercial Biomass Gasifier at State Central Heating and Cooling Plant. Feasibility Study. Sacramento, CA: Fuels Office Alternatives Division, California State Energy Commission.

Gross, J. R. 1978. "Food, Forest Wastes = Low Btu Fuel." Agricultural Engineering. Vol. 59 (No. 1): p. 30.

Jet Propulsion Laboratory. 1976. The Cost of Energy from Utility-Owned Solar Electric Systems: A Required Revenue Methodology for ERDA/EPRI Evaluation. Pasadena, CA: California Institute of Technology; JPL 5040-29.

Reed, T. B.; Jantzen, D. E. 1979. Generator Gas. Golden, CO: Solar Energy Research Institute; SERI/SP-33-140.

Reed, T. B.; Stevenson, W. A. 1975. Energy from Wood. Maine Wood Study Group.

Williams, R. O.; Gross, J. R. 1977. "An Assessment of the Gasification Characteristics of Some Agricultural and Forest Industry Wastes." Davis, CA: University of California. Manuscript from Department of Agricultural Engineering.

Witholder, Robert. 1978. Levelized and Energy Inflating Model: ECOST 1. Golden, CO: Solar Energy Research Institute.

Chapter 12

Gas Conditioning

R. Bennett
Mittlehouser Corp.

12.1 INTRODUCTION

Biomass gasifiers of current design produce a raw gas consisting chiefly of carbon monoxide and hydrogen, with minor amounts of methane, higher molecular weight hydrocarbons, sulfur compounds, tars, and oil. When burned as a boiler fuel, the raw gas requires little or no cleanup. If the product gas is intended for use as a synthesis gas (for example, in the production of methanol) it will require substantial cleanup. Further improvements in gasifier design may reduce and even eliminate the tar and oil problems associated with gas cleanup. These improvements may also greatly reduce the formation of undesirable hydrocarbons. However, system studies for near-term commercial production of synthetic fuels from biomass must be based on current technology.

Many options are available for cleaning raw gas from currently available biomass gasifiers. However, before detailed designs of commercial facilities can be made, some preliminary review of available technology is needed to evaluate methods for the separation of tars, oil mists, and undesirable hydrocarbons from the raw gas and to examine the costs and requirements of each technology.

A study on gas conditioning was performed by the Mittelhauser Corporation under contract to SERI. Specific objectives of the study were to:

- Survey the technology available for eliminating oil mists from a hot gas stream.
- Estimate tolerances of commercial methanol synthesis catalysts for CH_4, C_2H_2, C_2H_4, C_2H_6, H_2S, and Cl.
- Survey the technology available for separating CH_4 and higher hydrocarbons from a CO-H_2 mixture.
- Estimate incremental costs of upgrading a 500 ton/day raw pyrolysis-gas stream to a synthesis gas for a methanol plant; specifically, a Purox gasifier, ICI methanol process.

The study was based on the following general assumptions:

- The ambient atmospheric pressure was assumed to be 14.7 psia.
- The overall methanol synthesis facility was assumed to be a grass-roots plant in the northeastern United States. The gas cleanup facilities to be studied were assumed to be part of the larger complex; thus, electric power, cooling water, and steam would be available as needed by the cleanup facilities. It was also assumed that concentrated waste hydrocarbon gases could be used as fuel in the plant's auxiliary steam generation system. Wastewater from the gas cooling step was assumed to require treatment as part of the gas cleanup system.
- All costs were taken on a first quarter 1979 basis.
- Synthesis gas compression facilities were specifically excluded from the scope of the study. A qualitative assessment was made of the effect of different gas separation schemes and synthesis gas compression requirements.

- The methanol synthesis loop was assumed to be an Imperial Chemical Industries 50-atm process for manufacturing crude methanol.
- Generally, all plant units processing the main synthesis gas stream were designed to operate on a 90% stream factor. Spare parts were included to ensure this and to allow for on-line maintenance that could not be accomplished in a normal, once-per-year "turnaround."

Attention was focused on review and definition of the available technologies for oil mist elimination, gas separation, and gas cleanup. The designs developed here are felt to be reasonable, workable, and generally representative of the capital and operating requirements associated with the function each system performs. However, these designs are not optimized and should not be regarded as such. It is not possible to optimize a given section of a plant without considering fully all of the physical and economic interactions between that section and the remainder of the plant. Such considerations were outside the scope and time frame of the study, and were not made.

The study was based on 500 short tons per day of raw gas from a Union Carbide Corporation Purox gasifier fed with wood waste. The composition of the wood waste was based on that used in a study by Raphael Katzen Associates (1975). Yield data were based on published studies on the Purox process by Ralph M. Parsons Company (1978) and the City of Seattle (Mathematical Sciences 1974). These data were augmented by telephone conversations with Union Carbide technical personnel responsible for the Purox process.

Based on the sources cited above, the following assumptions were made:

- Oil yield on wood waste was assumed to be twice as high as on municipal refuse. The composition was assumed to be 94.6% carbon by weight; the balance hydrogen (Mathematical Sciences 1974).

  Based on conversations with Union Carbide personnel, it was assumed that the oil was entrained in the raw gas as droplets from 1 to 10 microns in diameter. No specific size was available; it was assumed that 99.99% of the droplets were equal to or less than 10 microns diameter. The gas was assumed to be available at 400 F, 3 psig, as this was consistent with pressures and temperatures found in the literature (Ralph M. Parsons 1978).
- The wood waste was arbitrarily assumed to contain 0.1 wt % sulfur. This results in a quenched gas sulfur content comparable to that from municipal refuse. According to information from Union Carbide personnel, this is a conservatively high estimate.
- The yield of water-soluble organics was assumed to be the same as for municipal refuse. The composition was taken from the Ralph M. Parsons (1978) study.
- The moisture content of the feed was assumed to be 25 wt %, the same as in the Raphael Katzen (1975) study.
- Oil recovered in the gas cooling section was assumed to be recycled to extinction in the gasifier.
- Apart from water-soluble organics, sulfur, and oil, the raw gas yield was assumed to be the same as in Parsons (1978).

Tables 12-1 and 12-2 show the assumed rate and composition of the biomass feed and the effluent raw gas, respectively. This raw gas was used as the basis for all work done in this study.

12.2 OIL MIST ELIMINATION

A brief review was made of the available technologies for removing oil droplets from a hot gas stream. For each technology a short description was prepared and currently commercial applications, expected efficiencies, advantages and disadvantages, rough utility requirements, and appropriate costs were reviewed and tabulated. The major effort was expended on the review of the applicable devices for oil mist removal currently on the market.

Table 12-1. GASIFIER FEED (WOOD WASTE) COMPOSITION

Component	Feed Rate (lb/h)	Weight Fraction
Carbon	11,509.17	0.38049
Hydrogen	1,427.42	0.04719
Oxygen	9,017.00	0.29810
Nitrogen	22.69	0.00075
Sulfur	30.25	0.00100
Moisture	7,562.07	0.25000
Ash	679.68	0.02247
	30,248.28	1.00000

Table 12-2. TYPICAL GASIFIER YIELD, INCLUDING EFFECT OF OIL RECYCLE
(Basis for Gas Conditioning Discussions)

Component	Mol Wt	Yield lb/h	Weight Fraction	lb/mol/h	Dry, Oil-Free Mole Fraction
H_2	2.016	526.62	0.012639	261.22	0.236668
CO	28.01	12,197.01	0.292726	435.45	0.394522
CO_2	44.01	11,498.28	0.275957	261.27	0.236713
CH_4	16.043	978.23	0.023477	60.98	0.055249
C_2H_2	26.04	198.43	0.004762	7.62	0.006904
C_2H_4	28.05	641.26	0.015390	22.86	0.020711
C_2H_6	30.07	98.61	0.002367	3.28	0.002972
C_3H_6	42.08	91.65	0.002200	2.18	0.001975
C_3H_8	44.09	18.75	0.000450	0.43	0.000390
C_4H_8	56.10	203.27	0.004878	3.62	0.003280
C_4H_{10}	58.12	105.26	0.002526	1.81	0.001640
C_5H_{12}	72.15	835.15	0.020043	11.58	0.010492
N_2+Ar	28.02	385.67	0.009256	13.76	0.012467
NH_3	17.03	25.11	0.000603	1.47	0.001332
H_2S	34.08	25.71	0.000617	0.75	0.000680
Acetic acid	60.05	174.84	0.004196	2.91	0.002636
Methanol	32.04	216.28	0.005191	6.75	0.006116
Ethanol	46.07	84.09	0.002018	1.83	0.001658
Acetone	58.08	84.09	0.002018	1.45	0.001314
MEK	72.10	16.94	0.000407	0.23	0.000208
Propionic acid	74.08	67.15	0.001612	0.91	0.000824
Butyric acid	88.10	16.94	0.000407	0.19	0.000172
Furfural	96.08	84.09	0.002018	0.88	0.000797
Phenol	94.11	16.94	0.000407	0.18	0.000163
Benzene	78.12	9.98	0.000240	0.13	0.000118
Total dry oil-free		28,600.35	0.686403	1,103.74	1.000000
Oil		3,364.82	0.080755		
Total dry		31,965.17	0.767158		
Water vapor	18.016	9,701.83	0.232842	538.51	0.487896
Total		41,667.00	1.000000		1.487896

There are five basic mechanisms for collection of oil droplets from a flowing gas stream:

- Gravitational sedimentation. This will be of little consequence for droplets in the 1-10 micron size range.
- Inertial impaction and interception. This is a very effective method for mist removal that relies on multiple changes of direction of gas flow to cause collisions between droplets and a solid barrier.
- Centrifugal deposition. This mechanism relies on imparting a circular vortex motion to the gas stream, causing oil droplets to be hurled outward against a wall by centrifugal force. This is not particularly effective for droplets smaller than 5 microns in diameter.
- Electrostatic precipitation. If an electrostatic charge is induced on the droplets, they can be removed from the gas stream by a potential gradient. This mechanism is effective on all droplet diameters and can achieve a high collection efficiency.
- Droplet growth. The enlargement of a droplet by condensation on it of additional liquid, or by collision with other droplets, allows the droplet to be more easily collected by centrifugal force or inertial impaction.

The following paragraphs examine the ways in which different scrubbing media and equipment might utilize these five mechanisms to remove oil mists from biomass pyrolysis gas.

12.2.1 Scrubbing Media

For removing oil droplets from the raw gas stream either oil or water or a combination of the two can be used as scrubbing media. Wet scrubbers use a liquid stream, either water or oil, to remove small liquid hydrocarbon droplets from a gas stream. The liquid droplets are captured by the liquid or by the scrubber mechanical structure and then washed off by the liquid. Table 12-3 outlines the salient features of oil and water scrubbing.

Table 12-3. FEATURES OF OIL AND WATER SCRUBBING

	Oil Scrubbing	Water Scrubbing
Disposal of purge liquid	Thermal oxidation with heat recovery or recycle after fractionation	Water treatment before discharge
Oil droplet removal considerations	Entrainment and saturation of gas stream with oil	Oil entrainment from H_2O
Makeup quality	Oil might require fractionation to achieve proper boiling range material	Condensate quality water
Metallurgy	Carbon steel equipment is probably adequate if no water condensation occurs	Water will be acidic due to contaminants in gas: stainless steel scrubber required
Source of scrubbing medium	Available if oil produced by process can be used; otherwise must be imported	Readily available

12.2.1.1 Oil Scrubbing

If a multicomponent oil is used as the scrubbing medium, the lower-boiling components of the oil tend to saturate the gas stream at the operating temperature and pressure of the scrubbing device. A small amount of oil is unavoidably entrained in the gas stream. These two characteristics of oil as a scrubbing medium significantly reduce its capability to remove oil droplets from a raw gas stream. A purge stream equal to the quantity of oil removal from the gas stream must be taken out of the scrubbing system to maintain a constant oil inventory. This purge stream can be fractionated to remove contaminants and recycled as scrubbing oil makeup or burned as a source of heat energy.

The scrubbing oil has the composition and physical properties of the oil removed from the Purox gas stream. To decide whether this oil is suitable for scrubbing, more physical property data on this oil are required. Ideally, the oil should have low viscosity at the system pumping temperature and a high boiling point to minimize vaporization losses. If the oil collected from the gas is not suitable as a scrubbing medium and cannot be upgraded by fractionation, scrubbing oil must be imported.

The oil scrubbing system can be made of carbon steel as long as there is no water condensation during removal of the oil droplets. If water condenses, it will collect the acidic components of the gas stream and corrode the carbon steel.

12.2.1.2 Water Scrubbing

If water is used as the scrubbing medium, the gas is saturated with water at the outlet temperature and pressure of the scrubbing system. Generally, water condenses from the gas stream and must be purged from the scrubbing system. Oil droplets removed from the gas stream by the water must be separated from the water phase. Furthermore, all the water-soluble components in the raw gas stream are present in the water. Consequently, the purge water would require treatment before discharge to make it environmentally acceptable.

Oil captured by the water in the scrubbing system may be reentrained in the gas stream. For example, in a plate column the raw gas may pick up the oil floating on the surface of the water.

The scrubbing water probably is corrosive to carbon steel due to the presence of organic and inorganic water-soluble acids. Consequently, the water scrubbing system might have to be stainless steel unless the surface were protected by passivation with H_2S or were coated with a corrosion-resistant material.

12.2.2 Oil Mist Elimination Devices

12.2.2.1 Plate Scrubbers

A plate scrubber is a vertical tower with one or more horizontal trays mounted on its inside surface. Gas enters at the bottom of the tower and must pass through perforations, valves, slots, or other openings in each plate before leaving the top of the scrubber. The scrubbing medium is introduced at the top plate and flows over each plate as it moves downward. In some designs, the gas passes through holes covered with caps. The caps act as impingement plates and are set below the liquid level on the plates. At low gas velocities, lightweight caps on alternate rows rise first while the heavyweight caps in the other rows remain in the closed position. All the caps are finally opened when the gas flow reaches the design condition.

The liquid flows across each tray and is kept in a froth by the gas, which exits each cap at high velocity. Fine droplets of liquid are generated that will absorb impurities from the gas stream. Also, adiabatic cooling and condensation or humidification of the gas stream occurs. Before the gas stream leaves the scrubber it passes through a mist eliminator to remove liquid droplets.

In oil mist separation devices that use wet scrubbing, collection efficiency increases with pressure drop. For plate scrubbers, gas pressure drops of as much as 6 to 15 in. of water can be achieved. Approximately 80% of droplets of 5-micron and larger diameters can be removed with a pressure drop of 10 in. of water. The oil droplet collection efficiency is set by the performance of the mist eliminators. If the water is used as the scrubbing medium, some of the oil removed from the gas is reentrained as the gas passes through the oil-water mixture. If oil is used as the scrubbing medium, product oil is removed, but scrubber oil is entrained and vaporized in the gas stream.

12.2.2.2 Packed Bed Scrubbers

Scrubbers contain packing such as rings or saddles. The gas-liquid contact may be cocurrent, countercurrent, or cross flow. The primary collection mechanisms in packed beds are inertial impaction and centrifugal deposition with subsequent drainage.

Collection efficiency for droplets larger than 0.3 micron rises as packing size decreases. Approximately 50% of 1.5-micron droplets can be removed by a column packed with 1-in. Berl saddles or Raschig rings. A 1/2-in. packing can achieve 50% removal of 0.7-micron droplets at a gas velocity of 30 fps.

Packed scrubbers are subject to plugging but can be shut down periodically to change the packing. Temperature limitations are of special importance when plastic packing is used, and corrosion can result when metallic packing is used. Packed columns have the same reentrainment problems as those described for plate columns.

12.2.2.3 Spray Scrubbers

A spray scrubber collects oil droplets or liquid droplets that have been atomized by spray nozzles. The properties of the droplets are determined by the configuration of the nozzle, the liquid to be atomized, and the pressure at the nozzle. Sprays leaving the nozzle are directed into a chamber shaped so that the gas passes through the atomized droplets. Horizontal and vertical gas flow paths have been used, as well as spray trajectories either cocurrent, countercurrent, or crossflow to the gas. If the tower is vertical, the gas flow must be slower than the terminal settling velocity of the droplets to prevent massive droplet entrainment.

Droplet collection in these units results from inertial impaction on the droplets generated by the spray. Droplet removal efficiency is a complex function of droplet size, gas velocity, liquid-to-gas ratio, and droplet trajectories. The optimal droplet diameter varies with fluid flow parameters.

Spray scrubbers utilizing gravitational settling can remove about 50% of 2-micron particles at moderate liquid-to-gas ratios. Gas phase pressure drop is usually very low. Spray scrubbers are almost immune to plugging on the gas flow side but are subject to severe problems on the liquid side. The circulating scrubber medium can erode and corrode nozzles, pumps, and piping. Nozzles are subject to plugging with circulating solids. The liquid-to-gas ratio depends on the removal efficiency required but can run as high as 30 to 100 gal per 1000 ft^3 of gas treated: thus, sprays generate a heavy loading of liquid, which must be collected.

12.2.2.4 Venturi Scrubbers

A venturi scrubber uses high gas velocities (200 to 400 fps) to atomize liquid into droplets and then accelerate the droplets to promote droplet collection. Liquid may be introduced in several ways without affecting collection efficiency. Usually the liquid is introduced at the entrance to the throat through several straight pipe nozzles directed radially inward.

Oil mist removal from the gas is achieved by coalescence with the generated droplets. Removal efficiency increases with throat velocity and liquid-to-gas ratios.

Venturi scrubbers are the smallest and simplest of all scrubbers. They do not plug easily but are subject to corrosion due to the high throat velocity. They can be built with adjustable throat openings to permit variation in pressure drop and collection efficiency. Liquid-to-gas ratios ranging from 5 to 20 gal per 1000 ft^3 have been used. It is important to note that all of the scrubbing liquid is entrained in the gas and must be removed by subsequent separation.

Ejector venturis are spray devices in which a high-pressure spray is used both to collect the droplets and to move the gas. High relative velocity between the liquid and the gas helps droplet separation.

12.2.2.5 Wet Scrubber Combinations

Combinations of wet scrubbers can be used for oil droplet removal. For example, a venturi scrubber can be used to remove the bulk of the oil droplets, followed by a plate scrubber to separate the entrained liquid from the gas.

12.2.2.6 Mist Eliminators

Beds of fibers called mist eliminators can be used in various configurations for collecting oil droplets. The fibers can be made from plastic, spun glass, fiberglass, or steel. Fibrous packings usually have a very high void fraction ranging from 97% to 99%. The fibers should be small in diameter for efficient operation but strong enough to support collected droplets without matting. A cocurrent, countercurrent, or cross flow arrangement can be used to flush any collected material from the fiber.

Collection in a mist eliminator is by inertial impaction as the gas flows through the fibers. Efficiency increases as fiber diameter decreases and as the gas velocity increases. Approximately 50% of 5- to 10-micron droplets can be removed by a knitted wire mesh made of 0.11-in. diameter wire.

Mist eliminators are susceptible to plugging, and they can be impractical where scaling persists. They also are especially sensitive to chemical, mechanical, and thermal attack.

12.2.2.7 Wet Electrostatic Precipitation (ESP)

Wet electrostatic precipitators operate by electrostatically charging the oil droplets as they pass through a corona developed by a negatively charged electrode. Each droplet in the gas stream is attracted to a grounded collection plate or to the inside walls of the pipes through which the gas flows. After collection, the liquid is washed down by additional liquid flowing countercurrently to the gas. The wet ESP is very efficient for collecting very small, submicron-sized droplets; electric power usage is negligible, and pressure drop across the ESP is very low, usually less than 1 in. of water.

Droplet collection is extremely efficient; essentially all droplets larger than 1-micron diameter can be collected. Disadvantages of wet ESP are high capital cost, poor performance when flow variations are encountered, and high maintenance requirements. For a wet ESP to operate satisfactorily, the gas must be cooled from 400 F to about 150 F. This often requires a wet scrubber ahead of the ESP to saturate the gas, with the ESP then used as a final cleanup device.

12.2.3 Similar Applications

The use of scrubbers to control various air pollution sources was studied in a survey carried out as part of the work reported in the Scrubber Handbook (APT, Inc. 1972). The only wet scrubbers reported to be used for oil mist removal are packed bed, mist eliminators, and spray towers. Wet ESP is used to remove entrained coal tar and coal tar mist from coke oven gas (COG) in COG processing plants. Table 12-4 summarizes the application of the various scrubbers.

Table 12-4. SURVEY OF SCRUBBER APPLICATIONS IN A VARIETY OF INSTALLATIONS

	Scrubber Type				
	Plate[a]	Packed	Mist Eliminators	Spray	Venturi
Calcining	6 (1)[b]	2 (1)	- (0)	13 (5)	21 (23)
Combustion	17 (3)	- (0)	- (0)	5 (2)	2 (2)
Crushing	6 (1)	- (0)	- (0)	- (0)	- (0)
Drying	39 (7)	- (0)	- (0)	10 (4)	18 (19)
Gas Removal	17 (3)	72 (33)	40 (2)	45 (18)	9 (10)
Liquid Mist Recovery	0 (0)	24 (11)	60 (3)	7 (3)	- (0)
Smelting	17 (3)	2 (1)	- (0)	20 (3)	50 (54)

[a]The table should be read vertically. For example, 39% of plate-type scrubbers are used to control discharges from drying processes.

[b]The numbers in parentheses refer to the number of separators reporting information to the survey.

12.2.4 Summary of Findings

Table 12-5 summarizes the findings of this survey. The major operating costs of wet scrubbers are power requirements for circulation of the scrubbing medium. For mist eliminators and ESP, power requirements are minimal. Capital costs of wet scrubbers can vary widely depending on design and operating conditions of the devices surveyed; mist eliminators are generally least expensive and ESP the most expensive.

12.3 METHANOL CATALYST TOLERANCE

Available information was reviewed and suppliers of commercial methanol synthesis catalysis were contacted by telephone to determine the catalyst tolerances to impurities found in the raw pyrolysis gas. A table was prepared containing the catalyst supplier, synthesis process, catalyst type, specific poison, and maximum recommended concentration. In addition to hydrocarbons, H_2S, COS, chlorides, nitrogen compounds, and HCN were investigated.

The study of methanol catalyst tolerances produced surprisingly sparse results. This is due at least partly to the fact that most manufacturers have little or no operating experience with synthesis gases derived from feedstocks such as coal, municipal solid waste, or biomass. Most present commercial methanol processes are based on a synthesis gas produced by steam re-forming natural gas, LPG, or naphtha. Therefore, the only hydrocarbon present to an appreciable extent is methane. Nitrogen is present primarily as N_2.

Little information is available about the potential catalyst poisoning capabilities of olefins, acetylene, HCN, NH_3, and NO_x. Available information is usually expressed in qualitative terms. The three catalyst suppliers contacted indicated that little is known about synthesis catalyst poisons in synthesis gases produced by the gasification of municipal solid waste, biomass, or coal.

Table 12-5. SUMMARY OF MIST ELIMINATION SURVEY RESULTS

Device	Plate Scrubber	Packed Scrubber	Spray Scrubber	Venturi	Mist Eliminator	Wet ESP
Pressure drop (in. H_2O)	10	0.24-0.5[d]	1 - 3	10 - 30	1 - 3	1
Droplet size (microns) at percentage removal	5 at 80	1.5 at 50	2 at 50	5	5 - 10 at 50	1 at 100
Circulation (gpm/1000 acfm)	2-50	2-50	30-100	5-20	3-5[c]	Variable
Capital cost	(a)					(b)
Operating cost	Power	Power	Power	Power	Minimal	Minimal
Maintenance cost	Nominal	Nominal	Nominal	Nominal	Nominal	High

[a]Plate scrubber is the most expensive wet scrubber; venturi is the least expensive wet scrubber.

[b]Wet ESP is the most expensive of all devices considered.

[c]Three to five gpm/ft^2 of mist eliminator cross-sectional area.

[d]Pressure drop per foot of packed height.

The catalyst suppliers were concerned primarily with sulfur and chlorine. When the sulfur and chlorine levels are lower than 50 ppm each, zinc oxide provides a satisfactory means of desulfurization and dechlorination. The literature commonly refers to complete sulfur removal by means of zinc oxide guard beds. Most methanol synthesis plants use zinc oxide as the final desulfurization step, just prior to steam re-forming. However, activated carbon must be used when acetylene is present, because the high temperatures (500-700 F) required for proper use of zinc oxide cause polymerization of the acetylene and plugging of the beds.

Methane and heavier paraffin hydrocarbons, together with nitrogen and water vapor, are inert. However, their presence in significant quantities reduces the conversion of CO and H_2 to methanol by lowering the partial pressures of these reactants.

Table 12-6 summarizes the information obtained from the study.

Table 12-6. METHANOL SYNTHESIS CATALYST POISON TOLERANCE

	Information Source					
Information	Supplier: United Catalysts Louisville, KY	Supplier: Haldor-Topsoe Houston, TX	Supplier: Katalco, Inc. Oak Brook, IL	Literature[a]		Literature[b]
Process	ICI (50 atm)	Haldor-Topsoe (50-150 atm)	ICI (50 atm)	Not specified		300-400 atm
Catalyst Type	C79-4 Cu-Zn Base	Cu-Zn-Cr Oxides		Cu-Based	Zn-Based	ZnO 570-750 F
Component:						
C_2H_2	Possibly poisonous	Unknown	Apparently not a problem		not poisonous in "small quantities"	Poison at more than 3 ppm
C_2H_4, higher olefins	Possibly poisonous	Unknown	Apparently not a problem			
CH_4, C_2H_6, higher paraffins	Inert	Inert	Inert			
Sulfur (as H_2S, COS, CS_2)	0.1 lb sulfur per ft^3 of catalyst[c]	Poison at more than 0.03 ppm	Poison at more than 0.5 ppm	Poison at 0.7 ppm	Reversible Poison	Poison at more than 3 ppm
Chlorides	0.035 lb per ft^3 catalyst[c]	Poison at more than 0.03 ppm	Poison at more than 0.2 ppm			
NH_3	Possible poison with liquid H_2O present	Unknown				
NO_x	Unknown	Possible poison			May cause amine formation	
HCN	Unknown	Possible poison	Possible poison			
Fe, Ni						Form carbonyls with CO, causing CH_4 formation over the catalyst

[a]Natta 1955.

[b]*Manufacturing Chemist* 1978.

[c]Catalyst is spent when this level is reached in the upper half of the bed.

12.4 GAS SEPARATION TECHNOLOGY

Using as a basis the Purox raw gas shown in Table 12-2, a rough estimate was made of the rate and composition of the gas, leaving out the oil mist elimination step. The primary objective of the gas separation study was to review various methods of removing unsaturated hydrocarbons from the synthesis gas. Our early review of available data indicated that paraffin hydrocarbons are not poisons to the catalyst. However, technology was incorporated in each design that would remove most of these paraffins from the synthesis gas to reduce the purge gas requirements in the methanol synthesis process.

Removal of sulfur compounds and chlorine from the synthesis gas is accomplished in facilities separate from the gas separation units. The design of these facilities is discussed in Section 12.5.2.2.

The three separation technologies reviewed were hydrogenation, re-forming, and cryogenic separation. Process alternatives to these technologies were examined qualitatively and are discussed in Section 12.5.2.9.

12.4.1 Hydrogenation

12.4.1.1 Design Basis

A block flow diagram of hydrogenation technology is shown in Fig. 12-1. The hydrogenation scheme consists of two principal sections: the first provides for the hydrogenation of olefins by the hydrogen in the clean raw gas, and the second is designed to remove CH_4 and heavier paraffins from the gas by oil absorption.

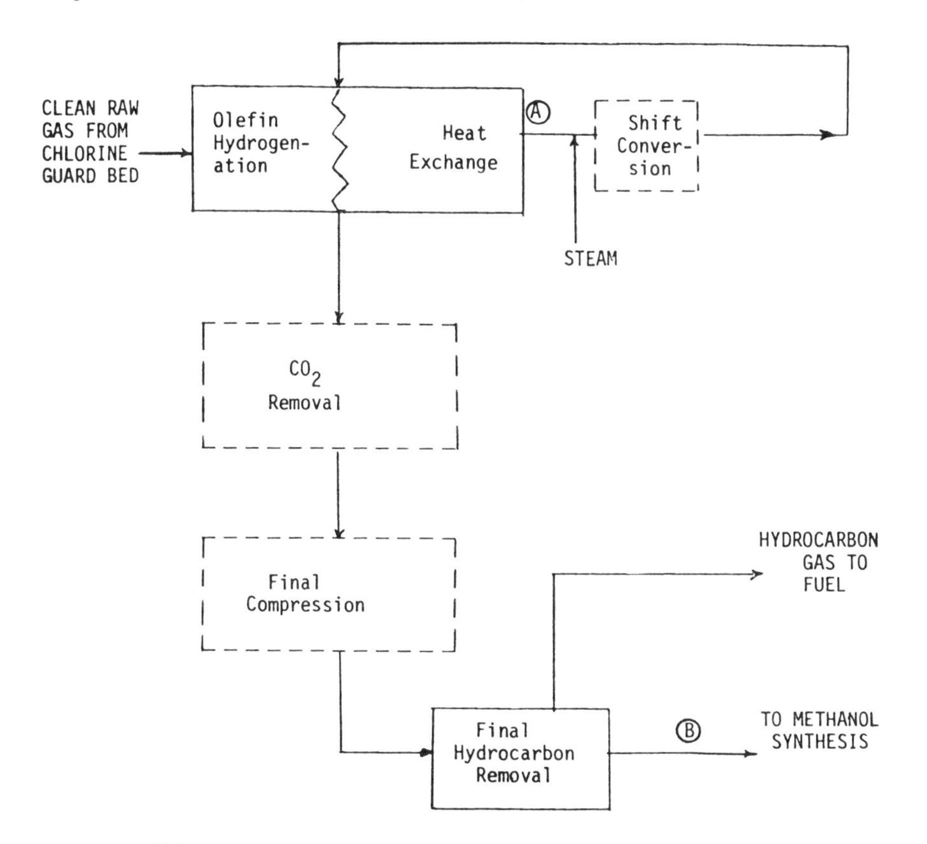

Notes:
Units in dotted outline are not part of the separation scheme; gas at point A contains no more than 100 ppmv olefins, wet basis; and gas at point B contains no more than 3.8 mole percent hydrocarbons, wet basis.

Figure 12-1. Hydrogenation Block Flow Diagram

Katalco Corporation is a leading supplier of methanol synthesis catalyst for the ICI 50-atm process. Their technical representatives recommended catalysts and process conditions for the hydrogenation section. A two-stage hydrogenation unit was selected, due to its reliability and ease of operation. The unit is designed to reduce the olefin content of the exit gas to 100 ppmv. Katalco supplied catalyst bed volumes, estimated reactor inlet temperature and temperature rises across each stage of hydrogenation, and gave catalyst prices and estimated catalyst lives. For the oil absorption section, a rough design was prepared based on published literature (Sherwood and Pigford 1952).

12.4.1.2 Process Description

In the following discussion, reference is made to Fig. 12-2, Process Flow Diagram for Gas Separation—Hydrogenation, and the associated material balance shown in Table 12-7. The battery limit of the hydrogenation technology is the outlet of the chlorine guard beds. At this point the sulfur and chlorine have essentially been completely removed from the gas, which is at 150 F and 121 psig.

The poison-free gas, containing about 3 vol % unsaturated hydrocarbons, is heated and hydrogenated in two steps. In the first step, the raw gas is heated to 300 F against the partially cooled shift converter effluent gas in exchanger E-9 and passed over a bed of palladium-on-allumina catalyst contained in reactors. The palladium catalyst selectively hydrogenates acetylene to ethylene. This prevents polymerization of the acetylene at the higher temperatures used for general olefin hydrogenation. The exit temperature of this bed, about 365 F, is higher than normally employed. However, the CO concentration in the gas tends to moderate the reaction. The hydrogen partial pressure in this reactor is about 31 psia, which is sufficient to carry the hydrogenation essentially to completion.

In the second step, the acetylene hydrogenation reactor effluent is heated from about 365 F to 550 F against higher temperature shift effluent in exchanger E-1. The gas is then passed over a bed of nickel-molybdenum catalyst in reactor R-4A&B. In this bed, hydrogenation of the remaining olefins takes place. The palladium catalyst used in the first hydrogenation step is poisoned by sulfur; therefore, sulfur has been removed from the gas prior to that step. However, in the absence of sulfur, cobalt-molybdenum catalyst, which ordinarily would be used in hydrogenating olefins, promotes the methanation reaction:

$$CO + 3H_2 \rightarrow CH_4 + H_2O$$

To prevent this loss of synthesis gas, the nickel-molybdenum catalyst has been used for second-stage hydrogenation.

Hydrogenated synthesis gas exits reactor R-4A&B at a temperature of approximately 750 F. Attemperated steam is then added as required for shift conversion, and the gas passes to the shift converter, which is not considered part of the gas separation scheme.

Both R-3 and R-4 are provided with full-capacity spares. In case excessive olefin breakthrough occurs, plugging of the catalyst by polymerized acetylene or poisoning of R-3 by sulfur breakthrough will result and each bed can be taken off line and the spare bed put in service.

Shift converter effluent is cooled in exchangers E-7, E-1, and E-9 by generating 665 psig saturated steam and preheating the hydrogenation reactor feed streams. Between E-1 and E-9 the shift effluent is cooled in exchanger E-8 by generating 50 psig saturated steam. This is done to keep the tube wall temperatures in E-9 sufficiently low to prevent polymerization of the olefins in the feed to R-3A&B. Final cooling is done in air cooled exchanger E-10, with condensate separation in vessel V-5. The cooled gas at 200 F then passes into the CO_2 removal unit, which is not considered part of the gas separation scheme.

The final part of the hydrogenation-gas separation process is an oil absorption unit. This unit follows final compression of the synthesis gas to about 750 psig. After being cooled to 100 F, the gas passes through the absorber, in which it flows countercurrently to a

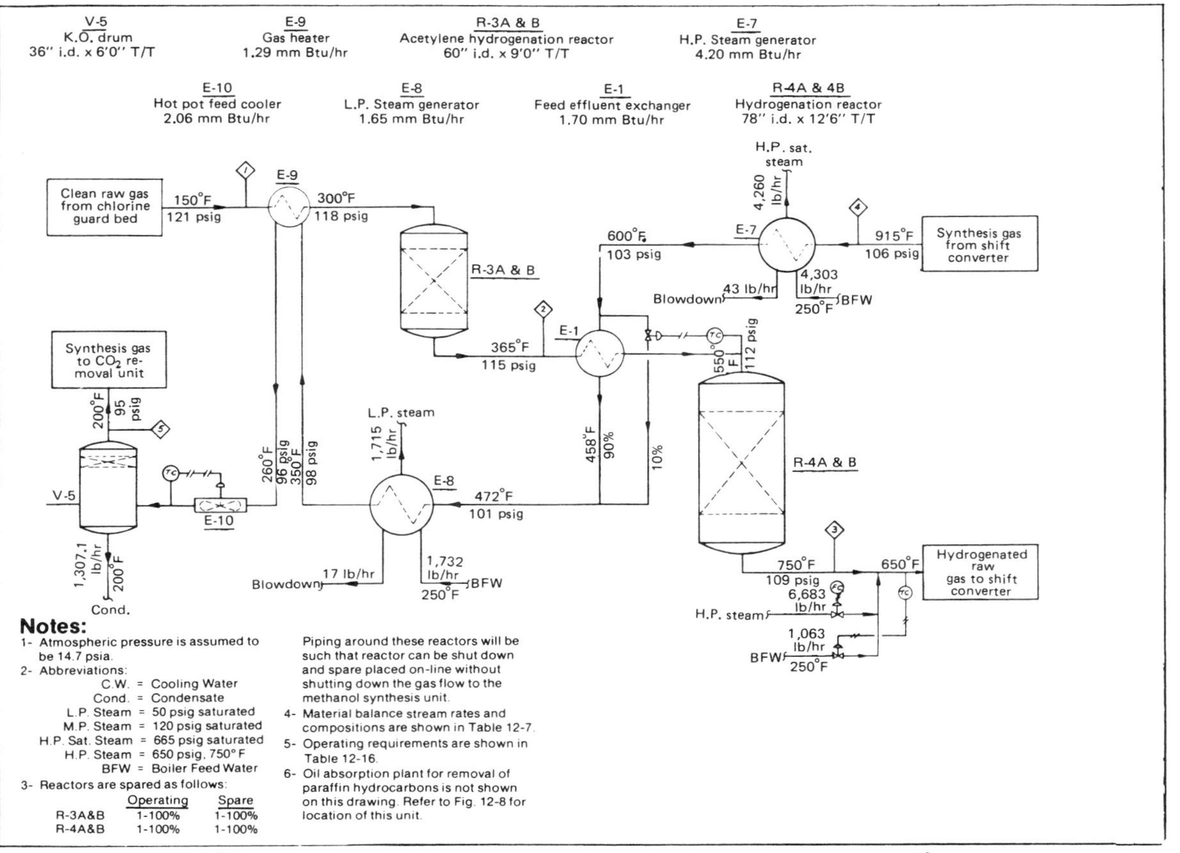

Notes:

1- Atmospheric pressure is assumed to be 14.7 psia.

2- Abbreviations:

C.W. = Cooling Water
Cond. = Condensate
L.P. Steam = 50 psig saturated
M.P. Steam = 120 psig saturated
H.P. Sat. Steam = 665 psig saturated
H.P. Steam = 650 psig, 750°F
BFW = Boiler Feed Water

3- Reactors are spared as follows:

	Operating	Spare
R-3A&B	1-100%	1-100%
R-4A&B	1-100%	1-100%

Piping around these reactors will be such that reactor can be shut down and spare placed on-line without shutting down the gas flow to the methanol synthesis unit.

4- Material balance stream rates and compositions are shown in Table 12-7.

5- Operating requirements are shown in Table 12-16.

6- Oil absorption plant for removal of paraffin hydrocarbons is not shown on this drawing. Refer to Fig. 12-8 for location of this unit.

Figure 12-2. Process Flow Diagram for Gas Separation—Hydrogenation

Table 12-7. **MATERIAL BALANCE FOR GAS SEPARATION—HYDROGENATION**
(Refer to Fig. 12-2)

Component (lb-mol/h)	Stream Number						
	1 Clean Gas from Chlorine Guard Bed	2 Acetylene Hydrogenation; Reactor Effluent	3 Hydrogenation Reactor Effluent	4 Shift Converter Effluent	5 Synthesis Gas to CO_2 Removal Unit	6 Synthesis Gas to Methanol Synthesis	7 Hydrocarbon Gas to Fuel
H_2	261.22	253.60	217.32	440.83	440.83	440.83	
CO	435.45	435.45	435.45	211.94	211.94	211.94	
CO_2	261.27	261.27	261.27	484.78	484.78	34.36	
CH_4	60.98	60.98	60.98	60.98	60.98	27.56	33.42
C_2H_2	7.62						
C_2H_4	22.86	30.48					
C_2H_6	3.28	3.28	33.76	33.76	33.76	0.44	33.32
C_3H_6	2.18	2.18					
C_3H_8	0.43	0.43	2.61	2.61	2.61		2.61
C_4H_8	3.62	3.62					
C_4H_{10}	1.81	1.81	5.43	5.43	5.43		5.43
C_5H_{12}	11.58	11.58	11.58	11.58	11.58		11.58
N_2 + Ar	13.76	13.76	13.76	13.76	13.76	13.76	
H_2S	>0.5 ppm[a]						
H_2O	14.66	14.66	14.66	221.13	148.58	1.03	
TOTAL	1,100.72	1,093.10	1,056.82	1,486.80	1,414.25	729.92	86.36

[a]Maximum value; less than 0.1 ppm expected.

stream of absorption oil of approximately 161 molecular weight. The oil removes paraffinic hydrocarbons from the gas. Rich oil from the base of the absorber is pumped to a steam stripper where the absorbed gases are distilled overhead and sent to the fuel gas system. Regenerated lean oil is pumped back to the absorber. The treated gas from the oil absorption unit contains approximately 4 mole % CH_4 and heavier hydrocarbons. It goes directly into methanol synthesis. The theoretical methanol make from this synthesis gas is 162,970 lb/day.

An overall rough energy balance for the hydrogenation technology is presented in Table 12-8. This balance excludes the oil absorption unit, as no energy balance was made for that unit.

Table 12-8. ENERGY BALANCE FOR HYDROGENATION
(10^6 Btu/h)[a]

Inputs	
Raw gas	156.37
BFW import	1.15
Steam import[b]	9.22
Electric power	0.06
Outputs	
Shifted gas to CO_2 removal	157.35
Steam export	6.97
Cooling losses	2.06
Blowdown	0.02
Condensate export	0.18
	166.58

[a]Energy quantities include sensible enthalpies and higher heating values relative to 60 F, 1 atm pressure with water in the liquid state. Oil absorption is not included in the energy balance.

[b]Steam includes attemperation water.

12.4.2 Re-forming

12.4.2.1 Design Basis

The re-forming technology for hydrocarbon separation is shown schematically in Fig. 12-3. KTI, Inc., a leading supplier of steam-hydrocarbon re-forming furnaces, prepared a process design package for the re-forming step shown in the figure. To prevent cracking of olefins in the re-forming furnace, with subsequent carbon laydown, KTI recommended the hydrogenation of all olefinic compounds upstream of the re-former. Therefore, the hydrogenation unit described in Section 12.4.1 was also incorporated in this gas separation scheme. The re-former converts only paraffin hydrocarbons to CO and H_2. Furthermore, KTI recommended that the shift converter be placed between the olefin hydrogenation section and the re-former. This was done to reduce the concentrations of CO entering the re-former and to prevent cracking of CO with subsequent carbon laydown.

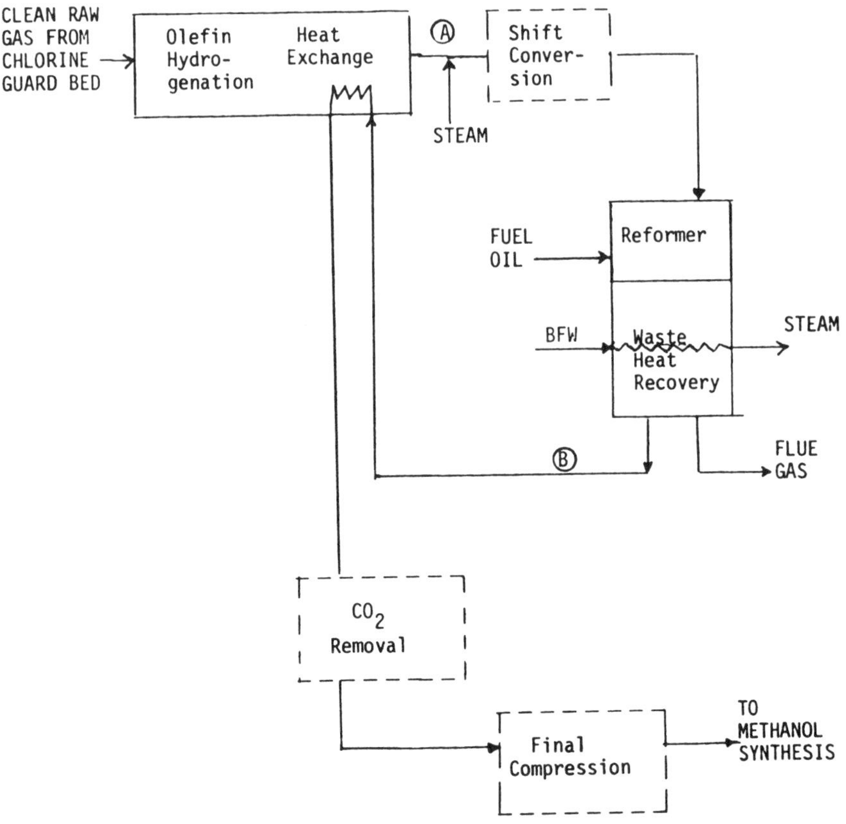

Notes:
Units in dotted outline are not part of the separation scheme, gas at point A contains no more than 100 ppmv olefins, wet basis; and gas at point B contains no more than 0.05 mole percent hydrocarbons, wet basis.

Figure 12-3. Re-forming Block Flow Diagram

KTI provided heat and material balance data for the re-former, shift converter, and waste heat recovery sections. Although the shift converter is not included in the "re-forming" scheme, knowledge of process conditions around it is required to specify steam requirements and to design the heat exchange trains for the scheme.

12.4.2.2 Process Description

Reference is made in the following discussion to Fig. 12-4, Gas Separation—Re-forming, and to the associated material balance shown in Table 12-9. The battery limits of re-forming technology are the same as for the hydrogenation technology just described.

Re-forming uses high temperatures and catalytic activity to crack higher paraffin hydrocarbons to CH_4 and to re-form the methane to CO and H_2:

$$C_2H_6 + H_2 \rightarrow 2CH_4 , \tag{12-2}$$

$$CH_4 + H_2O \rightarrow CO + 3H_2 . \tag{12-3}$$

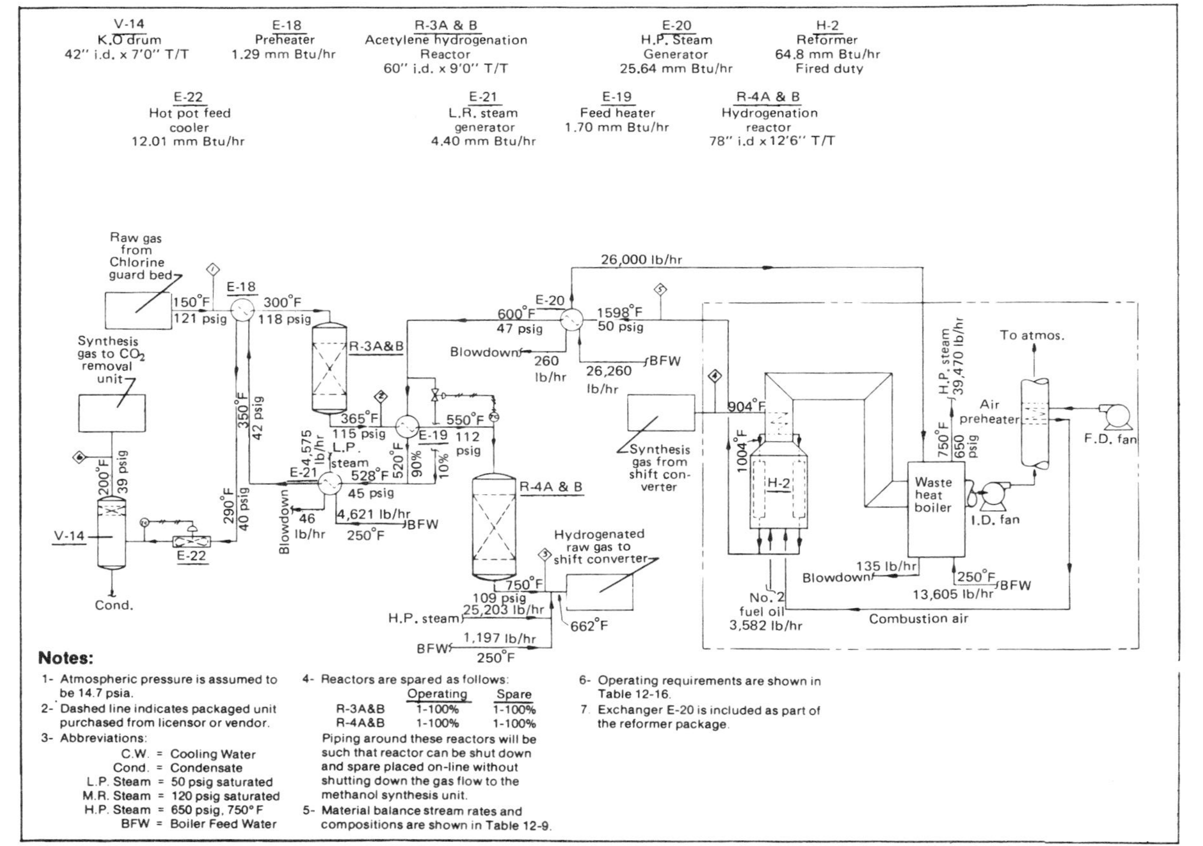

Figure 12-4. Process Flow Diagram for Gas Separation—Re-forming

Table 12-9. MATERIAL BALANCE FOR GAS SEPARATION - RE-FORMING
(Refer to Fig. 12-4)

Component (lb-mol/h)	Stream Number						
	1 Clean Gas from Chlorine Guard Bed	2 Acetylene Hydrogenation; Reactor Effluent	3 Hydrogenation Reactor Effluent	4 Shift Converter Effluent	5 Reformer Effluent	6 Synthesis Gas to CO_2 Removal Unit	7 Synthesis Gas to Methanol Synthesis
H_2	261.22	253.60	217.32	579.33	952.59	952.59	952.59
CO	435.45	435.45	435.45	73.44	457.37	457.37	457.37
CO_2	261.27	261.27	261.27	623.28	454.05	454.05	74.21
CH_4	60.98	60.98	60.98	60.98	1.26	1.26	1.26
C_2H_2	7.62						
C_2H_4	22.86	30.48					
C_2H_6	3.28	3.28	33.76	33.76			
C_3H_6	2.18	2.18					
C_3H_8	0.43	0.43	2.61	2.61			
C_4H_8	3.62	3.62					
C_4H_{10}	1.81	1.81	5.43	5.43			
C_5H_{12}	11.58	11.58	11.58	11.58			
N_2 + Ar	13.76	13.76	13.76	13.76	13.76	13.76	
H_2S	$<$0.5 ppm[a]						
H_2O	14.66	14.66	14.66	1,117.99	1,072.53	513.48	26.97
TOTAL	1,100.72	1,093.10	1,056.82	2,522.16	2,951.56	2,392.51	1,526.16

[a]Maximum value, less than 0.1 ppm expected.

Because of these high temperatures, all of the unsaturated hydrocarbons should be eliminated from the gas to prevent carbon laydown on the catalyst. In addition, high inlet concentrations of carbon monoxide can cause carbon laydown due to the Boudouard reaction:

$$2CO \rightarrow C + CO_2 \ . \tag{12-4}$$

Therefore, the gas must undergo shift conversion before it enters the re-forming furnace.

Clean raw gas from the chlorine guard beds is heated and hydrogenated exactly as in the hydrogenation technology discussed in Section 12.4.1. Effluent from reactor R-4A&B is quenched from 750 F to 662 F with attemperated steam and is fed to the shift converter. The shift converter is not considered part of the gas separation technology.

Shift converter effluent at 904 F passes to the re-forming furnace, H-2. The shift converter effluent, lean in CO, contains sufficient steam such that no additional steam injection is required before the furnace. The re-former feed gas is heated to 1004 F in the convection section of the furnace and fed to the catalyst beds, where the hydrocarbons are re-formed to CO and H_2. The re-former is fired with imported No. 2 fuel oil at a rate of 3,582 lb/h.

Waste heat is recovered from the re-forming furnace flue gases by generating high-pressure superheated steam at 650 psig, 750 F. The saturated steam produced by the high-pressure steam generator, E-20, is also superheated.

The re-former effluent, containing no hydrocarbons heavier than methane and containing hydrogen and carbon monoxide in the proper ratio for methanol synthesis, is cooled first by generating high-pressure saturated steam in exchanger E-20; it is cooled further against the hydrogenation reactor feed streams in exchangers E-19 and E-18, with an intermittent stage of low-pressure saturated steam generation in exchanger E-21.

The synthesis gas is finally cooled to 200 F in air-cooled exchanger E-22 before being fed to the CO_2 removal unit. This unit and downstream units are not considered to be part of the gas separation technology. An energy balance for re-forming technology is presented in Table 12-10.

The gas to the methanol synthesis reactor section will contain approximately 1 vol % inert gases, mainly nitrogen, on a dry basis. The theoretical methanol make as a result of employing this technology is 351,700 lb/day, more than twice the theoretical methanol make attributable to gas separation by hydrogenation.

12.4.3 Cryogenic Separation

12.4.3.1 Design Basis

Cryogenic technology for hydrocarbon separation is shown schematically in Fig. 12-5. The central technology is expansion-refrigeration, for which a package unit was supplied by Linde Division of Union Carbide Corp. However, this technology requires the following additional units ahead of the packaged unit to prepare the feed gas:

- bulk CO_2 removal,
- final CO_2 removal, and
- dehydration.

In addition, compressors are required before and after the cryogenic unit. The compressor ahead of the unit raises the gas pressure to 400 psig prior to expansion, while the second compressor restores the original pressure of approximately 103 psig. Neither compressor was included in the design, operating requirements, or costs of this scheme; however, the qualitative effects of differences in compression requirements among the gas separation schemes are discussed in Section 12.5.2.4.

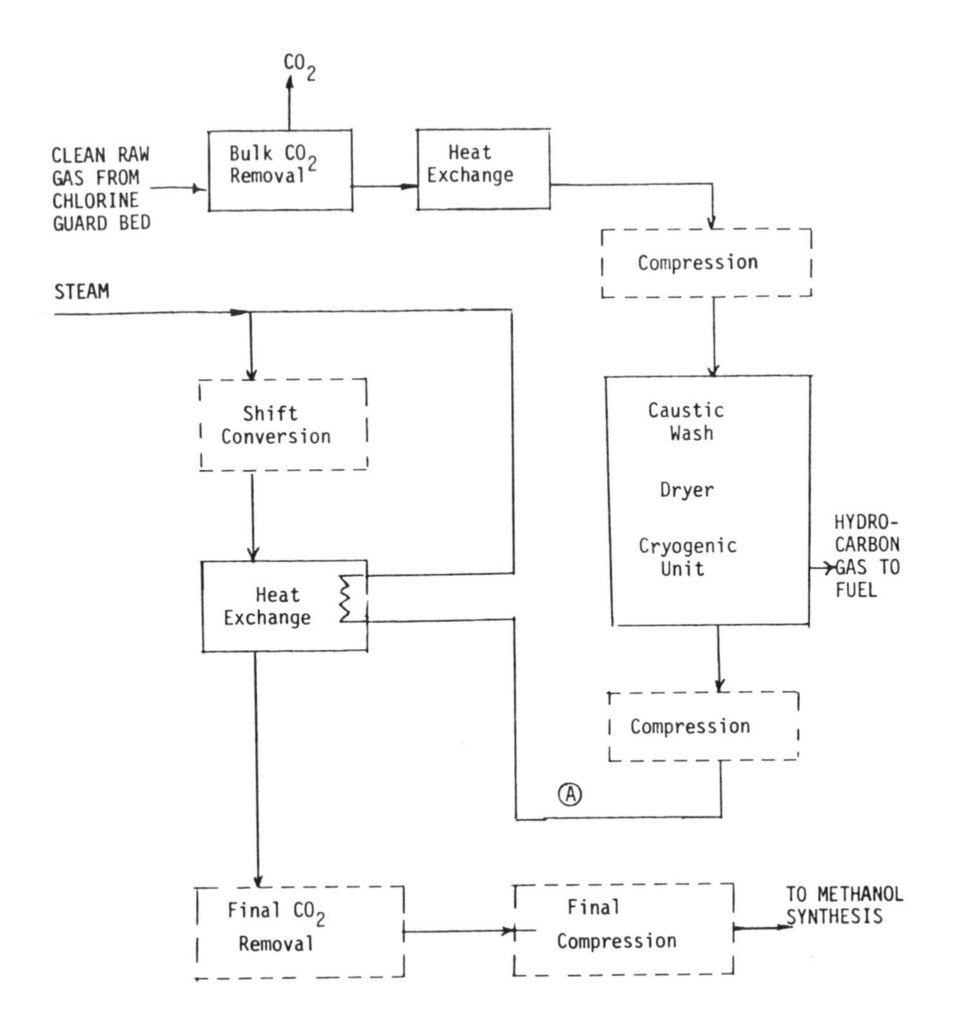

Notes:
Units in dotted outline are not part of the separation scheme; and gas at point A contains no more than 1.5 mole percent hydrocarbons, wet basis.

Figure 12-5. Cryogenic Separation Block Flow Diagram

Table 12-10. ENERGY BALANCE FOR RE-FORMING
(10^6 Btu/h)[a]

Inputs	
Raw gas	156.37
BFW import	8.47
Steam import[b]	34.22
Fuel oil import	64.83
Electric power	0.55
	264.44
Outputs	
Shifted gas to CO_2 removal	185.63
Steam export	58.51
Cooling losses	12.01
Blowdown	0.19
Condensate export	1.41
Re-former flue gas	5.75
	263.50

[a]Energy quantities include sensible enthalpies and higher heating values relative to 60 F, 1 atm pressure, with water in the liquid state.

[b]Steam includes attemperation water.

A hot potassium carbonate unit for bulk CO_2 removal was designed by Mittelhauser from published methods (Kohl and Riesenfeld 1974; Maddox and Burns 1967). For final CO_2 removal, caustic scrubbing was selected. Performance requirements and costs for a molecular-sieve dehydration unit were supplied by Linde.

12.4.3.2 Process Description

Reference is here made to Fig. 12-6, Gas Separation—Cryogenics, and to its material balance presented in Table 12-11.

Cryogenic separation technology requires that compounds which solidify or form hydrates at the low temperatures in the separation unit be removed from the gas before it enters the unit. Such compounds include H_2S, CO_2, HCl, and water. As in the previously described technologies, chlorine guard bed effluent is taken as the battery limits of the gas separation technology. Therefore, only CO_2 and water must be removed ahead of the cryogenic unit.

Clean raw synthesis gas is first heated to 200 F against CO_2 absorber overhead in exchanger E-12. It then enters the absorber, V-10, where about 96% of the CO_2 is absorbed by countercurrent stagewise contact with a hot aqueous potassium carbonate solution. Rich solution from the absorber flows to the stripper V-11, in which the CO_2 is liberated from the solution by reboiling with steam in exchanger E-11. The lean carbonate solution is pumped by P-8A&B back to the top of the absorber. The overhead from the top tray of the stripper is cooled against cooling water in a vertical tube bundle mounted in the top of the stripper. This process recovers water and potassium carbonate from the overhead;

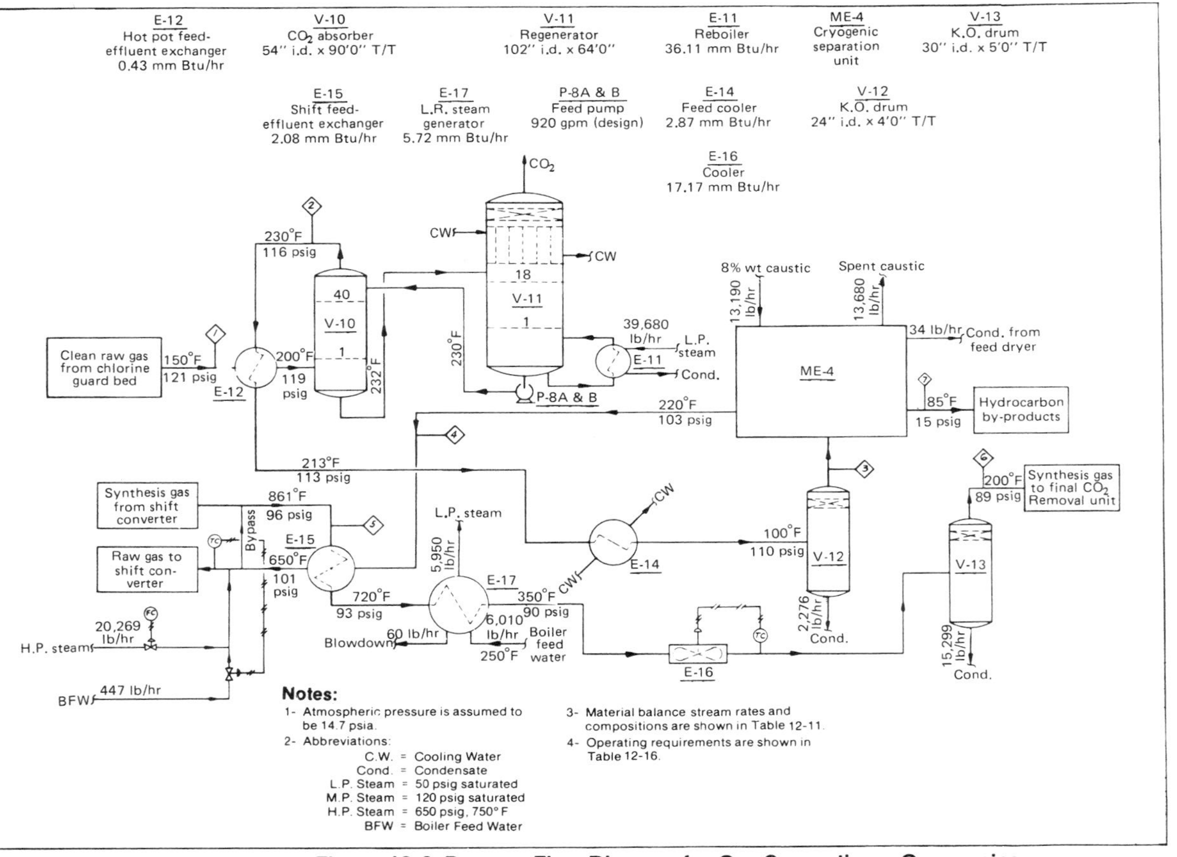

Figure 12-6. Process Flow Diagram for Gas Separation—Cryogenics

Table 12-11. MATERIAL BALANCE FOR GAS SEPARATION - CRYOGENIC SEPARATION
(Refer to Figure 12-6)

Component (lb-mol/h)	Stream Number							
	1 Chlorine Guard Bed Effluent	2 CO_2 Absorber Effluent	3 Cryogenic Unit Feed	4 Cryogenic Unit Effluent	5 Shift Converter Effluent	6 CO_2 Removal Unit Feed	7 Synthesis Gas to Methanol Synthesis	8 Hydro-Carbon Gas to Fuel
H_2	261.22	261.22	261.22	256.00	452.25	452.25	452.25	5.22
CO	435.45	435.45	435.45	413.68	217.43	217.43	217.43	21.77
CO_2	261.27	11.21	11.21		196.25	196.25	33.83	
CH_4	60.98	60.98	60.98	10.38	10.38	10.38	10.38	50.60
C_2H_2	7.62	7.62	7.62					7.62
C_2H_4	22.86	22.86	22.86					22.86
C_2H_6	3.28	3.28	3.28					3.28
C_3H_6	2.18	2.18	2.18					2.18
C_3H_8	0.43	0.43	0.43					0.43
C_4H_8	3.62	3.62	3.62					3.62
C_4H_{10}	1.81	1.81	1.81					1.81
C_5H_{12}	11.58	11.58	11.58					11.58
N_2 + Ar	13.76	13.76	13.76	12.00	12.00	12.00	12.00	1.76
H_2S	$<$0.5 ppm[a]							
H_2O	14.66	132.95	6.62		955.29	106.11	6.85	
TOTAL	1,100.72	968.95	842.62	692.06	1,843.60	994.42	705.74	132.73

[a]Maximum value; less than 0.1 ppm expected.

the cooled, CO_2-rich gas is vented to the atmosphere. Two atmospheric storage tanks, TK-1 and TK-2, have been included for fresh solution storage and to hold the liquid inventory of the system during planned maintenance shutdowns.

Gas from the CO_2 absorber, containing about 11,600 ppmv CO_2, is cooled to 100 F by heat exchange in E-12 and E-14. Condensate is separated in the knockout drum V-12. The cooled gas, at 110 psig, enters the cryogenic package.

The cryogenic unit relies on autorefrigeration of the gas by Joule-Thomson expansion to develop the low temperatures required for condensation of the hydrocarbons. First, the gas is compressed to 400 psig. Remaining carbon dioxide is removed from the gas by caustic scrubbing, and the gas is dehydrated by adsorption in a molecular sieve unit.

Then the gas is passed through a cold-box exchanger package and expanded to produce the desired separation. The separated hydrocarbon by-product leaves the unit at 15 psig and 85 F and is sent to the plant fuel system. A second compressor is required to compress the cleaned synthesis gas from 40 psig to 103 psig. In order to be consistent with other hydrocarbon separation technologies, neither the inlet nor the outlet gas compressor is considered part of the gas separation technology. Their costs were not included in the cost of the cryogenic package.

The synthesis gas leaving the cryogenic separation unit contains about 1.5 mole % CH_4. It is heated from about 220 F, the estimated compressor discharge temperature, to the shift converter feed temperature of 650 F in exchanger E-15 against the shift effluent gas. The shift converter effluent is further cooled to 200 F in exchangers E-17 and E-16, and condensate is separated in knockout drum V-13.

As in the other separation technologies, the final CO_2 removal unit is not considered to be part of the cryogenic gas separation technology. The synthesis gas delivered to the methanol synthesis loop contains from 1 to 2 vol % N_2. The theoretical methanol make is 167,190 lb/day, which is comparable to that from the hydrogenation technology. An energy balance for the cryogenic separation technology is presented in Table 12-12.

12.5 PYROLYSIS GAS CLEANUP

12.5.1 Design Basis

The gas cleanup facilities were designed to estimate the capital costs and operating requirements attributable to the upgrading of raw gas from a Purox gasifier to a quality suitable for feed to a methanol synthesis reactor. Design emphasis was placed on selecting units proven commercially in the same or similar service and on providing a conservative design wherever possible. Figure 12-7 shows schematically the various sections of the gas cleanup facilities.

The configuration of the gas cooling and oil mist elimination equipment was selected to match that used by Union Carbide Corp. at their Purox demonstration facility in South Charleston, W. Va. Union Carbide personnel reported satisfactory operation of these facilities during test runs. Design information provided by Union Carbide was used to size the raw gas spray cooler. The electrostatic precipitator performance data and costs were supplied by Koppers-Industrial Products, a leading manufacturer of tar-oil precipitation equipment.

Gravity settlers were designed for separating the raw gas scrubbing water from oil condensed in the scrubbers. No precise data were available on the ratio of oil removed in the scrubbing step to that removed by the precipitators; therefore, both the gravity settlers and the precipitators were designed to handle the entire plant net make of oil on a continuous basis.

Lastly, the final condenser, knockout drum, and all required pumps were designed and sized in-house based on a material and energy balance between the precipitation equipment and the battery limits of the section. The NH_3, CO_2, and H_2S contents of the gas leaving the knockout drum were estimated by using a computer program that predicts vapor-liquid equilibria in aqueous solutions of weak electrolytes.

Table 12-12. ENERGY BALANCE FOR CRYOGENIC SEPARATION (10^6 Btu/h)[a]

Inputs	
Raw gas	156.37
BFW	1.14
Steam[b]	63.53
Fuel	1.00
Net compression[c]	1.76
Electric power	0.87
	224.67
Outputs	
Shifted gas to CO_2 removal	89.13
Steam	6.85
Cooling (loss)	55.78
Blowdown	0.01
Condensate	2.23
Flue gas	0.52
Fuel gas	69.48
Acid gas	1.35
	225.35

[a]Energy quantities include sensible enthalpies and higher heating values relative to 60 F, 1 atm pressure, with water in the liquid state.

[b]Steam includes attemperation water.

[c]Compression horsepower less interstage and aftercooler duty, for units in the cryogenic separation package.

With the selected hydrogenation scheme, sulfur removal to 0.5 ppm or less is required ahead of the first hydrogenation reactor. A Stretford unit appeared to be the best available technology for removing most of the sulfur from the raw gas. The Stretford process operates well at low pressures and especially at low CO_2 partial pressure. The process produces a salable elemental sulfur product. Therefore, costs and utility requirements were estimated for a Stretford unit based on reducing the H_2S content of the gas from about 600 ppmv to 10 ppmv or less.

To protect the methanol synthesis and hydrogenation catalysts used in downstream processing, sulfur and chlorine guard beds of impregnated activated carbon were used. The unit was designed to reduce the H_2S content of the raw gas from 10 ppmv to less than 0.5 ppmv. The unit was located after the first stage of compression to take advantage of the somewhat elevated pressure and interstage cooling.

Although the amount of chlorine present in the raw gas was not quantified for this study, it was assumed that traces could be present due to the use of raw water in the plant. Katalco supplied data on the chlorine holding capacity of promoted alumina.

The hydrogenation and oil absorption designs developed for the review of gas separation technology were incorporated in the pyrolysis gas cleanup design without modification. The design work has previously been described in Section 12.2.

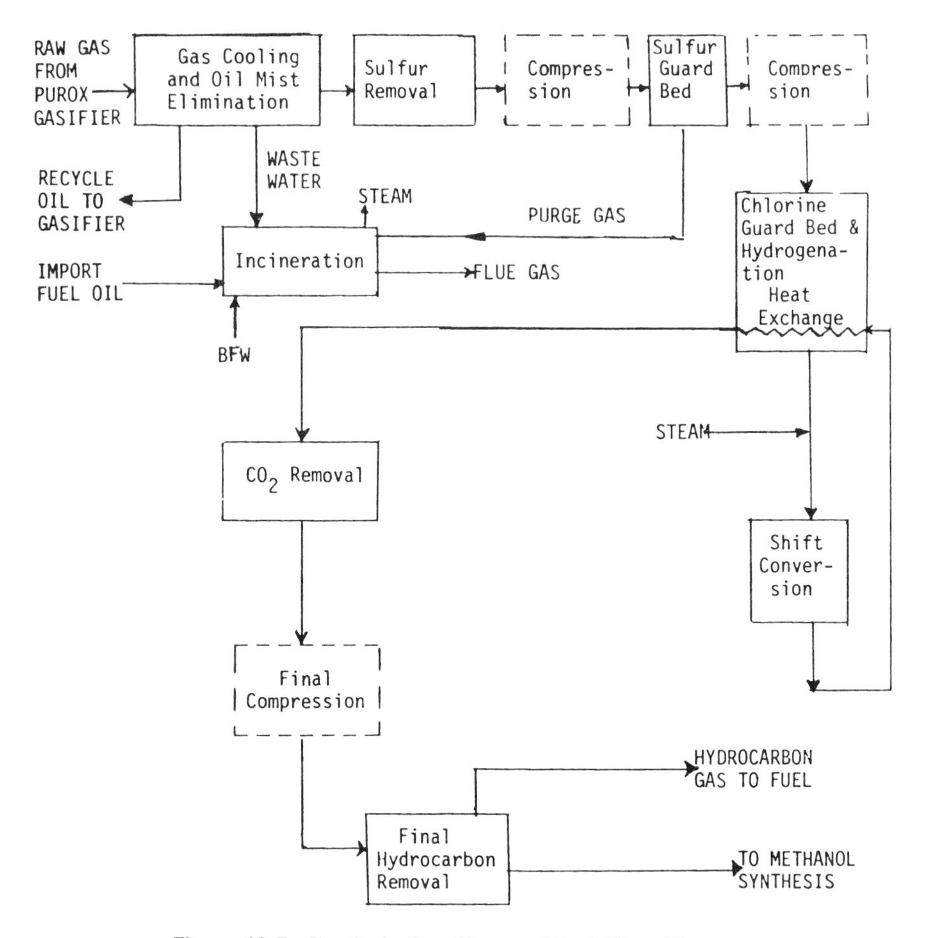

Figure 12-7. Pyrolysis Gas Cleanup Block Flow Diagram

The shift conversion unit was designed to produce a H_2 to CO ratio in the effluent of 2.08, as required for methanol synthesis. A maximum outlet temperature of about 950 F was used as a design basis. A 50 F temperature approach to equilibrium was assumed at the outlet of the reactor. The minimum inlet temperature was held at 650 F to provide optimum catalyst activity. A steam-to-dry-gas ratio of about 0.5 was used in the design.

Based on the above design data it was found that no shift bypass was required. Proper control of the unit can be maintained by attemperating the high-pressure superheated steam added to the feed.

Removal of carbon dioxide was done with a hot potassium carbonate unit because of the design data available in the open literature. The design specification for the product gas carbon dioxide content was 5% of the reactive components (H_2, CO, and CO_2) based on a previous study of methanol production from coal (McGeorge 1976).

The wastewater produced during raw gas cooling contains extremely high concentrations of water-soluble organic compounds. Conversations with Union Carbide technical personnel indicated that the water was extremely difficult to handle in a biological treatment system, although Union Carbide believed that their licensed Unox process might be able to treat this waste. The limited time-frame of the study prohibited obtaining performance data and cost estimates for a Unox system. Therefore, a rough design was prepared for an incinerator to burn the combustibles and evaporate the wastewater.

A waste gas stream is generated by the gas cleanup train; this is the regeneration gas stream from the activated carbon sulfur guard beds. This intermittent stream consists of either steam or nitrogen containing small amounts of sulfur and other gases. For this study, it was assumed that the regeneration gas would be incinerated in the same unit as the wastewater.

For waste heat recovery, a 50-psig steam generator was assumed to be included at the outlet of the incinerator.

Factored estimates of installed costs were prepared for each of the three gas separation technologies and for the gas cleanup train. First, equipment costs were estimated for each equipment item or vendor-supplied package.

The equipment costs were next factored into module costs, using factors developed in-house for installation labor and for materials such as piping, concrete, steel, instrumentation, electrical equipment, insulation, and paint. To the sum of direct materials and labor were added indirect charges such as payroll fringes, field expenses, tools, and equipment. Each of these factors was based on published data but was escalated separately to first quarter 1979 dollars using individual cost indices.

The modular costs were then combined to form factored cost estimates. To the sum of the modular costs were added allowances for process contingencies and offsites and for contractors' expenses and fees. Individual process cost contingencies were applied to each section of a given design rather than applying an overall contingency which might be high for some sections and low for others. In this way, the differences among technologies and their degrees of process risk were quantified individually.

Operating requirements for each section of each design were estimated from vendor-supplied information or from experience with the design or commercial operation of similar units. These operating requirements included utilities, operating and maintenance labor, and catalyst and chemical makeup requirements.

Utilities costs were estimated from the design requirements. In the gas separation technology review, steam requirements for shift conversion were included as a utility, even though the shift conversion unit itself was not considered part of the separation technology. This was done because it was found that the shift conversion section differed significantly for each of the three separation technologies studied.

The shift conversion section was not redesigned for each case, and a qualitative assessment of the effect of each separation technology on the costs of shift conversion is presented in Section 12.6.3. Based on the same sources, labor, chemical, and catalyst makeup requirements were also estimated for each section.

No costs were assigned to the operating requirements developed for this study. In-plant "transfer prices" of utilities can be estimated only by full consideration of the entire processing complex and its many interfaces with the subsystem under study; such a consideration was beyond the scope of this chapter. Also, labor rates are a strong function of the individual plant's location; only a generic location was used for this study. However, it is possible to make qualitative judgments among technologies based solely on the physical operating requirements themselves; such a discussion is presented in Section 12.6.

12.5.2 Process Description

This section describes a conceptual gas cleanup train designed to upgrade raw pyrolysis gas from a Purox biomass gasifier to methanol synthesis gas. Reference is made to Fig. 12-8 and to the material balance presented in Table 12-13.

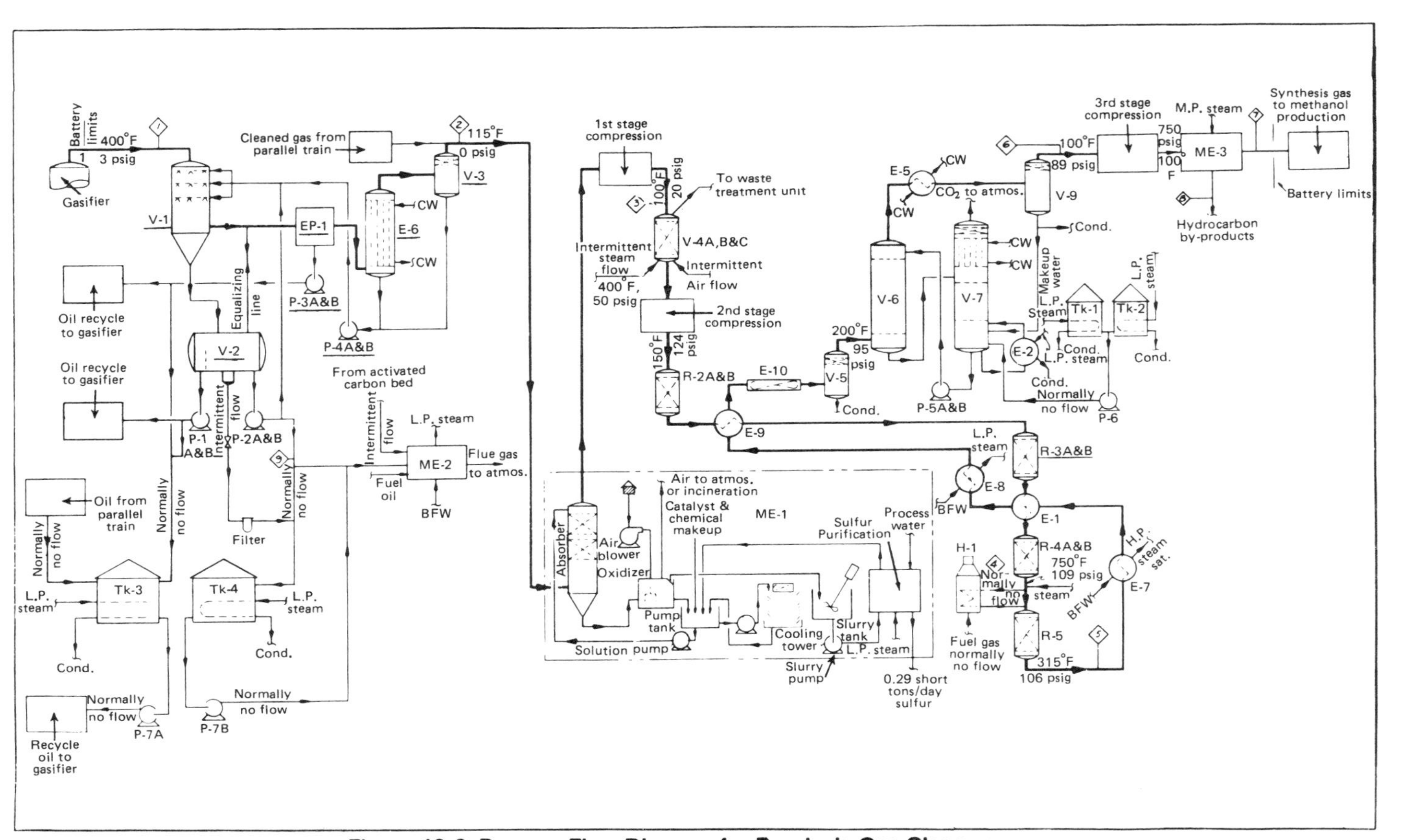

Figure 12-8. Process Flow Diagram for Pyrolysis Gas Cleanup

Equipment Identifcation for Figure 12-8

V-1 = scrubber
V-2 = decanter
P-1A&B = scrubber oil pump
Tk-3 = oil holding tank
P-2A&B = recirculation pump
Tk-4 = scrubber water tank
P-7A&B = transfer pump
EP-1 = electrostatic precir
P-3A&B = precipitator oil pump
P-4A&B = scrubber feed pump
E-6 = cleaned gas cooler
V-3 = cleaned gas K.O. drum
ME-2 = waste treatment unit (incineration)
V-4A,B&C = activated carbon bed
R-2A&B = chlorine guard bed
ME-1 = stretford unit
V-5 = K.O. drum
E-10 = hot pot feed cooler
E-9 = gas heater
E-5 = synthesis gas cooler
V-6 = hot pot absorber
P-5A&B = lean carbonate feed pump
V-7 = hot pot regenerator
E-8 = L.P. steam generator
H-1 = startup heater
V-9 = synthesis gas K.O. drum
E-2 = hot pot reboiler
R-3A&B = acetylene hydrogenation reactor
E-1 = feed effluent exchanger
R-4A&B = hydrogenation reactor
R-5 = shift converter
Tk-1 = rundown tank
Tk-2 = fresh solution tank
P-6 = makeup pump
E-7 = H.P. steam generator
ME-3 = final hydrocarbon removal unit

Notes for Figure 12-8

(1) Atmospheric pressure is assumed to be 14.7 psia.
(2) 1st, 2nd, and 3rd stage compression units are not included in battery limits.
(3) Equipment listed below is part of one of two parallel 50% capacity trains. All other equipment is for one 100% capacity train.

V-1	V-3	V-2
EP-1	P-3A&B	P-1A&B
E-6	P-4A&B	P-2A&B

(4) Dashed line indicates packaged unit purchased from licensor or vendor.
(5) Abbreviations:
CW = cooling water
Cond. = Condensate
L.P. steam = 50 psig saturated
M.P. steam = 120 psig saturated
H.P. sat. steam = 665 psig saturated
H.P. steam = 650 psig, 750°F
BFW = Boiler feed water
(6) Outlet temperature from 2nd stage compression is assumed compression discharge temperature. This temperature should be low enough to eliminate polymerization of olefins and subsequent plugging.
(7) Reactors and carbon drums are spared as follows:

	Operating (%)	Spare (%)
R-2A&B	1-100	1-100
V-4A,B&C	2-100	1-100
	One Regeneration One Absorption (%)	
R-3A&B	1-100	1-100
R-4A&B	1-100	1-100
R-5	1-100	

Piping around these reactors will be such that reactor can be shut down and spare placed on-line without shutting down the gas flow to the methanol synthesis unit.
(8) Tanks Tk-1, Tk-3, and Tk-4 are for holding liquid inventory during planned annual shutdown and cleanout of process facilities.
(9) Operating requirements are shown in Table 12-17.
(10) Material balance stream rates and compositions are shown in Table 12-13.

12.5.2.1 Gas Cooling and Mist Elimination

Raw pyrolysis gas leaving the Purox gasifier enters the gas cooling and mist elimination section at 400 F, 3 psig. At Union Carbide's recommendation, two parallel sets of gasifiers and gas cooling/mist elimination units, each producing 50% of total capacity, were assumed to be required to produce 500 ton/day of raw gas.

The raw gas is first scrubbed with water in a cocurrent spray tower, V-1, to remove entrained particulates and some of the oil produced in the gasifier. The gas is cooled to 150-180 F by adiabatic saturation. The water, with particulates and condensed oil, flows by gravity to the decanter V-2, where oil and water are separated by gravity settling,

Table 12-13. MATERIAL BALANCE FOR PYROLYSIS GAS CLEANUP
(Refer to Figure 12-8)

	Stream No.									
	1	2	3	4	5	6	7	8		9
Component (lb mol/h)	Gasifier Effluent[a]	Stret-ford Feed[b]	Stret-ford Effluent	Hydro-gen-ation Effluent	Shift Converter Effluent	Synthesis Gas to Compression	Synthesis Gas to Methanol Synthesis	Hydro-carbon Gas to Fuel	Component (lb/h)	Purge Water
H_2	261.22	261.22	261.22	217.32	440.83	440.83	440.83		NH_3-free	1.252
CO	435.45	435.45	435.45	435.45	211.94	211.94	211.94		NH_4^+	20.097
CO_2	261.27	261.27	261.27	261.27	484.78	34.36	34.36		CO_2-free	1.401
CH_4	60.98	60.98	60.98	60.98	60.98	60.98	27.56	33.42	HCO_3^-	65.841
C_2H_2	7.62	7.62	7.62						CO_3	0.515
C_2H_4	22.86	22.86	22.86						H_2S	0.009
C_2H_6	3.28	3.28	3.28	33.76	33.76	33.76	0.44	33.32	HS^-	0.092
C_3H_6	2.18	2.18	2.18						NH_2COO^-	0.902
C_3H_8	0.43	0.43	0.43	2.61	2.61	2.61		2.61	Methanol	216.280
C_4H_8	3.62	3.62	3.62						Ethanol	84.090
C_4H_{10}	1.81	1.81	1.81	5.43	5.43	5.43		5.43	Acetic acid	174.840
C_5H_{12}	11.58	11.58	11.58	11.58	11.58	11.58		11.58	Acetone	84.090
N_2+Ar	13.76	13.76	13.76	13.76	13.76	13.76	13.76		MEK	16.940
NH_3	1.47	0.10							Propionic acid	67.150
H_2S	0.75	0.75	10 ppm	<0.5 ppm[c]					Butyric acid	16.940
Acetic acid	2.91								Furfural	84.090
Methanol	6.75								Phenol	16.940
Ethanol	1.83								Benzene	10.160
Acetone	1.45								Oil	63.820
MEK	0.23								Water	7,521.860
Propianic acid	0.91									
Butyric acid	0.19									
Furfural	0.88									
Phenol	0.18									
Benzene	0.13									
Oil[d]										
H_2O	538.51	121.00	30.53	14.66	221.13	7.32	1.03			
Total	1,642.25	1,207.91	1,116.59	1,056.82	1,486.80	882.57	729.92	86.36		8,447.30

[a]Total for both trains.
[b]Flow used for design.
[c]Maximum value; less than 0.1 ppm expected.
[d]Oil flow at 3364.82 lb/h.

although the specific gravities of the oil and water are so close that settling is quite difficult. A boot is provided in the water section of the decanter into which solids can settle. This boot is blown down intermittently to the waste treating section of the plant through a cartridge filter.

The aqueous condensate is pumped by P-2A&B back to the scrubber V-1. Net condensate is withdrawn and pumped to the waste treatment section. As shown in Table 12-13, this condensate contains approximately 11% by weight of water-soluble organic compounds.

Oil recovered from the decanter is pumped by P-1A&B to the gasifier, in which it is assumed to be recycled to extinction.

It is anticipated that frequent maintenance may be required in the gas cooling and mist elimination section, particularly in the scrubbing and decanting equipment. Therefore, holding tanks TK-3 and TK-4 were provided to contain the liquid inventory of the system during shutdown and cleanout operations. Tank TK-3 has a capacity of one day's net oil make, while Tank TK-4 can hold one week's make of aqueous condensate to allow for shutdowns in the waste treatment section of the plant.

Gas leaving the scrubber V-1 is saturated with water at 150-180 F and 2.5 psig. Next it flows through a wet electrostatic precipitator EP-1 in which oil mist is recovered from the gas. The precipitator is designed to remove 99% of the oil mist and is sized to handle the entire gasifier net oil make. In addition, spare units are provided so that one unit may be cleaned without shutting down the entire gas cleanup train. This extremely conservative arrangement should provide maximum reliability in removing oil mists from the raw gas.

Oil collected in the precipitator is pumped by P-3A&B back to the gasifier, in which it is recycled to extinction.

The gas leaving the precipitator is next cooled to 115 F against cooling water in exchanger E-6, and condensate is separated from the gas in knockout drum V-3. Downstream of this point, the two parallel, 50% capacity trains are manifolded into a single 100% capacity train.

12.5.2.2 Sulfur Recovery

The cooled gas next flows through a Stretford unit ME-1, in which H_2S is scrubbed from the gas. Data from the Purox process operating on municipal solid waste have levels of organic sulfur in cooled, scrubbed gas of less than 1 ppmv, showing that the Stretford process is effective in high-efficiency sulfur removal.

The Stretford process is a licensed proprietary process of the Northwest Gas Board, United Kingdom. It operates by absorption of H_2S in a solution of sodium carbonate, sodium meta-vanadate, and anthraquinone disulfonic acid. Through a series of oxidation-reduction reactions, the H_2S is first converted to HS ion, then oxidized to elemental sulfur. The sulfur is released as a froth by air blowing through the solution. The froth is skimmed from the oxidation tank and processed in a melter to recover solution, producing about 0.29 ton/day of marketable elemental sulfur. The scrubbed gas leaving the Stretford unit contains no more than 10 ppmv of H_2S.

12.5.2.3 Guard Beds

The sweetened gas is next compressed to about 20 psig and passed over a bed of impregnated activated carbons for final sweetening. Three beds are used, V-4 A,B,C. At any time, one bed is on adsorption service, one is being regenerated by steam or nitrogen from the air separation plant associated with the Purox process, and the third is a spare. Placing the carbon unit between compression stages takes advantage of low interstage gas temperatures to greatly enhance adsorptive capacity. The gas leaving the carbon beds contains less than 0.5 ppmv of sulfur.

Regeneration gas from the carbon beds, an intermittent stream, is incinerated in the waste treatment unit of the plant.

The sweetened gas is compressed to about 124 psig and passed through a chlorine guard bed, V-2A&B, of promoted alumina, which will reduce the chlorine content of the outlet gas to less than 0.2 ppmv. A full-capacity spare is provided for the chlorine guard bed, allowing for shutdown of a bed and removal of the spent guard material without shutting down the gas cleanup train.

12.5.2.4 Compression

Compression has been excluded from consideration in this study; however, in designing the gas cleanup facilities, consideration was given to the placement of process units relative to compression and to the compression requirements.

In the processing scheme selected for design, most of the gas cleanup is done at pressures below 150 psig to minimize the requirements for compressing CO_2 to elevated pressures. Only oil absorption is done at methanol synthesis pressure. It has been found that the first- and second-stage compressors must be designed very carefully to minimize discharge temperatures. Excessive temperatures can cause polymerization olefins in the gas, and the larger polymers can plug valves. We have assumed that the discharge temperature from second-stage compression is held to 150 F.

12.5.2.5 Gas Separation

In the gas cleanup described here, hydrogenation and oil absorption technology were used. The processes and equipment involved are described in Section 12.4 and are not repeated here.

12.5.2.6 Shift Conversion

Hydrogenated pyrolysis gas must be shifted to provide the correct H_2 to CO ratio for methanol synthesis. This reaction,

$$CO + H_2O \rightarrow CO_2 + H_2 ,$$

is performed in the shift converter R-5. An iron-chrome high-temperature shift catalyst is used to shift approximately 50% of the CO in the feed gas to CO_2. In the current design, no bypass of gas around the shift reactor is required to obtain the desired ratio of H_2 to CO (2.08). Instead, the reaction is controlled by the addition of attemperated steam upstream of the reactor.

No spare is required for the shift converter because the spare guard beds and hydrogenation facilities upstream of the shift converter ensure that shift catalyst poisons never reach the shift reactor. The shift catalyst is a rugged catalyst that should be extremely long-lived and require very little maintenance. Catalyst changeouts can be done during planned maintenance shutdowns.

Shift effluent gas is cooled against hydrogenation reactor feeds and by raising steam, and it is then sent to CO_2 removal.

A startup heater, H-1, heats the shift and hydrogenation reactors during startups. Its duty has been set at 2.0×10^6 Btu/h, which is approximately 35% of the shift preheat duty required in normal service. The shift section can be "boot-strapped" to full throughput once a 20% gas flow has been established and the shift reaction is initiated.

12.5.2.7 CO_2 Removal

CO_2 produced in the shift conversion step is removed from the gas in a CO_2 removal unit similar to that detailed in Section 5.3.3. Hot potassium carbonate solution is used to absorb CO_2 from the synthesis gas in absorber V-6. The solution is regenerated in the stripper V-7. Two atmospheric storage tanks, TK-1 and TK-2, have been included for fresh solution storage and to hold the liquid inventory of the system during maintenance shutdowns.

12.5.2.8 Waste Treatment

Wastewater produced by the condensation of water in the gas cooling and mist elimination section is evaporated, and the combustible organics are incinerated in the waste treatment unit, ME-2. This unit is a thermal oxidizer with waste heat recovery capability. The wastewater is oxidized with supplemental fuel oil firing to raise the combustion chamber temperature to 1800 F. Thirty-percent excess air is used based on fuel oil heating value; the total heat release in the combustion chamber is 26.2×10^6 Btu/h. About 45% of the oxidizer heat release is recovered in a packaged water-tube steam generator. Fifty-psig saturated steam is generated in this equipment. The flue gas exits to the atmosphere at approximately 400 F.

12.5.2.9 Process Alternatives

There are a number of process alternatives for each section of the gas cleanup facilities. Although scope and time limitations did not permit a detailed examination of these alternatives, some qualitative assessments were made.

Process and equipment for gas cooling and oil mist elimination were discussed in Section 12.5.2.1. Although an electrostatic precipitator is more costly than other devices, it has been proved in performance at South Charleston. Union Carbide reported that a venturi scrubber had been tested for oil mist elimination at South Charleston but had not performed effectively.

A number of sulfur removal technologies other than the Stretford method were considered for the gas cleanup service. Two of them were MEA and solid iron oxide.

MEA (monoethanolamine) absorption removes hydrogen sulfide from the cooled raw gas by chemical absorption. After regeneration of the solvent by heating, the H_2S is released from solution and flows overhead from the regenerator. It is then converted to elemental sulfur in a Claus sulfur recovery unit. Based on in-house experience with such processes, we concluded that in the small size under consideration, the separate MEA-Claus installations would be more costly than a single Stretford unit.

Solid iron oxide has been used for many years to purify both natural and synthetic gases containing trace amounts of H_2S. Perry Gas Processors, Inc., a supplier of commercial iron oxide units, estimated that 15 vessels, each 90 in. by 20 ft high, would be required. The estimated life of the total inventory of the beds was 150 days. Because of its high anticipated capital and operating costs, such an installation was not considered for this application.

Zinc oxide is a widely used alternative for trace sulfur removal. Katalco, a leading supplier of zinc oxide, was contacted to ascertain the usefulness of zinc oxide in the gas cleanup train. The applicability of zinc oxide is affected by the choice of olefin hydrogenation scheme. For the selected scheme, sulfur had to be removed ahead of the acetylene hydrogenation reactor to prevent poisoning of the palladium catalyst. However, Katalco indicated that acetylene would polymerize in the zinc oxide beds. This makes the use of zinc oxide incompatible with two-stage hydrogenation.

Three alternatives for separation of hydrocarbon gases from the synthesis gas were considered, in addition to those studied in detail. The alternatives are single-stage hydrogenation, molecular sieve adsorption, and low-pressure refrigerated oil absorption. These alternatives were not examined in detail, but some observations are presented below.

Olefins and acetylene can be hydrogenated over a single-stage cobalt-molybdenum catalyst, rather than the two-stage scheme adopted in the study. In a single-stage hydrogenation, however, acetylene may crack and lay down carbon on the catalyst. The catalyst can be regenerated periodically by burning off the carbon with air. Use of the single-stage scheme would permit the use of high-temperature zinc oxide for trace sulfur removal, as the cobalt-molybdenum catalyst is not poisoned by sulfur. Therefore, the zinc oxide single-stage hydrogenation scheme may be attractive for this service. The two-stage hydrogenation scheme was selected for its anticipated operating simplicity; i.e., lack of a periodic burnoff of the catalyst beds.

An alternative to the removal of paraffin hydrocarbons by high-pressure oil absorption is their adsorption in a molecular sieve pressure-swing-adsorption (PSA) unit. This type of unit was briefly discussed with the supplier, Union Carbide Corp. A rough estimate of capital cost indicated that a PSA unit would be more costly than the oil absorption system. Furthermore, losses of H_2 and CO predicted by Union Carbide for the PSA unit were significantly higher than those predicted for an oil absorption unit. Therefore, this alternative was not considered further.

Another alternative in gas separation would be to use a single refrigerated oil-absorption step to remove both olefins and paraffin hydrocarbons ahead of shift conversion. Mittelhauser performed preliminary process simulations on such a system, using the SSI/100 computer program. With a lean oil temperature of -40 F, approximately 25% of the CH_4 and essentially all of the acetylene, olefins, and heavier paraffins can be removed in a reasonable-sized absorber. Unfortunately, scope and time constraints prevented completion of the process design work. The alternative, however, should be investigated further when conceptual commercial designs are undertaken.

Several process alternatives exist for removing CO_2 from synthesis gases. Hot potassium carbonate is often used as a chemical solvent for CO_2. Other commercial CO_2 removal processes use either physical solvents or mixtures of physical and chemical solvents. Descriptions of these processes are available in the literature.

One physical solvent process that may be attractive in synthesis gas cleanup is Allied Chemical Corporation's proprietary SELEXOL process. This method is effective at higher pressures; in the absence of sulfur compounds the solvent is regenerated by pressure letdown or air stripping, without the use of steam for reboiling. This can result in a considerable savings in operating cost when compared with the hot carbonate process.

In further research work, alternative CO_2 removal processes for synthesis gas cleanup should be compared.

Mittlehauser investigated the possibility of using Union Carbide's UNOX process for treating the highly concentrated wastewater from the gas cooling and mist elimination sections. Design data and cost estimates could not be obtained in time for inclusion of such a design in the study. However, since this approach has been used in at least one conceptual study, it should be investigated further.

12.5.2.10 Technology Assessment

The investigations of methanol synthesis catalyst tolerance presented here reveal that little is known about potential poisons other than sulfur and chlorine. Synthesis gas from the biomass scheme may contain many more chemical species than commercial methanol processes that produce a synthesis gas by steam re-forming of natural gas, LPG, or naphtha.

Katalco, United Catalysts, and Haldor Topsoe were uncertain about the effects of many of the trace compounds for which more exact tolerance levels are required. In addition, the concentrations of these compounds in the raw synthesis gas produced from biomass should be better defined. Nevertheless, the gas cleanup system as designed represents a conservative approach to removing known and suspected methanol catalyst poisons.

All of the units and equipment designed for the gas separation technologies and gas cleanup designs presented here have been employed commercially or in demonstration facilities in the petroleum or coal and gas processing industries. The gas cooling and mist elimination designs were identical to those proven in performance at Union Carbide's Purox demonstration facility at South Charleston, W. Va.

The Stretford process, a joint development effort by the Clayton Aniline Company, Ltd. and the North Western Gas Board, was designed initially for the desulfurization of coke oven gas. The process has been used for treating refinery gases, synthesis gas and natural gas, and has been commercially used in Europe and the United States.

The impregnated activated carbon and promoted alumina material used as sulfur and chlorine guards have been used commercially for treating natural gas and light hydrocarbon feed stocks. The two-stage hydrogenation catalysts have been used extensively in refinery service. The simpler, single-stage hydrogenation over a cobalt-molybdate catalyst discussed in Section 12.4.1 has been used in acetylene service. However, the catalyst would require some laboratory test runs under expected conditions to determine the rate of catalyst coking. Catalyst suppliers are equipped to perform such tests.

Shift and re-forming systems have been widely used in the refining and methanol synthesis industries for years. Shift catalysts have been specifically developed for the coal-to-SNG industry. Cryogenic separation systems have generally been used for the purification of hydrogen but have been commercially modified for the separation of hydrocarbons from synthesis gas streams.

Many systems are available for removing carbon dioxide from synthesis gas streams. A proprietary system licensed by Benfield is a catalyzed, hot potassium carbonate system similar to the one used in this study. It has been employed at the British Gas Corporation, Westfield, test facility to remove acid gases from town gas.

12.5.2.11 Overall Review

For removing hydrocarbon contaminants from methanol synthesis gas, it appears that cryogenic separation is less favored economically as compared with hydrogenation. No such conclusions should be drawn between re-forming and hydrogenation, however. These two technologies have too many differences that should be studied in detail in the context of an overall, commercial-scale methanol plant design.

12.6 COST ESTIMATES

12.6.1 Capital Costs

The cost estimates for three gas separation technologies are summarized in Table 12-14. The hydrogenation technology is least costly in capital. The cryogenic separation technology is by far the most costly because of the high cost of the cryogenic package relative to the re-former and the oil absorption plant in the other technologies, and also because of the added CO_2 removal step ahead of the cryogenic package.

A qualitative assessment of the effects of the three gas separation technologies on capital requirements for other gas cleanup units is presented in Table 12-15. This table shows that external costs are likely to be somewhat higher for re-forming and cryogenic separation than for hydrogenation. However, the table also illustrates the tremendous increase in potential methanol yield afforded by re-forming, a result of the conversion of the paraffin hydrocarbons to additional synthesis gas. For gas separation capital cost only, the total capital required per potential daily ton of methanol is $21,450 for hydrogenation but only $16,030 for re-forming.

12.6.2 Operating Costs

A summary of operating requirements for the three gas separation technologies is presented in Table 12-16. As discussed previously, the steam requirements for shift conversion have been included as part of the operating requirements to afford a more realistic view. The cryogenic separation technology is a heavy importer of steam, hydrogenation a moderate importer, and re-forming a net exporter of steam to the overall methanol-from-biomass plant. However, the re-forming technology requires a significant import of fuel oil with which to fire the re-forming furnace. The cost of this requirement at least partly offsets the value of the exported steam.

It should also be pointed out that the hydrogenation and cryogenic separation technologies supply a significant quantity of hydrocarbon fuel gas to the methanol-from-biomass plant, while the re-forming technology does not, having converted the hydrocarbons to synthesis gas instead.

Operation labor and maintenance expenses are lowest for hydrogenation and highest for cryogenic separation.

Table 12-14. CAPITAL REQUIRED FOR GAS SEPARATION

	Costs in Thousands of 1979 Dollars		
	Hydro-genation	Re-forming	Cryogenic
Equipment	384.7	1204.1	2150.8
Other materials[a]	259.1	195.9	444.1
Installation	192.1	159.7	342.3
Installed facilities, Field costs	835.9	1560.2	2937.2
Indirect charges[b]	303.8	303.0	709.1
Initial charge of catalyst and chemicals	70.8	106.8	45.7
Installed module	1210.5	1970.0	3692.0
Allowance for process contingencies[c]	174.5	180.1	399.3
Allowance for offsites[d]	114.0	189.9	364.6
Contractor's expenses and fee[e]	249.0	478.0	860.0
Total Capital Required	1748.0	2818.0	5315.9
Capital required per lb mol/h of synthesis gas	2.39	1.85	7.53
Capital required per potential daily ton of methanol	21.45	16.03	63.59
Estimated annual maintenance expense	36.3	59.1	110.7

[a]Includes piping, concrete, structural steel, instrumentation, electrical, insulation, and painting.

[b]Includes payroll fringes, field expenses, tools, and equipment.

[c]Calculated as a percentage of module cost net of catalyst and chemicals. The percentage varies depending on the type of service.

[d]Allowance for offsites, at 10% of net module costs, to enable connections, site preparation, retrofit adjustments, and required ductwork and controls.

[e]Covers home office construction services, design engineering, drafting, procurement, project management, and general indirect and overhead expenses. The fee is based on a fixed percentage of the module plus the allowance for contingencies and for offsites.

Table 12-15. POSSIBLE EXTERNAL CAPITAL EFFECTS OF GAS SEPARATION TECHNOLOGIES

	Hydrogenation	Re-forming	Cryogenic
Shift Conversion			
Total mol/h			
to reactor (% bypass)	0	0	49.5
Dry	1042.16	1042.16	342.57[a]
STM	444.64	1480.01	1151.54
Total	1486.80	2522.17	1494.11
Relative size	1.00	1.70	1.00
Projected cost	Base	Higher	Same
Final CO_2 Removal			
Feed mols			
Inerts	780.89	1424.98	692.06[b]
H_2O	148.58	513.48	106.11
CO_2	484.78	454.05	196.25
Total	1414.25	2392.51	994.42
Gas flow factor	1.00	1.69	0.70
mol CO_2 in effluent	34.36	74.21	33.83
mol/h CO_2 removed	450.42	379.84	162.42
SCF/min CO_2 removed	2849.0	2402.5	1027.3
Prorated gpm solution[c]	1140.0	961.0	410.9
Approximate reboiler duty	33.1 10^6	27.9 10^6	11.9 10^6
Projected cost	Base	Same	Lower
Compression			
Intermediate			
mol/h	-	-	842.62/692.06
P_{in} to P_{out} (psig)	-		100 to 400/40 to 103
Final			
mol/h (Dry)	815.25	1499.19	725.89
P_{in} to P_{out} (psig)	89 to 750	37 to 750	84 to 750
Projected cost	Base	Much Higher	Much Higher
Methanol Synthesis			
% Inerts in feed (dry mol%)	10.4	6.0	8.0
Potential methanol (ton/day)[d]	81.5	175.9	83.6

[a]1.5 vol % CH_4

[b]Includes CH_4 at 10.38 mol/h

[c]Based on lean loading of 2.5 and rich loading of 5.0 SCF CO_2/gal

[d]Mol/h CO 24 32.04/2000

Table 12-16. OPERATING REQUIREMENTS FOR GAS SEPARATION

	Hydrogenation		Re-forming		Cryogenic	
Cooling water (gpm)	ME-3	455			V-11	2030
					E-14	164
					ME-4	57
		455				2251
Steam export (lb/h)						
50 psig, sat.	E-8	1715	E-21	4575	E-11	(39680)
					E-17	5950
					TK-1& TK-2	(40)
		1715		4575		(33770)
120 psig, sat.	ME-3	(20240)				
		(20240)				
650 psig, 750 F[a]	Shift	(5683)	Shift	(25203)	Shift	(20269)
	E-7	4260	Re-former	39470		
		(2423)		14267		(20269)
BFW import (gpm, 250 F)						
	Shift	2.3	Shift	2.5	Shift	1.0
	E-7	9.1	Re-former	84.6	E-17	12.7
	E-8	3.7	E-21	9.8		
		15.1		96.9		13.7
Electric power (kWh/h)	E-10	18.6	E-22	119.3	E-16	119.3
	ME-3	343.0	Re-former	42.5	P-8A&B	132.4
		361.6		161.8		251.7
Fuel oil (lb/h)			Re-former	3582		
Fuel gas (SCFM)[b]					ME-4	20
Catalyst [ft^3 (life, yr)]	R-3	80(1)	R-3	80(1)	Dryer	59(3)
	R-4	211(3)	R-4	211(3)		
Chemicals (lb/day)					K_2CO_3	8.6
					NaOH	25,321
					Condensate	265,875
Operating labor (man-hour/day)		6		12		24

[a]Does not include desuperheating water.
[b]Fuel gas at 1000 Btu/SCF (HHV).
[c]Molecular Sieve.

12.6.3 Incremental Costs of Gas Cleanup

Installed costs for gas cleanup are shown in Table 12-17.

The capital requirements of gas cleanup amount to about $127,000 per potential daily ton of methanol.

The major capital cost items in pyrolysis gas cleanup are the electrostatic precipitators, at about $2.2 million, installed cost, and the CO_2 removal unit at about $1.8 million, installed cost. We believe that these costs may be lessened somewhat by selecting different processing facilities. However, selection of alternatives must be made by thorough comparisons on consistent bases that fully account for the differing effect of each alternative on the overall methanol plant.

Operating requirements for gas cleanup are summarized in Table 12-18. The most significant requirements in terms of potential cost are the steam imports and the fuel requirements for the Waste Treatment Unit, ME-2.

Table 12-17. CAPITAL REQUIRED FOR PYROLYSIS GAS CLEANUP

	Costs in Thousands of Dollars
Equipment	3,144.1
Other materials[a]	1,300.2
Installation	1,052.1
Installed facilities, field costs	5,496.4
Indirect charges[b]	1,831.3
Initial charges of catalyst and chemicals	218.8
Installed module	7,546.5
Allowance for process contingencies[c]	511.9
Allowance for offsites[d]	736.9
Contractor's expenses and fee[e]	1,526.0
Total Capital Required	10,321.3
Estimated annual maintenance expense	331.1

[a]Includes piping, concrete, structural steel, instrumentation, electrical, insulation, and painting.

[b]Includes payroll fringes, field expenses, tools, and equipment.

[c]Calculated as a percentage of module cost net of catalyst and chemicals. The percentage varies depending on the type of service.

[d]Allowance for offsites, at 10% of net module costs, to enable connections, site preparation, retrofit adjustments, and required ductwork and controls.

[e]Covers home office construction services, design engineering, drafting, procurement, project management, and general indirect and overhead expenses. The fee is based on a fixed percentage of the module plus the allowance for contingencies and for offsites.

Table 12-18. OPERATING REQUIREMENTS FOR GAS CLEANUP

Cooling water (gpm)		
	V-7	1960
	E-5	259
	E-6	507
	M-3	455
Steam export (lb/h)		
50 psig, sat.	TK-1	(20)
	TK-2	(20)
	TK-3	(40)
	TK-4	(350)
	ME-1	(50)
	E-2	(36400)
	E-8	1715
	ME-2	12180
		(22985)
50 psig, 400 F	V-4ABC	(3750)
		(3750)
120 psig, sat.	ME-3	(20240)
		(20240)
665 psig, sat.	E-7	4260
		4260
650 psig, 750 F	R-5	6683
		6683
BFW import (gpm, 250 F)		
	R-5	2.3
	E-7	9.1
	E-8	3.7
	ME-2	26.1
		41.2
Condensate import (gpm)	ME-1	.12
		.12
Electric power (kW)		
	EP-1	21.3
	E-10	18.6
	P-1A&B	.3
	P-2A&B	8.9
	P-3A&B	.3
	P-4A&B	3.4
	P-5A&B	161.2
	P-6	1.9
	P-7A&B	.8
	ME-1	14.9
	ME-2	5.2
	ME-3	343.0
		579.8

Fuel oil (lb/h)		
	ME-2	920
		920
Fuel gas (SCFM)		
	H-1	40
		40
Utility air (SCFM)	V-4ABC	37.5
		37.5
Catalyst [ft^3 (life, yrs)/vessel]	V-4ABC	375(4)
	R-2A&B	123(2)
	R-3A&B	80(1)
	R-4A&B	211(3)
	R-5	564(4)
Chemicals (lb/day)		
	K_2CO_3	8.6
	ADA	0.8
	V_2O_5	0.8
	Na_2CO_3	25
Operating labor (Man-hours/day)		
Gas cooling and mist elimination		12
Sulfur removal		6
Guard beds		3
Gas separation		
Hydrogenation		6
Oil wash		3
Shift conversion		3
Final CO_2 removal		6
Waste treatment		3
		42

12.7 CONCLUSIONS AND RECOMMENDATIONS

The major conclusions that can be drawn from the study are as follows:

- Raw pyrolysis gas from the gasification of wood waste in a Purox gasifier can be upgraded to a synthesis gas which (so far as is now known to methanol catalyst suppliers) is of acceptable quality for commercial methanol synthesis. This upgrading is technically feasible with commercially available equipment.
- Several alternatives can be defined for a number of gas cleanup unit operations. At least some of these alternatives should be studied in more detail with a view to reducing the overall cost of gas cleanup.
- Among alternatives for separation of hydrocarbons from methanol synthesis gas, hydrogenation of olefins followed by oil absorption of paraffins and catalytic reforming appear to be more attractive than cryogenic separation.

Problems and uncertainties in the current literature include:

- Detailed characterizations of raw gas from gasification of wood waste in a Purox gasifier are not yet available. Such characterizations from commercial scale equipment are required to properly design downstream processing facilities, especially those in which performance is controlled by minor components such as HCN, COS, CS_2, NH_3, tars, oils, and water-soluble organic compounds.
- Detailed studies are needed of the long-term effects of compounds known or suspected to be present in biomass pyrolysis gas, on commercially available methanol synthesis catalysts.
- The biological treatability of Purox wastewaters from biomass gasification needs definition. Basic parameters for the design of biological treatment systems can be developed only from such treatability studies.
- The problem of scaleup to commercial methanol plant sizes must be addressed. At the 80-175 ton/day size addressed in this study, methanol production from biomass may or may not be economical. The relationship of product methanol cost to plant size must be quantified, together with problems associated with scaleup of plant facilities.

Based on the conclusions developed in this study, and the problems and uncertainties identified thereby, some aspects of a comprehensive research program may be defined. These research needs may be broadly classified as system level, subsystem level, and component studies.

On the system level, the following research programs should be undertaken:

- A conceptual commercial design should be made of a complete grass-roots plant to convert biomass to methanol. The suggested scale of the plant is 1500 tons per day, which is a reasonable scale for a large, single-train, methanol synthesis process. This design would identify and address system level problems associated with siting, construction, and operation of such a facility and would help to quantify the expected cost of methanol from such a plant.
- Comparative commercial scale designs of methanol plants using different biomass gasifiers should be performed to identify the most promising gasification processes for further commercial development. Particularly interesting would be comparison of air versus oxygen-blown gasifiers and atmospheric versus pressurized gasifiers.
- Sensitivity studies should be performed on commercial-scale designs to examine the effect of variations in design and of economic parameters on the cost of methanol from biomass. These parameters include feedstock and fuel costs and overall plant size.

On the subsystem level, the following research activities are recommended:

- Studies of alternative wastewater treatment method.
- Study of optimal location of compression facilities.
- Study of process alternatives for CO_2 and sulfur removal for commercial-scale facilities.

Component studies that should be performed are as follows:

- An experimental program to characterize thoroughly the types and quantities of trace components, such as nitrogenated compounds, water soluble organics, and sulfur compounds, produced by developing biomass gasifiers. Included in this program are a correlation of these component production rates with gasifier conditions and development of a method for predicting the production of such components.
- Scaleup and operational studies of the biomass gasifiers themselves, with an objective of determining the optimal size of a commercial gasifier.

- Biological treatability studies on wastewaters produced from biomass gasification.
- Laboratory studies of long-term tolerances of commercially available methanol synthesis catalysts to various compounds produced in biomass gasification.

12.8 REFERENCES

Abernathy, M. W. 1977. "Design Horizontal Gravity Settlers." Hydrocarbon Processing. Sept.

APT, Inc. 1972. Economic Feasibility Study - Fuel Grade Methanol from Coal. Wilmington, DE: E. I. DuPont de Nemours; TID-27156.

Calvert, S. 1977. "How to Choose a Particulate Scrubber." Chemical Engineering. Vol. 77 (No. 19).

Guthrie, K. M. 1974. Process Plant Estimating, Evaluation, and Control. Solana Beach, CA: Craftsman Book Company of America.

Kohl, Arthur L.; Riesenfeld, Fred C. 1974. Gas Purification. Second Edition. Houston, TX: Gulf Publishing Company.

Maddox, R. N.; Burns, M. D. 1967. "Lease Gas Sweetening." Part 6, Oil and Gas Journal. Vol. 65 (Oct. 9); Part 7, Oil and Gas Journal. Vol. 65 (Nov. 13).

Manufacturing Chemist. 1958. Feb. p. 63.

Mathematical Sciences Northwest, Inc. 1974. Feasibility Study - Conversion of Solid Waste to Methanol or Ammonia. NTIS Publication No. PB-255-249.

McGeorge, A. 1976. Economic Feasibility Study - Fuel Grade Methanol from Coal. Wilmington, DE: E. I. DuPont de Nemours, TID-27156.

Natta, G. 1955. "Synthesis of Methanol." Chapter 8 of Catalysis, P. Emmett, editor. New York: Reinhold Publishing Corp.

Neveril, R. B. et al. 1978. "Capital and Operating Costs of Selected Air Pollution Control Systems - I." J. of the Air Pollution Control Association. Vol. 28 (No. 8).

Popper, H., editor. 1970. Modern Cost Engineering Techniques. New York: McGraw-Hill.

Raab, M. 1976. "Caustic Scrubbers Can be Designed for Exacting Needs." Oil and Gas Journal. Vol. 74 (Oct. 11).

Ralph M. Parsons Company. 1978. Engineering and Economic Analysis of Waste-to-Energy Systems. Palo Alto, CA: Report No. EPA-600/7-78-086.

Raphael Katzen Associates. 1975. Chemicals from Wood Waste. NTIS Publication No. PB-262-489.

Sherwood, T. K; Pigford, R. L. 1952. Absorption and Extraction. Second Edition. New York: McGraw-Hill.

Chapter 13

Production of Fuels and Chemicals from Synthesis Gas

E. I. Wan, J. A. Simmins, T. D. Nguyen
Science Applications, Inc.

13.1 INTRODUCTION

This chapter reviews the chemistry of synthesis gas (CO and H_2 mixtures) reactions and the state-of-the-art process technologies suitable for converting biomass-derived synthesis gas to various fuels and chemicals. The review includes three major product areas:

- alcohols,
- hydrocarbon fuels and gasoline, and
- ammonia.

The section on alcohols discusses the synthesis of methanol and higher alcohols. An in-depth process evaluation and economic comparison of methanol technology is presented. The section on hydrocarbon fuels and gasoline evaluates conventional Fischer-Tropsch synthesis in terms of the various hydrocarbon fuels expected from a chain-growth process. A recent advancement in gasoline production from methanol is also presented. The final section discusses the technology of producing ammonia from synthesis gas.

13.2 FUNDAMENTAL ASPECTS OF SYNGAS CHEMISTRY

13.2.1 Thermodynamics

Reactions between hydrogen and carbon monoxide to form hydrocarbons, alcohol, and other chemicals are favored thermodynamically at lower temperatures, less than 700 C. These reactions were discovered over 75 years ago by Sabatier and Senderens. Some selected reactions are listed in Table 13-1, which also shows the approximate temperatures at which the Gibbs free energy for each reaction becomes zero and, hence, below which the reactions are favored (Stull et al. 1969). Figure 13-1 shows the temperature dependence of the equilibrium constants for most of the reactions in Table 13-1.

Several features in Fig. 13-1 are worth noting. Methane is favored at the highest temperatures, above 600 C. At lower temperatures, generally below 350 C, the formation of higher alkanes is favored at the expense of methane. Indeed, the insertion of a methylene group into a general straight chain hydrocarbon (see the reaction labeled "alkane + CH_2") is favored at temperatures below 380 C. Although not listed, branched chain hydrocarbons are favored thermodynamically at the expense of straight chains. Also, the formation of alkanes from hydrogen and carbon monoxide is favored as compared with olefins and alcohols. Within any one homologous series of alcohols and olefins, the longer chains (higher homologues) are favored.

Thus it is clear from thermodynamics alone that in nonspecific catalytic synthesis, such as the Fischer-Tropsch synthesis, a substantial amount of the products are heavy hydrocarbons. Conversions of methanol into gasoline (Mobil process) or higher alcohols (Union Carbide process) are strongly favored thermodynamically.

Table 13-1. SELECTED SYNTHESIS GAS CONVERSIONS

Reaction	Approximate T (°C) at Which $\Delta F = 0$[a]	ΔH[a] (kcal/mol syngas)	Percent of Heating Value of Syngas Lost[c]
Methane: $CO + 3H_2 = CH_4 + H_2O$	690	-12.3	18.2
Ethane: $2CO + 5H_2 = C_2H_6 + 2H_2O$	510	-11.9	17.5
Propane: $3CO + 4H_2 = C_3H_8 + 3H_2O$	470	-11.9	17.5
Nonane: $9CO + 19H_2 = C_9H_{20} + 9H_2O$	410	-12.0	17.8
Decane: $10CO + 19H_2 = C_{10}H_{22} + 10H_2O$	410	-12.0	17.8
Alkane + CH_2 $R\text{-}R' + CO + 2H_2 = RCH_2R' + H_2O$	380	-12.0	17.8
Ethylene: $2CO + 5H_2 = C_2H_4 + 2H_2$	380	-8.4	12.4
Methanol: $CO + 2H_2 = CH_3OH$	140	-10.3[b]	15.2[b]
Ethanol: $2CO + 4H_2 = C_2H_5OH + H_2O$	300	-11.8[b]	17.4[b]

[a]In standard gas states unless otherwise noted.

[b]Alcohols in liquid state.

[c]Syngas heating value is approximately 67.8 kcal/mol.

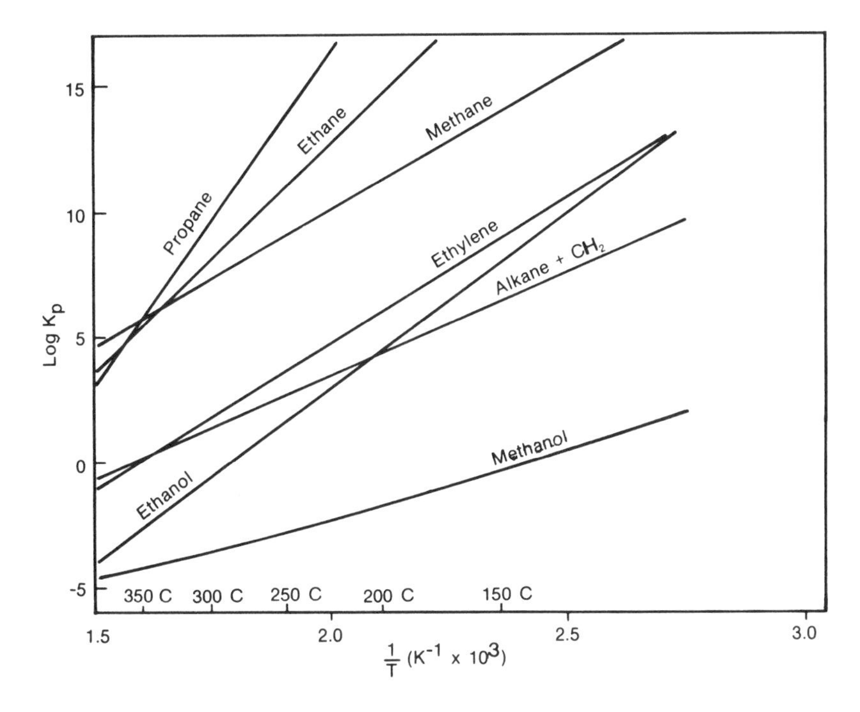

Figure 13-1. Temperature Dependence of the Equilibrium Constants for Reactions for the Synthesis of Hydrocarbons and Alcohols from Carbon Monoxide and Hydrogen

Table 13-1 (columns 3 and 4) also indicates another thermodynamic result that is important to energy efficiency in the synthesis of fuels: the heating value of the synthesis gas is degraded by conversion to other fuels, especially condensable fuels. Column 3 lists the enthalpy change for the reaction divided by the sum of the moles of the reactants. The last column is the ratio of this value to the heat of combustion of CO or H_2 (which are nearly the same on a molar basis) and represents the heat of combustion of the synthesis gas.

Syngas heating value is degraded least by its conversion to ethylene or methanol. For any hydrocarbon fuel the loss in chemical energy is less than 20% of the synthesis gas heating value. However, when the entire conversion process is considered, the net energy yield is still lower since energy is needed to operate the conversion process (e.g., energy is required for compression, gas cleaning, purification of product, etc.).

13.2.2 Kinetics and Mechanisms

Although the conversion of synthesis gas to hydrocarbons and alcohols is favored thermodynamically at temperatures below 350 C, such conversions do not proceed at a practical rate without the use of catalysts. Because of this rate limitation, an equilibrium distribution of products is never achieved in a practical reactor. Indeed, one of the most successful and widely used conversions is the synthesis of methanol which, according to Fig. 13-1, is the conversion least favored thermodynamically. To some extent the unfavorable thermodynamics are overcome by using high pressures, 50 to 200 atm in the case of methanol.

Some catalysts used for synthesis gas conversions are highly specific, favoring almost exclusively the formation of a single product. An example is the ZnO or ZnO-CuO catalysts used for methanol synthesis. Other catalysts may be less specific, especially Fischer-Tropsch catalysts and mixed oxides of Cr, Mo, Cu, Zn, alkaline earth, or alkali metals used to form higher alcohols. The Fischer-Tropsch catalysts include Group VIII metals, especially iron, cobalt, nickel, and ruthenium. Nickel catalysts, especially Raney nickel, are fairly specific for the synthesis of methane. Catalysts with other forms of nickel, and certain iron and cobalt catalysts favor polymerization of carbon atoms to form higher hydrocarbons. The catalysts containing iron and cobalt have been used in commercial Fischer-Tropsch processes to convert coal to liquid fuels.

The mechanisms of the catalytic conversions are not fully understood. For example, at least three possible mechanisms have been proposed for Fischer-Tropsch synthesis. Each mechanism has supporting but not conclusive evidence. Instead, evidence suggests that each catalyst type has a unique reaction mechanism. For these reasons, further consideration of kinetics and mechanism is deferred to discussions of the individual synthesis gas conversion processes.

Certain common features may be noted, however. First, all the conversion reactions are exothermic (column 3 of Table 13-1). Hence, reactors must be designed with provisions for removing the heat of reaction. Too high a temperature reduces the extent of equilibrium conversion and can destroy catalytic activity (e.g., by sintering).

Another common feature is that all the presently known, commercial, conversion catalysts can be poisoned by H_2S and other sulfur-containing compounds. This is especially troublesome for coal conversion but may not be too serious a problem for biomass conversion. Extensive research is underway to find catalysts less sensitive to sulfur.

13.3 ALCOHOLS

The significant alcohol synthesis technologies from CO/H_2 can be divided into two major categories: methanol synthesis and higher alcohol synthesis. A summary discussion of each alcohol synthesis technology is presented in the following sections. Detailed process technology and economic data on methanol production via biomass gasification are described.

13.3.1 Methanol Synthesis

The synthesis of methanol dates from the 1920s, when methanol was produced together with other hydrocarbon liquids by Fischer-Tropsch synthesis (Nelleo 1951). Later studies at Badische Anilin und Soda Fabrik (BASF), W. Germany led to the development of methanol catalysts and were the foundation for modern methanol synthesis technologies. This area significantly advanced in 1966 when Imperial Chemical Industries (ICI) commercialized the first low-pressure catalyst (Strelzoff 1971).

The major reactions in the synthesis of methanol are:

$$CO + 2H_2 \longrightarrow CH_3OH, \tag{13-1}$$

$$CO_2 + 3H_2 \longrightarrow CH_3OH + H_2O. \tag{13-2}$$

Table 13-2 shows that the equilibrium conversion of CO and H_2 to methanol from synthesis gas is favored by high pressure and low temperature.

Although the thermochemistry of methanol synthesis is well understood and is supported by ample data, the kinetics of the heterogeneously catalyzed reaction are still the objective of substantial research. The catalyzed reaction has been modeled in various ways, with the rate-determining step ranging from absorption of the reactants to desorption of the products.

Table 13-2. EFFECT OF PRESSURE AND TEMPERATURE UPON THE EQUILIBRIUM CONCENTRATION OF METHANOL FORMED FROM A SYNTHESIS GAS WITH A HYDROGEN-CARBON MONOXIDE MOLE RATIO OF 4:1[a]

	Mole Percent in Product Gas at Temperature (°C)					
Pressure (atm)	240	280	300	340	380	400
50	26.0	13.9	8.7	2.88	0.94	0.57
100	31.7	25.7	20.4	9.95	3.95	2.43
150	32.8	30.1	27.0	17.3	8.47	5.56
200	35.1	31.8	30.1	23.0	13.3	9.44
300	33.3	32.8	32.1	28.6	21.4	17.0

[a]From Strelzoff (1971).

Natta and his coworkers made detailed rate measurements on powdered catalysts, one consisting of ZnO and Cr_2O_3 and the other oxides of Zn, Cu, and Cr in the ratio of 2:1:1 (Natta 1955). An overall rate expression was developed for reaction:

$$r_1 = \frac{\gamma_{CO}\,\rho_{CO}\,\gamma^2_{H_2}\,\rho^2_{H_2} - \dfrac{\gamma_{CH_3OH}\,\rho_{CH_3OH}}{k_1}}{\left(A + B\,\gamma_{CO}\,\rho_{CO} + C\,\gamma_{H_2}\,\rho_{H_2} + D\,\gamma_{CH_3OH}\,\rho_{CH_3OH}\right)^3}, \quad (13\text{-}3)$$

where

r_1 is g-moles methanol/g catalyst/h,
k_1 is the equilibrium constant for reaction 13-1,
γ_i is the activity coefficient of component i, and
ρ_i is the partial pressure of component i.

The constants A, B, C, and D are characteristic of the catalyst and vary with temperature in the form:

$$\ln A = \alpha + \beta/RT.$$

Methanol synthesis catalysts are easily poisoned by sulfur-containing contaminants in the synthesis gas. Zinc catalysts can maintain their activity in gases with a sulfur content as high as 10 ppm. Copper catalysts are more sensitive; the sulfur level must be less than 0.2 ppm to avoid loss of activity (Catalytica Assoc. 1978). This may not be too important for the biomass synthesis of methanol, since most biomass materials contain little sulfur. Small amounts of sulfur can be removed by a zinc oxide guard bed or by activated carbon placed ahead of the catalyst beds.

13.3.1.1 Current Methanol Synthesis Processes

The current methanol synthesis technology is divided into three categories: the older, high-pressure technology; the newer, lower pressure technology; and a liquid-phase methanol synthesis process, presently under development.

High-pressure process. This process, representing a large fraction of the methanol production capacity at the present time, was used exclusively through 1966, when Imperial Chemical Industries introduced its low-pressure process (Strelzoff 1971). The high-pressure process operates at 300-350 atm (4400-5100 psig) and 300-400 C (570-750 F). The catalyst used is a mixed oxide of zinc and chromium in a fixed-bed reactor. The product stream is cooled to condense and remove the methanol, and the unconverted synthesis gas is recycled to the reactor. Because the activity of the Zn/Cr catalyst is low at lower temperatures, a high temperature is required to achieve reasonable reaction rates. The high temperature results in equilibrium limitations on the synthesis reaction, requiring high pressures to drive the equilibrium.

Low-pressure process. The low-pressure process, originally introduced by ICI in 1966, is now available in a number of variations. All of these processes use a copper-based catalyst and require the feed to be free of sulfur and chlorine to maintain catalyst activity.

The original ICI process operated at temperatures below 300 C (570 F) and at a pressure of 50 atm (750 psig) (Strelzoff 1971). The use of the more active Cu/Zn/Cr catalyst requires very pure synthesis gas.

The growing use of methane steam re-forming produces an extremely pure feed gas, giving the sensitive copper-based catalyst a long life. In other respects, the low-pressure process is similar to the high-pressure process, requiring methanol condensation and synthesis gas recycle. Table 13-3 lists several low-pressure processes currently available with operating conditions and reactor designs. A high-pressure process is included for comparison. This table also identifies one of the major problems in methanol synthesis, the high heat of reaction. This results in a temperature increase in the reactor catalyst bed, magnifying the equilibrium limitations on the conversion. ICI uses cold-gas injection similar to the system used in the high-pressure process. Lurgi has introduced a tube-in-shell reactor design to closely control catalyst temperature, while Topsøe employs a radial-flow converter with a copper-based catalyst capable of operation to 350 C.

Table 13-3. TYPICAL METHANOL SYNTHESIS PROCESSES IN CURRENT USE

Vendor	Catalyst	Pressure (atm)	Temperature (°C)	Reactor Type	Cooling
ICI	Cu/Zn/Al	50-100	220-290	Single fixed-bed	Multiple gas quench
Lurgi	Supported Cu	30-50	235-280	Tube in shell	Steam generation
Topsøe	Cu/Zn/Cr	50-100	220-350	Radial flow	—
Vulcan-Cincinnati	Zn/Cr	300-350	300-400	Multiple bed	Cold-shot quench, plus external gas cooling

Liquid-phase methanol synthesis process. This process is in the developmental stage, but deserves some comment since published economic analyses forecast reduced costs for the product methanol (Shewin and Blum 1976). This process addresses one of the major problems in methanol synthesis: efficient removal of the reaction heat. A minimal temperature rise is achieved by fluidizing the catalyst in an inert liquid phase which is circulated outside of the reactor where the heat is removed. This close temperature control results in increased conversion to methanol. Problem areas can be the breakdown of the catalyst particles into easily lost, fine particles; inhibition of the catalyst by the fluid; and insufficient solubility of the synthesis gas in the fluid. Comparison of the economics of this process with the ICI process project a cost advantage of approximately 15% in the methanol produced (Shewin and Blum 1976).

The efficiency of methanol production generally has been based on the thermodynamic first law: the combustion enthalpy of the products divided by the energy of the feedstocks plus energy losses. The reported efficiency values are summarized here (SAI 1978):

Process Description	Efficiency (%)
Large-scale natural gas methanol plant using ICI low-pressure process	50-60
Vulcan-Cincinnati high-pressure process	63-69
Large coal gasification plant using ICI low-pressure process	41-75
Wood biomass gasification and ICI low-pressure process	30-47

13.3.1.2 Alternative Biomass to Methanol Processes

The basic process for producing fuel grade methanol from biomass feedstocks employs a thermal gasification step as shown in Fig. 13-2. The major processing steps are described here.

Gas purification. The partially purified syngas from the biomass gasifier(s) is compressed to about 100 psig and treated in a two-stage system to remove carbon dioxide. In the first stage, a hot potassium carbonate system is used to reduce the carbon dioxide content to about 300 ppm; in the second stage, this is reduced to about 50 ppm, with methylethanolamine as the scrubbing agent. The net product is a gas that is essentially a mixture of carbon monoxide and hydrogen.

Shift reaction. After purification, the gas is compressed to 400 psig for shift conversion. Here, a portion of the carbon monoxide reacts with water vapor to form additional hydrogen, to the extent that the final gas contains the required 2:1 hydrogen to carbon monoxide molar ratio. The shift reaction also produces carbon dioxide, which must be removed from the gas prior to the methanol reaction. This is done in a second hot potassium carbonate absorption system, which removes about 97% of the carbon dioxide formed during shift.

Methanol synthesis and purification. The synthesis gas, containing a 2:1 hydrogen to carbon monoxide ratio, is compressed to 1,500 psig and fed into the methanol synthesis reactor. Approximately 95% of the gas is converted to methanol, the balance being lost in a purge stream fed to the boiler. The product then passes to a distillation train for separation of the light ends and higher alcohols. A fuel grade methanol product is produced. The mixture of light ends and higher alcohols is used as a fuel in the boiler.

13.3.1.3 Methanol Production Economics

Capital cost of methanol plants. Table 13-4 lists capital costs for methanol plants utilizing different processes and both conventional and unconventional feedstocks. Similar information for methanol plants utilizing biomass feedstocks is listed in Table 13-5. The capital cost data are summarized graphically in Figs. 13-3 and 13-4, including the data from Stull (1969). Costs of plants have been brought to a common basis for comparison.

Figures 13-3 and 13-4 show that the capital costs increase with the 0.8 power of plant capacity. For the same plant size, the capital costs of a residual oil-based plant average about 75% more than the costs of a natural gas-based plant. The capital costs of plants utilizing coal or lignite are even higher, about 150% more than the natural gas-based plant. The estimates for the capital costs of plants utilizing biomass or refuse fall into a range that overlaps the costs for coal-based plants.

Table 13-6 summarized capital costs at various scales based on "best estimate" cost lines. Where there is no entry, a plant of that size is not regarded as feasible at the present time because of raw material supply considerations (wood and refuse); or because the small size would preclude profit (natural gas, residual oil, lignite, and coal).

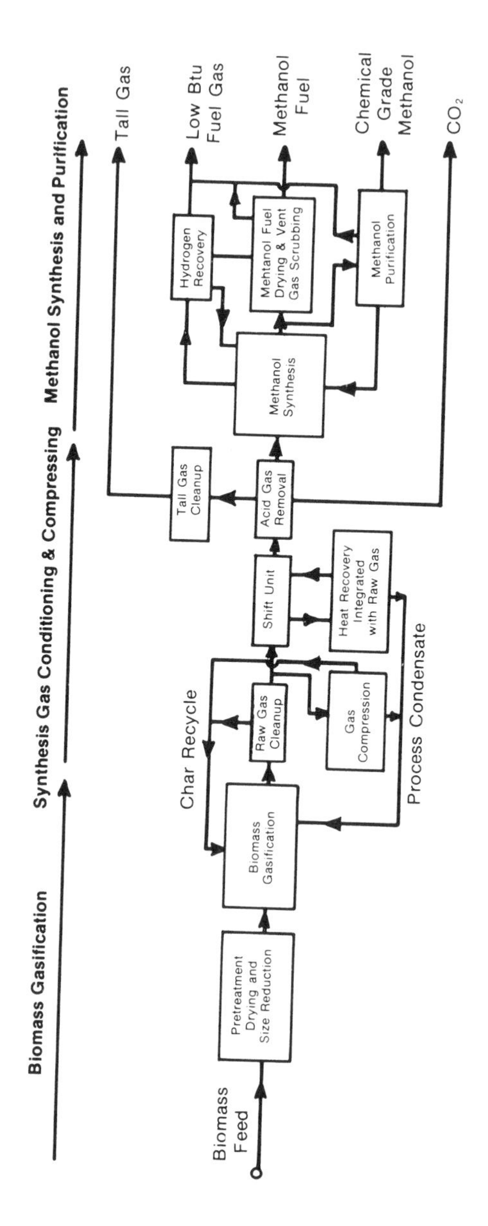

Figure 13-2. Simplified Biomass to Methanol Process Flow Diagram

Table 13-4. SUMMARY COMPARISON OF CURRENT METHANOL PRODUCTION COSTS—CONVENTIONAL PRODUCTION TECHNOLOGIES AND FEEDSTOCKS (1980)[a]

Source	Plant Size (ton MeOH/day)	Feedstock		Re-forming Oxidation or Gasification Process	Methanol Synthesis Process	Capital Cost[a] (million $)	Annual Operation & Maintenance Cost (million $)	Feedstock Cost	Unit Production Cost[b]		
		Type	Throughput per/day						($/gal MeOH)	($/ton MeOH)	($/MBtu)
Exxon Res. & Eng. Co. (1977)	2000	Natural gas	70,900 MBtu	Steam methane re-forming	Low pressure	149	13.5	$3.15/MBtu	0.64	191.7	9.9
Exxon Res. & Eng. Co.	2000	Residual oil	79,100 MBtu	Partial oxidation	Low pressure	242	15.9	$2.35/MBtu $15.0/barrel	0.74	221.6	11.4
Exxon Res. & Eng. Co.	2000	Illinois coal	3436 ton	Koppers Totzek	Low pressure	355	20.4	$21.8/ton $0.96/MBtu	0.74	225	11.4
Exxon Res. & Eng. Co.	2000	Illinois coal	3212 ton	Improved process (Texaco or Koppers-Shell)	Low pressure	315	17.8	$21.8/ton $0.96/MBtu	0.65	196	10
Badger Plants, Inc. (1978)	58,300	Coal	63,000 ton	Slagging gasifier	Lurgi low pressure	3800	593	$31/ton	0.23	69	3.7
Ralph M. Parsons (1977)	16,392	Illinois coal	24,566 ton	Foster Wheeler gasification	Chem Systems low pressure	2100	114	$31/ton $1.26/MBtu	0.41	123	6.4
Ralph M. Parsons	16,392	Illinois coal	22,918 ton	British Gas Council/ Lurgi gasification	Chem Systems low pressure	1900	110	$31/ton $1.26/MBtu	0.39	117	6.1
Ralph M. Parsons	16,392	Illinois coal	24,574 ton	Koppers-Totzek	Chem Systems low pressure	2900	163	$31/ton $1.26/MBtu	0.53	159	8.3
Ralph M. Parsons	16,392	Illinois Coal	22,100 ton	Texaco gasification	Chem Systems low pressure	2400	134	$31/ton $1.26/MBtu	0.44	134	7.0
Wilson et al. (1977)	245	Refuse 25.8% moisture	1500 ton	Purox (Union Carbide)	Low pressure	126	16	$-14/ton[b]	0.72	217	10
Mathematical Sciences Northwest (1974)	275	Refuse 25% moisture	1500 ton	Purox (Union Carbide)	ICI low pressure	31	3.1	$-6.4/ton[b]	0.42	127	6.5

[a]Costs have been extrapolated to 1980 dollars by using the Chemical Engineering Cost Index with appropriate extrapolation.
[b]Negative numbers mean that the methanol producer receives money by taking the feedstock (which is refuse in this case). This money comes from the refuse and drop charges.

Table 13-5. SUMMARY COMPARISON OF PROJECTED METHANOL PRODUCTION COSTS BIOMASS FEEDSTOCKS ($1980)[a]

Source	Plant Size (ton MeOH /day)	Feedstock		Gasification Process	Methanol Synthesis Process	Mass Conversion Efficiency dry ton (feedstock/ ton MeOH)	Capital Cost[a] (million $)	Annual Operation & Maintenance Cost (million $)	Feedstock Cost ($/dry ton)	Unit Production Cost[a]		
		Type	Throughput (dry ton per/day)							($/gal) MeOH)	($/ton MeOH)	($/ MBtu)
Reed, T. (1976)	300	Wood (dried)	900	Purox	Available commercial process	3.0	45	5.0	30.3	0.58	173	8.9
Intergroup (1978)	1000	Wood 35% moisture	2380	Purox	Available commercial process	2.3	223	16	37	0.76	229	11.8
Mackay and Sutherland (Canada) (1976)	1000	Wood (dried)	3160	Purox	ICI medium pressure	3.2	223	13.8	46	0.96	290	15
Mitre (Blake and Salo 1977)	1340	Wood 50% moisture	3400	Purox	ICI low pressure	2.5	130	21	45	0.66	199	10
Mitre (Blake and Salo 1977)	335	Wood 50% moisture	850	Purox	ICI low pressure	2.5	46	8.9	45	0.84	253	13
Raphael Katzen Associates (1975)	500	Wood waste 50% moisture	1500	Moore-Canada	Vulcan Cincinnati intermediate pressure	3.0	90	7	48	1.35	404	20.7
Raphael Katzen Associates	2000	Wood waste 50% moisture	6000	Moore-Canada	Vulcan Cin. I.P.	3.0	237	N/A	48	1.02	304.0	15.6
SRI (1978)	666	Wood 50% moisture	1000	Oxygen blow gasification	not specified	3.0	100.8	9.0	19.1	0.51	154	7.96
SRI	1990	Wood 50% moisture	3000	Oxygen blow gasification	not specified	3.0	268.7	29.4	19.1, 38.2	0.50, 0.62	150, 185	7.77, 9.53

[a]Costs have been extrapolated to 1980 dollars by using the Chemical Engineering Cost Index with appropriate extrapolation.

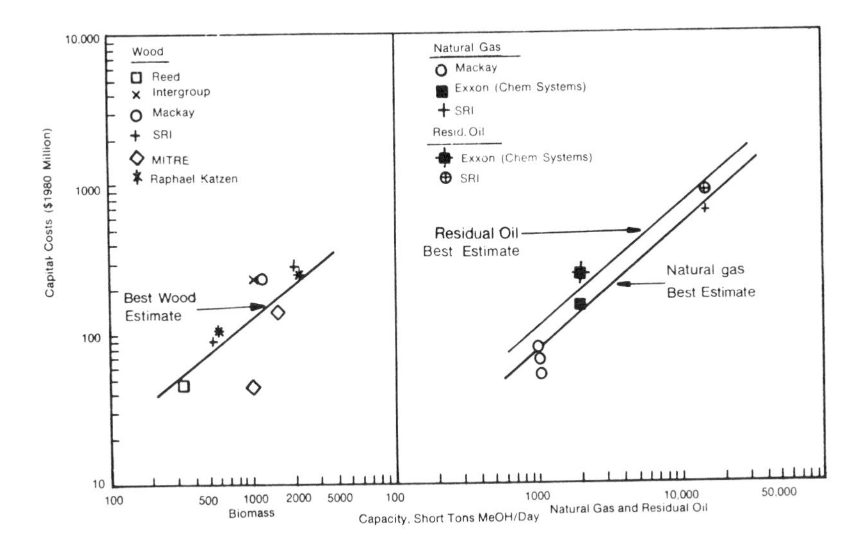

Figure 13-3. Capital Costs of Methanol Plant: Biomass and Natural Gas, Residual Oil

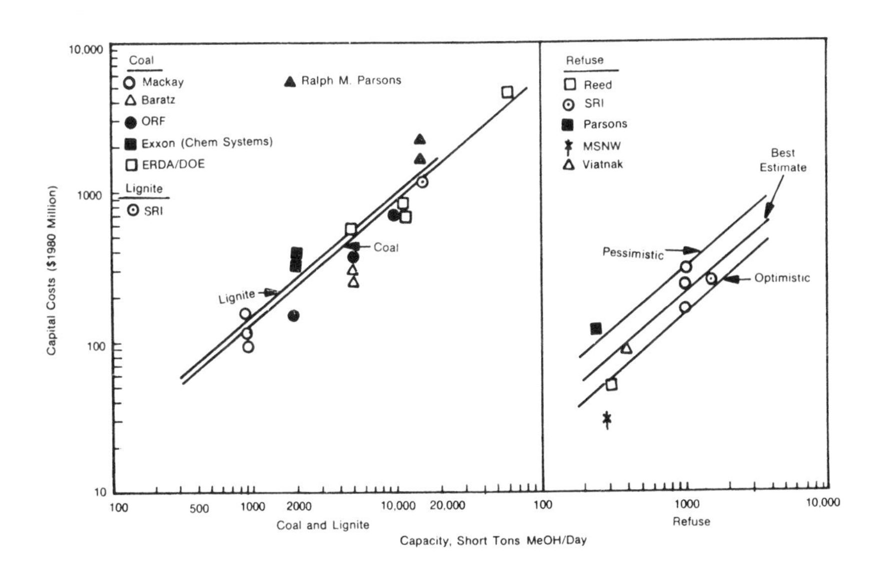

Figure 13-4. Capital Costs of Methanol Plants: Refuse and Coal, Lignite

Table 13-6. CAPITAL COSTS OF METHANOL PLANTS

(Millions of 1980 Dollars)

	Methanol Plant Capacity (ton/day)					
Feedstock	200	500	1000	2000	5000	60,000
Wood	40	80	120	220	—	—
Natural gas	—	—	78	136	283	—
Oil residue	—	—	112	190	397	—
Coal	—	—	137	238	495	3800
Lignite	—	—	146	255	531	—
Refuse	54	112	195	—	—	—

Operation and maintenance costs (O&M). O&M costs include the costs of utilities, chemicals and catalysts, labor, and maintenance and are listed in Tables 13-4 and 13-5. The O&M costs for coal plants are lower than for the SRI wood to methanol plant in the same size range. O&M costs for residual oil and natural gas plants are the lowest of all of the energy sources.

Methanol production costs. As shown in Table 13-5, typical estimated production costs of methanol from biomass range from $0.58/gal to $1.35/gal; the range for the non-biomass technologies is from $0.42/gal to $0.72/gal.

13.3.1.4 Alternative Methanol Process from Biomass-Methane Hybrid Feedstocks

Analysis of the current technology for large-scale methanol production from biomass by thermochemical gasification indicates that methanol production costs are significantly affected by plant size, feedstock cost, hybrid feedstock potential, and future technological improvements in gasification and in methanol synthesis. Aside from probable long-term technological breakthroughs in biomass gasification and methanol synthesis, the near-term commercialization of biomass to methanol processes appears to suffer from the lack of cheap biomass resources for large-scale conversion and from the high investment costs required for plant construction. However, several new concepts could alleviate the resource constraint and reduce the methanol production costs. Of particular interest is production of methanol by mixing the synthesis gases obtained from biomass gasification and from the re-forming of methane. A recent study estimated that methanol production from wood biomass could become economically competitive if it were based on the use of a biomass-methane hybrid feedstock (Intergroup 1978). The following sections discuss some of the technical and economic issues of such a system.

Technical advantages of biomass-methane hybrid methanol system. The current proposed technology for large-scale methanol production from renewable biomass feedstocks employs a thermal gasification process with no additional feedstocks after the gasification step (see Fig. 13-2). The synthesis gas thus produced is cleaned, compressed, and shifted to obtain the required stoichiometric ratio of hydrogen to carbon monoxide. Adjustment of the hydrogen-carbon monoxide ratio through use of the water-gas reaction lowers the mass conversion efficiency of the overall process.

Re-forming of methane produces a synthesis gas rich in hydrogen. Combination of these two synthesis gas streams in the proper proportion would allow adjustment of the hydrogen-carbon monoxide ratio without shift conversion while maximizing the amount of carbon available for conversion to methanol. Depending on the gasification process, this methane-hybrid system can increase methanol outputs per unit biomass feedstock to about two to five times the level achievable by biomass gasification alone. In addition, the methane-hybrid methanol system, through elimination of the shift conversion, would reduce the CO_2 scrubbing requirements. Figure 13-5 is a simplified flow diagram of alternative biomass-methane hybrid methanol processes. Table 13-7 compares mass conversion efficiencies of a biomass and a methane-hybrid methanol system.

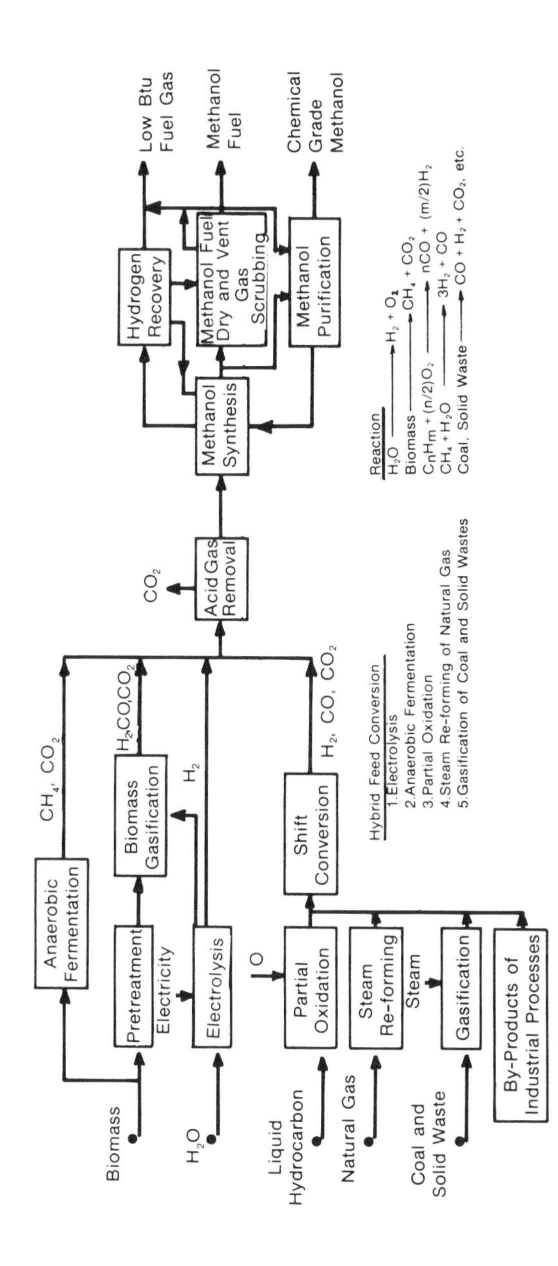

Figure 13-5. Biomass to Methanol Process Flow Diagram Using Hybrid Feedstock

Table 13-7. COMPARISON OF THE RELATIVE MASS CONVERSION EFFICIENCY OF A SIMPLE BIOMASS AND A BIOMASS-METHANE HYBRID TO METHANOL SYSTEM

(Tons of methanol per ton of dry wood feed)[a]

Gasification Process	Simple Gasification[b]	Methane Hybrid[c]
Purox	0.464	2.09
Moore-Canada	0.505	1.86
Koppers-Totzek	0.497	1.80
Wellman-Galusha	0.462	1.91

[a]Estimated yields do not include fuels used for process energy or removed from the methanol reactor purge stream.

[b]Gasification plus shift conversion to $H_2/CO = 2$.

[c]Gasification plus appropriate steam re-formed methane addition to adjust $H_2/CO = 2$.

Economic advantages of biomass-methane hybrid methanol systems. For near-term development, a biomass-methane hybrid methanol system could be constructed in two stages (within five years) without a significant capital cost penalty:

- Stage One would provide (within three years) a natural gas methanol plant with a deliberately under-capacity steam re-forming process.
- Stage Two would add the biomass gasifiers within one to three years from the start of initial methanol production.

The biomass-methane hybrid system also can be viewed as a mechanism for retrofitting existing methanol from natural gas plants. In this capacity, biomass gasifiers could replace as much as 30% of the natural gas feedstock. Therefore, the only capital cost requirement would be that associated with the biomass gasifier. It is estimated that such retrofitting arrangements would not only prolong the natural gas supply in several regions but would also reduce the methanol production cost.

For longer term development (1990 and after), methanol production from biomass can benefit economically from consideration of several methane gas production alternatives including (1) anaerobic digestion of biomass, MSW, sewage sludge, or peat; (2) synthetic natural gas (SNG) production from fossil fuels; and (3) SNG from petroleum sources. Other possible biomass hybrid feedstocks also are being investigated. One is to augment the biomass gas stream with hydrogen only, although this hydrogen hybrid is estimated to be less cost effective than the methane-hybrid systems. In the long term, the hydrogen source could be generated from electrolysis of water, closed-cycle thermochemical decomposition, and hydrogen from fossil fuels.

All of the alternative biomass hybrid feedstocks appear to offer considerable technological and economic advantages over a simple, conventional biomass-to-methanol process.

13.3.2 Higher Alcohol Synthesis

The use of fuel-grade alcohols is not severely restricted by a requirement for high product purity. In this case, the less selective catalytic processes described below may be considered for liquid fuel production from syngas.

13.3.2.1 Mixed Alcohols Using Alkali Metal Oxide Catalysts

In the intensive efforts to find a suitable methanol catalyst during the past several decades it was discovered that several metal-containing catalysts could be used to produce mixtures of alcohols at high temperatures and pressures. These catalysts include metal pairs such as $Cu-Cr_2O_3$ and $MnO-Cr_2O_3$. It has long been recognized that these catalyst components can be added to the ZnO methanol catalyst if it is desirable to make higher alcohols. Catalysts for synthesizing higher alcohols can be prepared from Cu, Zn, Mn, Mo, and a combination of an alkali or alkaline earth oxide with a metal oxide of acid character; e.g., chromates, manganates, molybdates. Previous test results of alcohol synthesis catalysts are summarized in Table 13-8. It is noted that as the alkali ion concentration increases, the yield of methanol decreases and the yield of higher alcohols increases. For example, with a catalyst of composition $Cr_2O_3/MnO/Rb_2O$ = 1:0.85:0.42, the synthesis gas consisting of CO and H_2 at 400 C and 200 atm was converted to liquid products consisting of 42% methanol, 38% higher alcohol (mostly ethanol), and 15% aldehydes and acetals.

The prospect of using a unique catalyst for the simultaneous production of methanol and ethanol appears to be attractive at this time. Further research and development work in these areas is highly desirable.

Table 13-8. METHANOL SYNTHESIS OVER ALKALI METAL OXIDE CATALYSTS AT 400 C AND 200 atm

Catalyst	Product Ratio (wt %)	% Yield (g/h)	% Methanol in Product	Other Liquid Compounds in Product
$Cr_2O_3/MnO/Li_2O$	1:0.93:0.10	47	76.9	21.7
$Cr_2O_3/MnO/Na_2O$	1:0.93:0.08	43	63.9	32.9
$Cr_2O_3/MnO/K_2O$	1:0.93:0.12	39	60.8	38.4
$Cr_2O_3/MnO/Rb_2O$	1:0.93:0.00	62	80.5	13.0
	1:0.93:0.06	61	75.5	23.1
	1:0.93:0.13	62	67.2	33.1
	1:0.93:0.25	53	49.7	46.0
	1:0.85:0.42	50	42.0	54.0
$Cr_2O_3/MnO/Cs_2O$	1:0.93:0.11	53	82.1	18.8

13.3.2.2 The "Oxo" Process

The production of aldehydes via the hydroformylation of olefins had been accomplished on the laboratory scale in the early portion of this century. Reviews of the application of this reaction (Wender et al. 1957; Gates et al. 1979) attribute to Otto Roelen of Ruhrchemie AG in Germany the discovery of catalyst composition and reaction conditions at which the reaction of the following type could occur. The reactions are members of a general class and so they are referred to as "oxo" synthesis reactions:

$$RCH{=}CH_2 + CO + H_2 \longrightarrow \begin{cases} RCH(CHO)CH_3 \\ RCH_2CH_2CHO \end{cases}$$

Reactants in the stoichiometric ratio were mixed with a cobalt catalyst at a pressure of 100 atmospheres and at temperatures between 50 and 150 C. Carbonyls [e.g., $Co_2(CO)_8$] of cobalt, iron, nickel, and rhodium have been found to be active catalysts for this reaction. If the oxo reaction is followed by hydrogenation of the aldehyde, the overall result is production of an alcohol containing one carbon atom in excess of that of the original olefin.

Wender et al. (1957) report that in the hydroformylation of ethylene the free energy change of reaction varies from -13,900 cal/mole at 200 C to -14,460 cal/mole at 25 C. Since the reaction takes place in the liquid phase, the effect of total pressure on the equilibrium yield may be expected to be small. These considerations indicate that formation of aldehydes via the oxo synthesis is thermodynamically favored at temperatures below 200 C. The reaction is highly exothermic (-28 to -35 kcal/mole) and efficient heat removal is required for control of temperature. Gates, Katzer, and Schuit (1979) report that the following rate form may be representative of the kinetics of the oxo reaction:

$$r = \frac{k\ C_C C_{RCH=CH_2}\ P_{H_2}}{P_{CO} + K\ P_{H_2}}$$

where C_C is the catalyst concentration and the pressure independent rate for 1:1 H_2/CO ratios is notable.

In the oxo reaction, olefins are reacted in the liquid phase with hydrogen and carbon monoxide in the presence of a dissolved catalyst. In addition to favorable thermodynamic effects, increased gas partial pressures of carbon monoxide and hydrogen increase the reaction rate through a proportional increase in liquid phase concentrations, and prevents decomposition of the cobalt carbonyl catalyst complex.

Presently, BASF Aktiengesellshaft in Germany and the Union Carbide Corporation in the United States operate large-scale oxo processes. In the BASF process, linear or branched chain olefins in the C_9-C_{17} range are converted to aldehydes which are hydrogenated to the corresponding alcohol. A dissolved cobalt catalyst is employed and is recycled without significant material loss. The oxo synthesis is conducted at temperatures in the 150-190 C range and at pressures in the 100-200 atmosphere range. The product alcohols are employed in the production of sulfated washing and wetting agents. In the Union Carbide process, n- and iso-butraldehyde are produced through hydroformylation of propylene at pressures of 7-20 atmospheres and at at a temperature of approximately 100 C. An organo-metallic complex of rhodium is employed to obtain a product containing an excess of the normal isomer. Separation columns are employed to provide product streams of high purity in each of the two isomers.

13.4 HYDROCARBON FUELS AND GASOLINE

In addition to methanol, the greatest development in syngas utilization since the early 1920s has been the synthesis of liquid hydrocarbon fuels. Because of the flexibility in composition of these fuels, the restrictions of product selectivity are not as severe as those in the synthesis of methanol or ammonia. For industrial application, the most desirable liquid hydrocarbon fuels are gasoline, jet fuels, diesel fuels, and gas turbine fuels.

The most highly developed technology for producing liquid hydrocarbons from syngas is the SASOL technology based on Fischer-Tropsch reactions. This technology uses promoted iron catalysts and operates at medium pressures (10-30 atm). The product distribution is broad, including light hydrocarbons as well as waxes and a considerable percentage of oxygenated compounds. High selectivity to specific fuels of the type described in Section 13.3 for alcohols is not achieved. The presence of large amounts of olefins and only a small fraction of aromatics makes the SASOL product undesirable for either gasoline or jet fuel without considerable upgrading. This problem is typical of all synthesis efforts based on the conventional Fischer-Tropsch type of catalysis.

The only novel approach to the synthesis of hydrocarbon fuels from CO and H_2 has been pioneered by Mobil Oil Company over the last five years. Instead of relying on direct synthesis of fuels, the Mobil approach first synthesizes methanol and then proceeds through dimethyl ether as an intermediate to the desired hydrocarbons. By utilizing a novel catalyst, the Mobil technology can achieve a high selectivity for products of interest for gasoline manufacture, including a high yield of aromatics and no oxygenated products. More recently, the Mobil efforts have included attempts to start directly with synthesis gas.

The Fischer-Tropsch and Mobil gasoline technologies are discussed in the following sections.

13.4.1 Fischer-Tropsch Synthesis

13.4.1.1 Catalysts, Product Distribution, and Kinetics

The Fischer-Tropsch reaction is the nonspecific catalytic conversion of hydrogen and carbon monoxide to a mixture of hydrocarbons, alcohols, and other oxygenated hydrocarbons. Since its discovery over 75 years ago, a great deal of research has focused on the activities of a variety of catalysts, catalyst preparation, tailoring catalysts for specific products, and the reaction mechanism and kinetics. The early work was done by Fischer and Tropsch in the 1920s. They demonstrated synthesis at atmospheric pressure and showed that the Group VIII metals have the highest catalytic activities. Later research in Germany emphasized cobalt at low pressures and led to the production of Fischer-Tropsch fuels in Germany during World War II. Further research led to iron catalysts and synthesis at medium pressures (10 to 20 atm). This technology was utilized in the SASOL plant in South Africa, built in 1955. The SASOL plant is the only commercial Fischer-Tropsch plant in the world today.

Some of the main characteristics of Fischer-Tropsch synthesis using the major catalyst types are listed in Table 13-9. Ruthenium is the most active catalyst but is expensive and produces mostly high molecular weight products unsuitable for use as liquid fuels. Nickel, although a very active catalyst, produces primarily gaseous hydrocarbons. Cobalt, although it is active and produces a good mix of liquid products, is expensive. Iron has slightly less activity than cobalt but is much less expensive.

Table 13-9. SYNTHESIS CHARACTERISTICS OF FISCHER-TROPSCH CATALYSTS[a]

Catalyst	Temperature (°C)	Pressure (atm)	Products
Ruthenium	20	200	Hydrocarbons, high melting waxes
Cobalt	200	1 to 10	Paraffins, olefins
Nickel	200	1 to 10	Mainly methane but some paraffins and olefins
Iron	250	20	Paraffins, some olefins, and oxygenated hydrocarbons
Zinc Oxide	250-300	100-300	Methanol

[a]From Stull (1969) and Strelzoff (1971).

Most research in the United States since World War II has dealt with iron catalysts, and most of the research has been conducted at the Bureau of Mines. Typical, commercially available iron catalysts include fused iron oxide (magnetite) with 0.4 to 0.6% K_2O, 2 to 3% Al_2O_3, and 0.2 to 0.4% SiO_2 as promoters. The catalyst may be activated by reduction with H_2 at 450 C. A typical precipitated iron catalyst contains 55.4% Fe, 12.1% Cu, and 0.6% K as K_2CO_3. This catalyst is activated by treatment with H_2 at 250 C. Another catalyst is prepared by precipitating Fe_2O_3 on Al_2O_3 followed by reduction with H_2 at 450 C. Still another may be formed from lathe turnings of carbon steel (Univ. of Connecticut 1978).

Recent research at the Bureau of Mines and the University of Connecticut (1978) has shown that nitrided and carburized-nitrided fused iron catalysts improved yields of middle distillates and reduced yields of waxes and olefins. As shown in Fig. 13-6, this is achieved only at the expense of a modest increase in the yield of methane. Approximately 50 to 60% of the synthesis gas is converted to liquid products (Shultz et al. 1957).

For iron catalysts the best synthesis conditions appear to be approximately 250 C and 20 atm pressure. The H_2:CO ratio should be in the range from 1:1 to 2:1. Lower ratios sup-

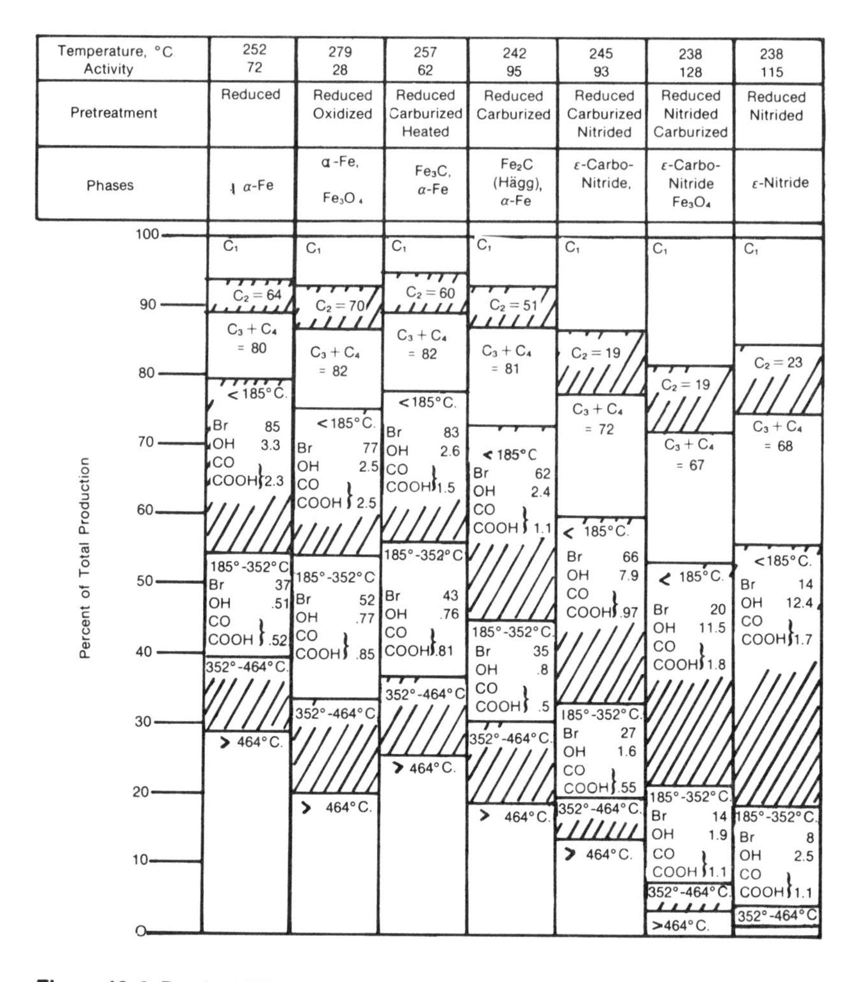

Figure 13-6. Product Distribution for Selected Fused Iron Catalysts Synthesis with $1H_2 + 1CO_2$ Gas at 300 psig. (From *Encyclopedia of Chem. Tech.* 1964)

Notations in blocks: Br, Bromine number of fraction; OH, CO, and COOH, weight-percentages of these groups.

press the formation of methane but may tend to coke the catalyst. A 1:1 ratio corresponds to the actual usage ratio of H_2 and CO in the reaction (Univ. of Connecticut 1978). Although the actual reaction is

$$2H_2 + CO = [-CH_2-] + H_2O\ , \tag{13-4}$$

part of the water produced is reconverted to hydrogen by the water-gas shift reaction at these conditions,

$$H_2O + CO = H_2 + CO_2\ . \tag{13-5}$$

Hence, the overall net reaction is approximately:

$$1.5H_2 + 1.5CO = [-CH_2-] + 0.5CO_2 + 0.5H_2O\ . \tag{13-6}$$

It is well known that water inhibits the formation of hydrocarbons, and CO_2 has been reported to be a mild inhibiter. However, experience at SASOL has shown that the presence of CO_2 reduces the selectivity of iron catalysts for CH_4 production (Dry 1976). Basic constituents in the catalyst have a similar effect.

Sulfur also is known to have an inhibiting effect. However, recent research by Exxon Research and Engineering has indicated that iron catalysts, as well as cobalt catalysts, can be made sulfur-resistant to some extent by alkali metals (Madon et al. 1977). In the same work, it was reported that sulfided (and therefore sulfur-resistant) $Co\text{-}Mo/Al_2O_3$, $Ni\text{-}W/Al_2O_3$, and KOH-promoted MoS_2 were found to be active Fischer-Tropsch catalysts.

The rate of hydrocarbon production over iron catalysts is described by the expression:

$$r = \frac{k_o e^{-E/RT} \, (P_{CO}) \, (P_{H_2})}{(P_{CO}) + a \, (P_{H_2O})} , \qquad (13\text{-}7)$$

where the inhibiting effect of water vapor is exhibited. Work at the U.S. Bureau of Mines and the University of Connecticut indicates that the activation energy is in the range of 19 to 20 kcal/mol (Univ. of Connecticut 1978). Lower values, 15 kcal at low temperatures and 6 kcal at high temperatures, were reported for the SASOL process by Dry (1976).

13.4.1.2 SASOL Process

A block diagram of the SASOL plant is shown in Fig. 13-7. Coal is gasified and the product is rigorously cleaned to contain only CO, H_2, and CH_4. The gas is divided into two streams, the larger being fed to a fixed bed reactor (Fe-Cu catalyst) operating at 230 C and 360 psig. The tail gas, which is stripped of low-boiling hydrocarbons and CO_2 by a Rectisol unit, is then combined with the remainder of the feed gas. This gas is re-formed over a nickel catalyst with steam and O_2 to produce additional synthesis gas. A fluidized bed reactor (fused iron catalyst), operating at 325 C and 330 psig, converts this gas. Both reactor units contain internal recycle streams. Typical compositions of the fresh feed to the reactors are given below.

Volume Percent

	H_2	CO	CO_2	CH_4	N_2
Fixed bed	54	32	1	13	0
Fluid bed	62	22	7	5	4

As mentioned above, the only current commercial Fischer-Tropsch plant is that at Sasolbury, South Africa. The starting material is coal and the products include a high Btu gas, gasoline, diesel oil, waxes, and chemicals. Of the coal converted, approximately 40% is gasoline and 20% is diesel fuel. As of the early 1960s, only about 18% of the coal fed to the plant was converted to liquid products. Approximately 42% is used to provide power and process steam. The plant is commercially successful only because of a very unusual economic situation in South Africa. In a recent study by Air Products, fuels from Fischer-Tropsch were determined not to be competitive in the United States with methanol fuel synthesized from CO and H_2 (Drissel 1977).

Numerous other Fischer-Tropsch processes have been proposed but none have become commercial. Most of them have unique ways of removing the heat of reaction and controlling the reactor temperature. The U.S. Bureau of Mines has developed two processes: the hot gas recycle, in which all the heat is removed in the gas; and a recycled catalyst-oil slurry, in which the heat is removed by the oil. This latter process is similar to the Chem Systems three-phase process for methanol synthesis.

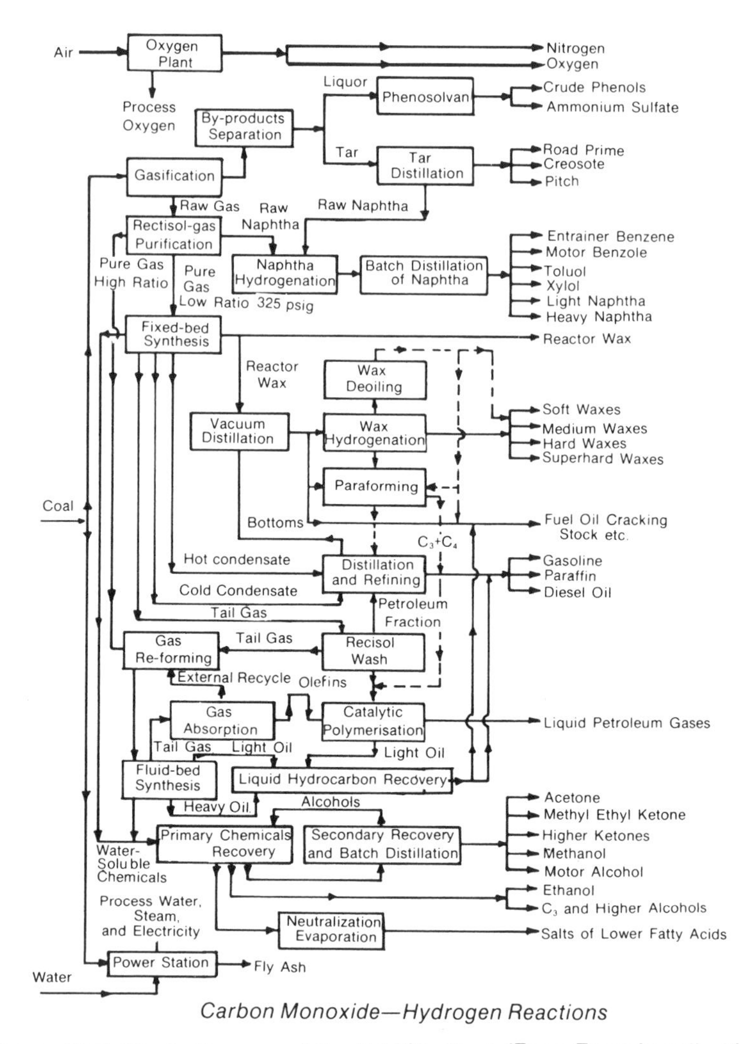

Figure 13-7. Block Diagram of the SASOL Plant (From *Encyclopedia of Chem. Tech.* 1964)

Fischer-Tropsch synthesis appears to be well adapted for biomass conversion. Especially intriguing is the possibility of performing the conversion with little or no chemical cleaning of the gas from the gasifier. Gasified biomass typically has an H_2:CO ratio of approximately unity, which is correct for iron catalysts. The gas contains appreciable CO_2, which is beneficial for the production of liquid products. Finally, the gases from most biomass materials contain little sulfur, which is important if presently available catalysts are to be used. In spite of this adaptability, Fischer-Tropsch conversion may not be able to compete economically with conversion to methanol.

13.4.1.3 Fischer-Tropsch Liquid Fuels Costs

The literature cost data (Table 13-10) are from R. M. Parsons' study (1977) performed for the Electric Power Research Institute.

Table 13-10. FISCHER-TROPSCH COSTS[a]

(1980 Dollars)

Plant size:	5573 tons or 223,545 MBtu per day
Feedstock:	22,918 tons Illinois coal/day at $31/ton or $1.26/MBtu
Gasifier:	British Gas Council/Lurgi Slagger
Capital Cost:	$2,200M
Annual Operation and Maintenance:	$122M
Production Cost:	$6.7/MBtu or $269/ton

[a]From Ralph M. Parsons (1977).

R. M. Parsons' researchers judged that their cost estimates for the Fischer-Tropsch process are less accurate than their cost estimates for methanol production with the British Gas Council/Lurgi Slagging, Koppers-Totzek, and Texaco gasifiers. This is due to technological uncertainties. However, the capital, operating, and production costs per MBtu are within the same range as those from coal-to-methanol plants. The difference in capital costs between the British Gas Council/Lurgi Slagging methanol plant and the Fischer-Tropsch liquids plant arises from the higher capital cost of the Fischer-Tropsch synthesis unit (about twice the cost of the methanol synthesis unit).

13.4.2 Mobil Gasoline Technology

A class of crystalline zeolite catalysts recently has been discovered which can induce transformation of short chain aliphatic hydrocarbons to mixtures of higher aliphatics, olefins, and alkyl-substituted aromatics. Moreover, the catalysts and associated conversion processes can be tailored to give a mixture, in high yield, which shows promise as a direct substitute for high-octane gasoline. The most publicized process of this kind is the Mobil gasoline from methanol process (Voltz et al. 1976). In this process, industrial-grade methanol is converted to hydrocarbons consisting mainly (greater than 75%) of a gasoline grade material with small amounts of LPG (C_3 and C_4) and fuel gas (C_1 and C_2). The overall gasoline yield can be increased to over 90% by alkylating the C_3 and C_4 olefins with the isobutane produced by the process. Figure 13-8 depicts the Mobil methanol-to-gasoline process flow scheme using a fixed bed reactor system.

The Mobil methanol-to-gasoline process offers a new route for the conversion of biomass to high-octane gasoline and other desirable products. The raw gasoline product is 30 to 50% aromatics, 45 to 55% isoparaffins, and the balance olefins, with an unleaded research octane number of over 90. Therefore, the gasoline product from the Mobil process could be used alone or it could be blended with petroleum-derived gasoline.

13.4.2.1 Reaction Path and Potential Product Characteristics

The reaction path of the methanol to gasoline process appears to be represented by the following mechanisms:

$$2CH_3OH \xrightarrow{-H_2O} CH_3OCH_3 \xrightarrow{-H_2O} \underset{\text{olefins}}{C_2\text{-}C_5} \longrightarrow \begin{matrix}\text{paraffins}\\ \text{aromatics}\\ \text{cycloparaffins}\end{matrix}$$

Two versions of this process were explored by Mobil. In the first, denoted the fixed bed process, the conversion is carried out in two stages, each employing a separate catalyst. In the first stage, methanol is dehydrated to an equilibrium mixture of methanol and dimethyl ether. In the second stage, this mixture is passed over a proprietary "conversion catalyst" to form the desired gasoline mixture. Olefins appear as intermediates. In the second version, a mixture of both catalysts is used in a fluidized bed reactor.

Figure 13-9 shows the effect of space velocity on product distribution in methanol conversion to gasoline products.

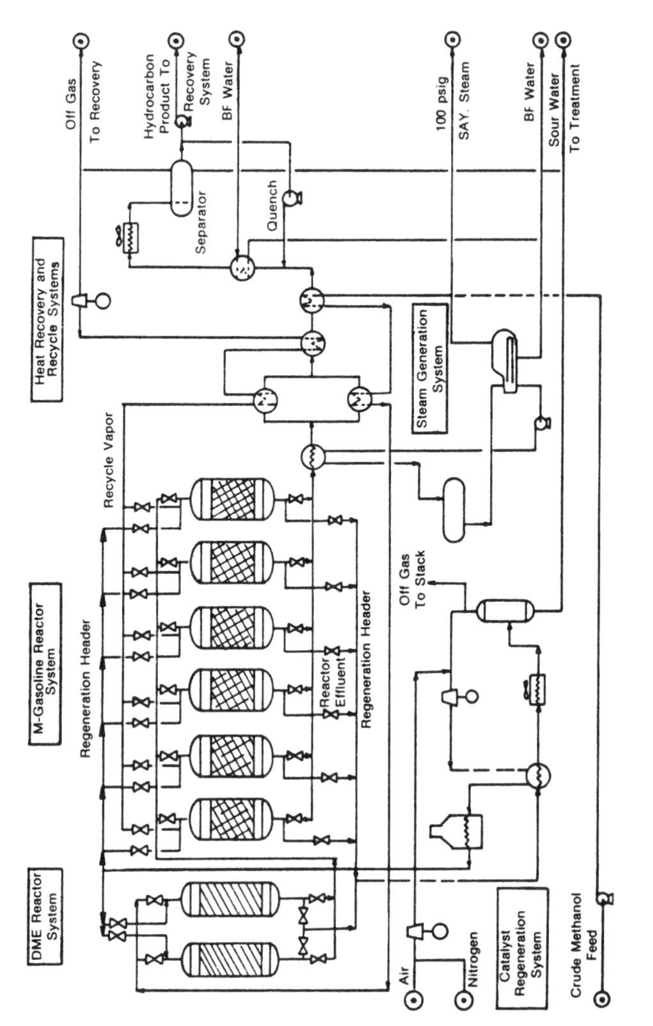

Figure 13-8. Mobil Methanol to Gasoline Process Flow Scheme — Fixed Bed Option (From Intergroup Consulting Economists 1978)

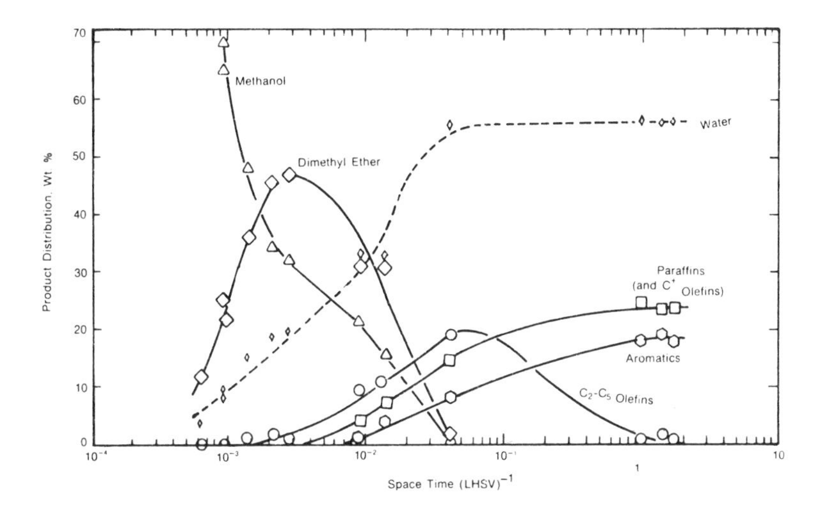

Figure 13-9. Product Distributions as a Function of Space-Time

Since the reaction path in the Mobil process indicates that the primary hydrocarbon products are light olefins, it is possible, at a low oxygenate conversion per pass, to produce ethylene and/or propylene. Laboratory work (Wise et al. 1977) has shown that, with catalyst and process modifications, it is possible to increase the level of the more desirable ethylene to about 30% at approximately 48% oxygenate conversion.

13.4.2.2 Alternative Gasoline Conversion Processes

Figure 13-10 shows an integrated process for converting biomass to high-octane gasoline via the fluidized bed reactor version of the Mobil methanol-to-gasoline process.

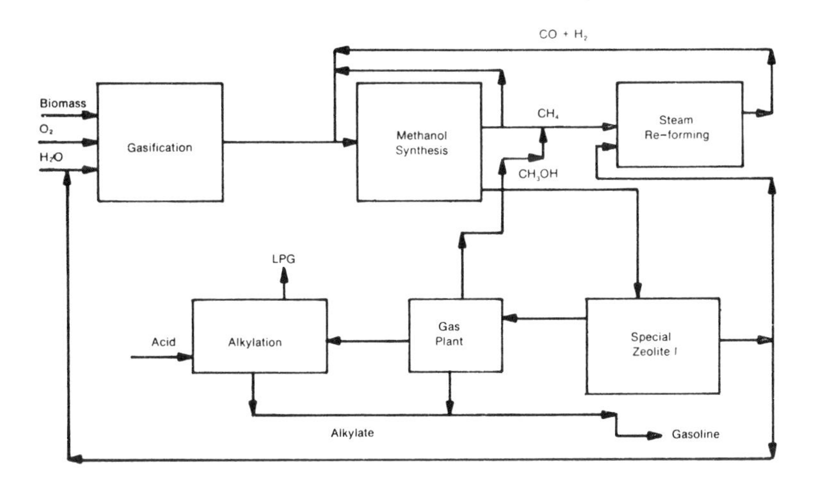

Figure 13-10. Integrated Simplified Flow Diagram of a Biomass-to-Gasoline Process

Figure 13-10 shows that biomass, oxygen, and steam are suitably reacted to produce a synthesis gas that is admixed with auxiliary synthesis gas. The synthesis gas mixture is converted to methanol via a methanol synthesis loop. The unreacted portion of the synthesis gas may be separated into a stream comprised of methane and a stream comprised of carbon monoxide and hydrogen, or it may be processed further without separation. In either case, a methane stream is steam re-formed to produce auxiliary synthesis gas. The organic portion of the product is primarily methanol and is converted to gasoline via a special zeolite catalytic process. The products from this conversion include water, which is recycled either to biomass gasification or to steam re-forming, or both, and a hydrocarbon product comprised of C_5+ aromatic gasoline and C_4- hydrocarbons.

In addition to the Mobil methanol-to-gasoline process, there are several other process alternatives for converting biomass-derived synthesis gas to gasoline products. They are described here.

A two-stage conversion of synthesis gas to dimethyl ether followed by conversion to gasoline products (U.S. patent 4,011,275). This two-stage process for the conversion of synthesis gas (mixed CO and H_2) to gasoline involves (1) contacting synthesis gas with a modified methanol synthesis catalyst to produce a mixture of dimethyl ether and methanol; and (2) contacting the first-stage product, in its entirety, with a crystalline aluminosilicate catalyst to convert it to high-octane gasoline.

Conversion of synthesis gas having a smaller hydrogen to carbon monoxide ratio than that required for methanol stoichiometry is achieved by passing it over a zinc-chromium acid or copper-zinc-alumina-acid modified methanol synthesis catalyst. The product is a mixture of methanol and dimethyl ether. The mixture is then converted to hydrocarbons in a second stage, using zeolite catalysts operating at 700 F and a space velocity of 1 LHSV (liquid hourly space velocity) to produce a stream consisting primarily of C_5 aromatic hydrocarbons.

Conversion of synthesis gas to methanol followed by carbonylation (U.S. patent 4,039,600). This process involves reacting carbon monoxide and hydrogen at about 450 to 750 F, in contact with a methanol synthesis catalyst, to yield a gas stream of methanol and carbon monoxide. This mixture then is reacted at about 300 to 800 F in contact with a carbonylation catalyst, to form methanol and acetic acid. With a zeolite catalyst at about 500 to 1200 F, this mixture is converted to aromatic hydrocarbons in the gasoline-boiling-range. The particular admixture produced by the combination of methanol synthesis followed by carbonylation is convertible to a product unexpectedly higher in aromatic hydrocarbons than that predicted from a consideration of the conversion obtainable from individual reactants.

13.4.2.3 Economics of Gasoline Production

Costs of Mobil's methanol to gasoline process. Mobil Oil's process requires approximately 2.4 gal of methanol per gal of synthetic gasoline. The conversion cost from methanol escalated to 1980 is $0.063 per gallon of gasoline (Voltz et al. 1976). This cost does not include the cost of producing the methanol feed. The total cost of each gallon of gasoline is thus the cost of manufacturing 2.4 gal of methanol plus $0.063. Assuming 0.13 MBtu per gallon of synthetic gasoline, the production cost can be determined for this product. The calculations are shown in Table 13-11 for various sources and costs of methanol. On a Btu basis it is apparent that synthetic gasoline is about 23% more expensive than synthetic methanol.

Table 13-12 shows the projected cost of producing gasoline via the Mobil process and is compared with estimates of production costs of obtaining gasoline from a synthetic crude oil produced from both coal and shale raw materials.

The alternative conversion process schemes differ from the basic Mobil methanol to gasoline process in that, while the synthesis gas produced from biomass is basically deficient in hydrogen for methanol synthesis, such synthesis gas may be ideally suited for conversion to gasoline or other products without the expense of going through an intermediate conventional methanol production stage. These process schemes could not only increase the carbon utilization efficiency from biomass resources but could also eliminate costly unit processes such as water-gas shift conversion and methanol purification.

Table 13-11. TYPICAL PRODUCTION COSTS OF MOBIL'S SYNTHETIC GASOLINE IN 1980

(Methanol costs from Tables 13-4 and 13-5)

Source	Feedstock	Cost of Methanol	Cost of Gasoline	
		($/MBtu)	($/gal)	($/MBtu)
Exxon	Coal	11.5	1.84	14.10
Badger	Coal	3.7	0.62	4.80
Ralph M. Parsons	Coal	8.3	1.34	10.30
Ralph M. Parsons	Refuse	10.0	1.79	13.80
Intergroup Consulting Economists	Wood	11.8	1.89	14.50
Raphael Katzen Associates	Wood	15.6	2.51	19.30

Table 13-12. SYNTHETIC GASOLINE COSTS IN 1980 DOLLARS

Sources	Feedstock	Process	Production Cost ($/MBtu)	Percent of Methanol Cost (from Coal)
Lawrence Livermore Laboratory (Intergroup Consulting Economists 1978)	Coal	Refining syncrude	5.1	75 (MeOH from Koppers-Totzek gasifier)
Amax Inc. (Intergroup Consulting Economists 1978)	Coal	Refining syncrude	8.0	115
Exxon Research & Engineering Co. (1977)	Shale	Refining syncrude	2.6	89 (MeOH from Lurgi gasifier)
	Coal	Refining syncrude	3.8	123 (MeOH from Lurgi gasifier)
Mobil (Voltz et al. 1976)	Coal	Conversion of methanol	Varies with methanol costs 8-13	123

13.5 AMMONIA

13.5.1 Thermodynamic and Kinetic Considerations

Ammonia is produced in large scale by passing hydrogen and nitrogen over an iron-based catalyst at elevated pressure and moderate temperature. The overall chemical reaction is expressed as:

$$N_2 + 3H_2 \xrightarrow{\text{catalyst}} 2NH_3$$

The equilibrium among N_2, H_2, and NH_3 is shown in Table 13-13 for the percentage of ammonia at equilibrium to 200 atm pressure. These data show the very beneficial effect of pressure on ammonia conversion at equilibrium and the opposite effect of increase in temperature.

The chemical processes involved in ammonia synthesis are fairly complicated, as are many heterogeneous catalytic reactions. At the conditions used in industrial ammonia synthesis, it appears that this step is the chemisorption of nitrogen onto a surface covered mainly by nitrogen atoms. The equation most widely used over the years to correlate

Table 13-13. PERCENTAGES OF AMMONIA AT EQUILIBRIUM[a]

Temperature, (°C)	$\frac{P_{NH_3}}{P^{1/2}_{N_2} \times P^{3/2}_{H_2}}$	Ammonia in gas mixture (%) at pressures (atm)			
		1	30	100	200
200	0.660	15.3	67.6	80.6	85.8
300	0.070	2.18	31.8	52.1	62.8
400	0.0138	0.44	10.7	25.1	36.3
500	0.0040	0.129	3.62	10.4	17.6
600	0.00151	0.049	1.43	4.47	8.25
700	0.00069	0.0223	0.66	2.14	4.11
800	0.00036	0.0117	0.35	1.15	2.24
900	0.000212	0.0069	0.21	0.68	1.34
1000	0.000136	0.0044	0.13	0.44	0.87

[a]From Slack and James (1977)

ammonia synthesis rate data is the Tempkin-Pyzhev equation (Slack and James 1977) as shown in Table 13-14. In these equations, ω is the net reaction rate; k_1 and k_{-1} are the rate constants for synthesis and decomposition, respectively; and α is a constant.

Table 13-14. KINETIC EXPRESSIONS FOR AMMONIA SYNTHESIS[a]

$$\omega = k_1 P_{N_2} \left[\frac{P_{H_2}{}^3}{P_{NH_3}{}^2}\right] - k_{-1} \left[\frac{P_{NH_3}{}^2}{P_{H_2}{}^3}\right]^{1-\alpha}$$

$$\omega = \frac{k_{-1}\left(a_{N_2} K^2 - a_{NH_3}{}^2 / a_{H_2}{}^3\right)}{NN}$$

$$NN = [1 + K_3\, a_{NH_3} / a_{H_2}{}^{\omega}]^{2\alpha}$$

$$k_{-1} = k^{o}_{-1} \exp(-E_{-1}/RT)$$

$$k_3 = k^{o}_3 \exp(-E_3/RT)$$

$$= \frac{k P_{N_2}{}^{1-\alpha}\left[1 - \frac{1}{K} \frac{P_{NH_3}{}^3}{P_{N_2} P_{H_2}{}^3}\right]}{\left[\frac{1}{P_{H_2}} + \frac{1}{K} \frac{P_{NH_3}{}^2}{P_{N_2} P_{H_2}{}^3}\right]^{\alpha} \left[\frac{1}{P_{H_2}} + 1\right]^{1-\alpha}}$$

$$\omega_+ = k' P_{H_2}{}^{\alpha} P_{N_2}{}^{1-\alpha}$$

[a]From Slack and James (1977)

The heart of any ammonia plant is the synthesis catalyst. The main constituents of the ammonia catalyst are FeO and Fe_2O_3. Modern catalysts differ from the early ones mainly in the amount of metallic oxides added as promoters. These metallic oxides may include the oxide of aluminum, calcium, potassium, silicon, and magnesium.

13.5.2 Ammonia Synthesis Processes

At the present time, ammonia synthesis processes may be classified according to synthesis loop pressures as high pressure (500-800 atm), medium pressure (240-350 atm), and low pressure (100-190 atm). A flowsheet for the production of ammonia by a typical process, but starting with clean synthesis gas from a wood biomass gasifier, is shown in Fig. 13-11. The major processing steps are described here.

13.5.2.1 CO Shift

The synthesis gas is preheated to 550 F prior to entering the first-stage shift reactor. The gas is quenched with condensate to 400 F before it enters the second-stage shift.

13.5.2.2 Carbon Dioxide Absorption

The synthesis gas is then passed through the regenerator reboiler of a Benfield type CO_2 scrubbing system. The condensate from the reboiler passes to a degasser, where the process condensate is returned to the waste heat boiler as makeup. The synthesis gas then passes through the absorber where the CO_2 is absorbed at high pressure with the Benfield solution. The Benfield process is basically a promoted hot carbonate process.

The CO_2-enriched Benfield solution from the bottom of the absorber passes to a turbine, where its pressure is reduced, and then to the regenerator. The rich solution at low pressure is stripped free of CO_2 in the regenerator, and the Benfield solution then is recycled to the absorber.

13.5.2.3 Methanation

The synthesis gas from the Benfield system is methanated to remove the remaining carbon monoxide and carbon dioxide. The gas is preheated to 500 F by heat exchange with the gasifier exit stream. The effluent from the methanator is cooled in a water-cooled condenser to remove most of the water from the synthesis gas. The balance of the water is removed by means of a refrigerated condenser. Final traces of CO_2 and water are removed by means of a molecular sieve.

13.5.2.4 Ammonia Synthesis Loop

The makeup synthesis gas is compressed in a multiple-stage reciprocating compressor and pumped into the synthesis loop. The ammonia converter consists of a multiple-bed cold gas quench reactor, where the product of gas-ammonia mixtures is separated through a series of heat exchangers and condensers. The unconverted synthesis gas is recycled to the ammonia converter via a recycle compressor.

13.5.3 Economics of Ammonia Production

13.5.3.1 Capital Costs for Ammonia Plants

The capital costs from Tables 13-15 and 13-16 are summarized in Fig. 13-12. Wood-fed ammonia plants show a cost versus plant size exponent of 0.8, based on SRI data (Schooley et al. 1978), and 0.6 based on Mitre data (Blake and Salo 1972). Therefore, an average "best" estimate of 0.7 was assumed for this type of ammonia plant. This cost line is placed between the SRI cost line (high) and Mitre cost line (low) in Fig. 13-12.

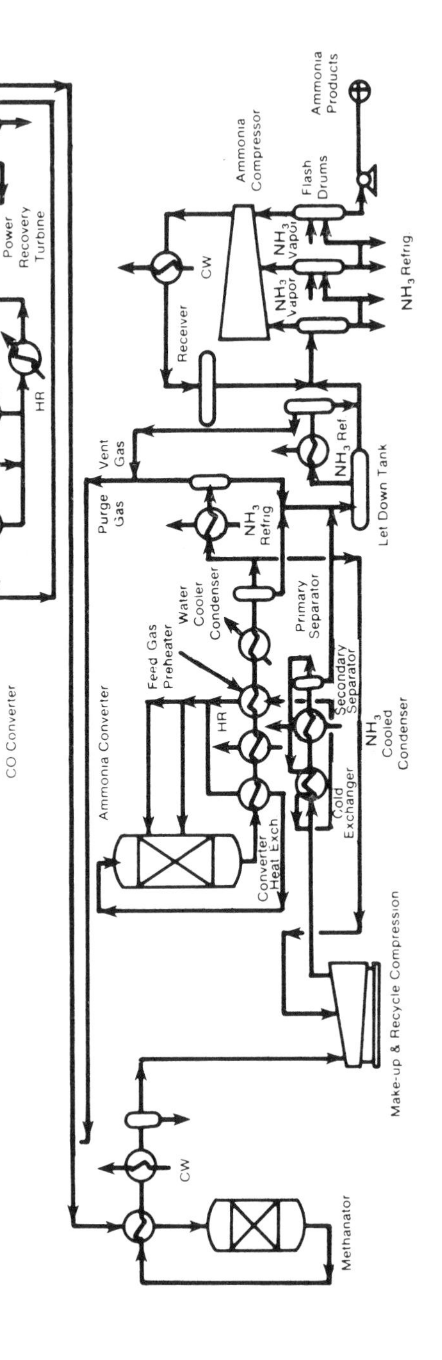

Figure 13-11. Ammonia Process

Table 13-15. SUMMARY COMPARISON OF CURRENT AMMONIA PRODUCTION COSTS ($ 1980) FROM NON-BIOMASS FEEDSTOCKS[a]

Source	Plant Size (ton NH_3 /day)	Feedstock		Re-forming, Oxidation or Gasification Process	Capital Cost (million $)	Annual Operation & Maintenance Cost (million $)	Feedstock Cost	Unit Production Cost ($/ton)
		Type	Throughput per/day					
Exxon Research and Engineering Co. (1977)	2000	Natural gas	70,000 MBtu	Steam methane re-forming	193.2	10.7	$3.15/MBtu	210
Exxon Research and Engineering Co. (1977)	2000	Residual oil	12,424 gal	Partial oxidation	292.2	19.2	$15/barrel $2.35/MBtu	248
Exxon Research and Engineering Co. (1977)	2000	Illinois coal	3545 ton	Koppers Totzek	400.5	22.98	$21.8/ton $0.96/MBtu	248
Exxon Research and Engineering Co. (1977)	2000	Illinois coal	3315 ton	Improved process (Texaco or Koppers-Shell)	367.2	20.72	$21.8/ton $0.96/MBtu	227
Ralph M. Parsons (Wilson et al. 1977)	350	Refuse 25.8% moisture	1500	Purox (Union Carbide)	140	17.9	$-14/ton[b]	134
Mathematical Sciences Northwest (1974)	335	Refuse 25% moisture	1500	Purox (Union Carbide)	39.7	3.9	$-6.4/ton[b]	134

[a]Costs have been extrapolated to 1980 dollars by using the Chemical Engineering Cost Index with appropriate extrapolation.

[b]Negative numbers mean that the plant makes money by disposing of the refuse and collecting a fee.

Table 13-16. SUMMARY COMPARISON OF AMMONIA PRODUCTION COSTS ($ 1980) FROM BIOMASS FEEDSTOCKS[a]

Source	Plant Size (ton MeOH /day)	Feedstock		Gasification Process	Mass Conversion Efficiency dry ton (feedstock/ ton MeOH)	Capital Cost (million $)	Annual Operation & Maintenance Cost (million $)	Feedstock Cost ($/dry ton)	Unit Production ($/ton MeOH)
		Type	Throughput (dry ton per/day)						
Mitre (Blake and Salo 1977)	492	Wood 50% moisture	850	Purox gasification	1.7	53.8	9.4	45	154
Mitre (Blake and Salo 1977)	1970	Wood 35% moisture	3400	Purox gasification	1.7	132.9	21.4	45	120
McKee Corp. (1978)	400	Brava (Bamboo)	1270	Thermex gasification	3.2	64 (in Nicaragua)	33.2	19.6	213 (in Nicaragua)
SRI (Schooley et al. 1978)	500	Wood 50% moisture	1000	Oxygen blown gasification	2.0	110.1	9.6	19.1	300
SRI (Schooley et al. 1978)	1542	Wood 50% moisture	3000	Oxygen blown gasification	2.0	267.3	20.6	19.1 38.2	249 287

[a]Costs have been extrapolated to 1980 dollars by using the Chemical Engineering Cost Index with appropriate extrapolation.

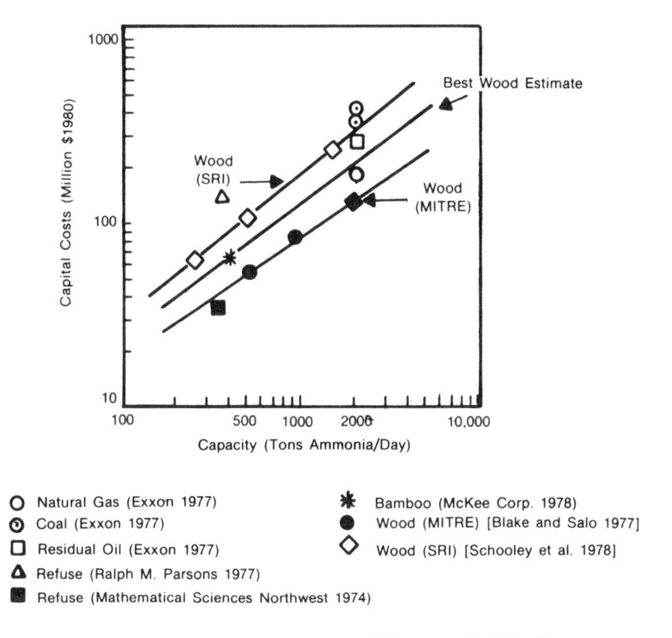

Figure 13-12. Capital costs of Ammonia Plants

Coal plants cost twice as much as methane-steam re-forming plants, while residual plants cost 50% more than re-forming plants, based on Exxon (1977) data. Mathematical Sciences Northwest's (1974) estimated cost for a refuse plant appears low, while Ralph Parsons' (1977) cost appears to be in the high range. To be conservative, Ralph M. Parsons' data are assumed to be more representative in light of the required equipment for shredding refuse and reclaiming metals. In addition, the Parsons data are more recent than those from Mathematical Sciences Northwest, and the costs of prototype equipment such as the gasifiers may be more current than those used by Mathematical Sciences Northwest.

The capital requirements for biomass-to-ammonia plants are slightly lower than those for coal, residual oil, and refuse plants. Table 13-17 summarizes capital costs at various scales based on "best estimate" cost lines. The cost lines for new biomass plants were assumed to have an exponent of 0.7.

Table 13-17. CAPITAL COSTS OF AMMONIA PLANTS

(Millions of 1980 dollars)

	Ammonia Plant Capacity (ton/day)				
Feedstock	200	500	1000	2000	5000
Wood	40	80	130	220	—
Natural gas	—	—	120	193	366
Oil residual	—	—	180	292	554
Coal	—	—	246	400	760
Refuse	95	180	—	—	—

13.5.3.2 Operation and maintenance costs

Tables 13-15 and 13-16 also show the annual estimated plant operating and maintenance (O&M) costs. These include utilities (power and water), chemicals, labor, overhead, and maintenance. No capital depreciation charges or base feedstock costs are included. The costs show no definite pattern, although it is apparent that for plants in the 1500-2000 tons of ammonia per day range, annual O&M costs are quoted around $20 million for the residual oil, coal, and wood plants. Steam re-forming of methane results in the lowest O&M cost.

13.5.3.3 Ammonia production costs

Tables 13-15 and 13-16 show that the estimated production costs of biomass-based ammonia range from $120 to $300/ton, while ammonia costs from other feedstocks range from $134 to $248/ton. The differential range is narrower than for methanol, indicating that ammonia may be able to penetrate the market more rapidily than methanol.

13.6 PROSPECTS FOR FUTURE RESEARCH AND DEVELOPMENT

A profitable potential exists for converting biomass-derived synthesis gases to fuels and chemicals through any of several thermochemical processes. The prospects for each type of process are summarized here.

13.6.1 Specialty Chemical Production

Ammonia production from biomass by current technology is both technically and economically attractive. The margin of this attractiveness should be enhanced by future technological improvements in biomass gasification for the production of hydrogen. The synthesis of other nitrogen-containing compounds from synthesis gas and simple organic molecules also should be explored. Such compounds might include aromatic isocyanates and simple amines.

13.6.2 Alcohol Fuels

In the near term, methanol is one of the most promising liquid fuels to be produced from biomass feedstocks. This can be realized by using a methanol hybrid production system with biomass and either methane or hydrogen feedstocks. The technical and economic advantages of such systems appear to allow biomass to compete with large-scale methanol production from coal and natural gas.

In the long term, new technologies may play a significant role in improving the methanol-from-biomass production economics and also may provide conversion process alternatives for the production of higher alcohols and gasoline products. The new technologies include improved methanol synthesis processes, direct higher alcohol synthesis, conversion of methanol to gasoline (Mobil processes), and improvements in biomass gasification technology to produce a more easily used synthesis gas.

13.6.3 Fischer-Tropsch Products

Several aspects of the current, commercial Fischer-Tropsch process limit the potential application of this technology to biomass feedstocks. This process results in higher costs of liquid fuels than would be true for the Mobil gasoline process or methanol synthesis. However, opportunities exist for integrating an alcohol fuel, chemicals, and hydrocarbon fuel production by a Fischer-Tropsch synthesis. The economic attractiveness of an integrated production system depends upon the market potential of various products and by-products. Also, carefully integrating process design and optimizing products for biomass feedstocks may be beneficial.

13.6.4 Gasoline Products

The new technology developed by Mobil for synthesizing gasoline from methanol and for direct synthesis of gasoline from synthesis gas may be economically attractive. Conceptual processes should be evaluated, especially those that include new biomass gasification techniques tailoring the synthesis gas composition to specific process requirements.

13.7 REFERENCES

Badger Plants, Inc. 1978. Conceptual Design of a Coal to Methanol Commercial Plant. FE-2416-24.

Blake, D.; Salo, D. 1977. Systems Descriptions and Engineering Costs for Solar-Related Technologies. Volume IX. Mitre Report No. MTR-7485.

Catalytica Associates, Inc. 1978. Evaluation of Sulfur-Tolerant Catalytic Processes for Producing Peak-Shaving Alcohol Fuels. Palo Alto, CA: Electric Power Research Institute; Report AF-687, EPRI TPS 76-649.

Drissel, G. M. 1977. "Economics of Ethylene Production via Pyrolysis of Coal-Based Fischer-Tropsch Hydrocarbons." Synthetic Fuels Processing. Pelofski, A. H., editor. Marcel Dekker, Inc.

Dry, M. E. 1976. "Advances in Fischer-Tropsch Chemistry." Industrial and Engineering Chemistry Production, Research and Development. Vol. 15: p. 282.

Exxon Research and Engineering Co. 1977. Production Economics for Hydrogen, Ammonia, and Methanol During the Period 1980-2000. Prepared for Brookhaven National Laboratory, Long Island, NY.

Gates, B. H.; Katzer, J. R.; Schuit, G. C. A. 1979. Chemistry of Catalytic Processes. New York: McGraw-Hill Book Co.

Gulf Oil Corporation. 1977. 1977 Petrochemical Handbook. Hydrocarbon Processing Reprint. Houston, TX: Gulf Publishing Co.

Intergroup Consulting Economists Ltd. 1978. Liquid Fuels from Renewable Resources; Feasibility Study. Prepared for the Government of Canada.

Kirk, R. E.; Orhmer, D. F., eds. 1964. Encyclopedia of Chemical Technology. Vol. 4, Second Edition. New York: Wiley Interscience.

Mackay, D.; Boocock, D.; Sutherland, R. 1977. The Production of Synthetic Liquid Fuels for Ontario. Toronto, Canada: University of Toronto.

Mackay, D.; Sutherland, R. 1976. Methanol in Ontario, a Preliminary Report Prepared for the Ministry of Energy. Toronto, Canada: University of Toronto.

Madon, R. J.; Backer, E. R.; Taylor, W. F. 1977. Development of Improved Fischer-Tropsch Catalysts for Production of Liquid Fuels. Final report for Contract No. E (46-1)-8003 for U.S. DOE; Morgantown Energy Research Center.

Mathematical Sciences Northwest, Inc. 1974. Feasibility Study: Conversion of Solid Waste to Methanol or Ammonia. Prepared for the City of Seattle; MNSW 74-243-1.

McKee Corporation. 1978. "Ammonia Production from Brava Case, an Economic Feasibility Study." Symposium Papers, Energy from Biomass and Wastes. Washington, D.C.; August 1978. Cleveland, OH: The McKee Corporation.

Natta, G. 1955. "Synthesis of Methanol." Chapter 8 in Catalysis, Vol. 3. New York: Reinhold.

Nelleo, Carl R. 1951. Chemistry of Organic Compounds. London: W. B. Saunders Company.

Ralph M. Parsons. 1977. Screening Evaluation: Synthetic Liquid Fuels Manufacture. Palo Alto, CA: Electric Power Research Institute; EPRI AF-523.

Raphael Katzen Associates. 1975. Chemicals from Wood Wastes. Madison, WI: U.S. Department of Agriculture; Forest Products Laboratory; R.K.A. N441.

Reed, T. B. 1976. A Survey of Our Primary Energy Sources and the Fuel We Can Make from Them." Conference on Capturing the Sun Through Bioconversion; Washington, D.C.; March 1976.

Schooley, F. A. et al. 1978. (Stanford Research Institute). Mission Analysis for the Federal Fuels from Biomass Program. Quarterly Progress Report to DOE. Second Annual Symposium on Fuels from Biomass; Rensselaer Polytechnic Institute; June 20, 1978.

Science Applications, Inc. 1978. Evaluation of the State-of-the-Art of Biomass Based Methanol Processes. McLean, VA: Science Applications, Inc.

Shewin, M.; Blum, D. 1976. Liquid Phase Methanol. Palo Alto, CA: Electric Power Research Institute; Report AF-202, Project 317-1, Annual Report.

Shultz, J. F.; Ableson, M.; Shaw, L.; Anderson, R. B. 1957. "Fischer-Tropsch Synthesis, Nitrides and Carbonitrides of Iron as Catalysts." Industrial and Engineering Chemistry. Vol. 49: pp. 2055-2060.

Slack, A. V.; James, G. R. 1977. Ammonia. Part III. Marcel Dekker, Inc.

Strelzoff, S. 1971. "Methanol: Its Technology and Economics." Methanol Technology and Economics, CEP Symposium. Vol. 66 (No. 98).

Stull, D. et al. 1969. The Chemical Thermodynamics of Organic Compounds. New York: John Wiley and Sons.

University of Connecticut. 1978. Liquids from CO and H_2 for Energy Storage. Palo Alto, CA: Electric Power Research Institute; EPRI AF-689.

Voltz, S. E. et al. 1976. Development Studies on Conversion of Methanol and Related Oxygenates to Gasoline. Mobil Research and Development Corporation; FE-1773-25.

Wilson, E. et al. 1977. Engineering and Economic Analysis of Waste to Energy Systems. Pasadena, CA: Ralph M. Parsons Co.; Report No. 5495-1.

Wise, J. J. et al. 1977. New Technology for the Conversion of Natural Gas to Gasoline and Petrochemicals. Presented at the Third Iranian Congress of Chemical Engineering; Nov. 6-10, 1977.

Chapter 14

Governmental Aids to Commercialization of Air Gasification

T. B. Reed, C. Bendersky, W. Montano
Pyros, Inc.

14.1 INTRODUCTION

The bulk of this report has been a technical evaluation of the past, present, and future of the gasification of biomass. The larger, social, commercial, and governmental issues of implementing gasification technology have not been of major concern. Nevertheless, it is recognized that successful engineering is only the first step in developing a new technology and that the speed of implementation will depend more on institutional factors than technical factors.

In dealing with scientists, engineers, manufacturers, and potential consumers while writing this survey, there have been many discussions on the role of government in implementing gasification. This chapter is a short summary of some of the problems and suggested solutions that are "in the air." At the Air Gasifier Workshop, "Retrofit '79," held February 2, 1979 in Seattle, Charles Bendersky of Pyros, Inc., suggested that the attendees, primarily small, struggling manufacturers trying to sell a new/old technology (gasification) should be in an ideal position to criticize present policies and suggest new ones.

With this in mind, a letter was sent to the 105 attendees asking for their comments and suggestions. Mr. Bendersky and his staff have summarized the replies in the table and the three attachments of Section 14.4. The comments are many and varied, but are summarized with SERI's comments in the following sections.

14.2 BARRIERS TO NEW ENERGY SOURCES

We are now entering a unique historical period in that for the first time our future will be shaped and limited more by the resources of the planet than by our ingenuity in using those resources. Until now, the patterns of our lifestyle have grown from small beginnings at a rate limited by technology, often creating new needs and desires in the process. Presently, our planet's limited energy resources support four billion people. In a terribly short time, substitute and renewable energy resources must be found to maintain this support. These substitutes will have to be developed in a few years. We need "instant" new technologies to satisfy established markets, and they must be developed in competition with well-developed, often well-subsidized technologies.

"For every man trying to open a new path, there are a thousand standing at the crossroads pointing to the old ways." This saying underlines the difficulties that will be encountered as we try to leave the old, comfortable ways of cheap oil and forge new, more costly and less convenient energy sources. It is difficult to conceive of the innumerable barriers unconsciously erected against change. Who will take the first steps?

One of the most formidable of the unconscious barriers is that the United States has had a "cheap energy" policy for most of this century. Probably a wise policy in the early days, this has taken the form of subsidies to energy industries ($77 billion in tax credits to the petroleum industry for exploration and production, according to a recent Battelle report), regulated prices on natural gas, etc., all leading to a high rate of consumption. These same policies, still in effect, make it very difficult for any other energy source to compete.

Another barrier that everyone faces in developing new energy sources is lack of capital. Not only was fossil fuel once cheap, but the equipment to burn it was also much cheaper than that required to burn wood or coal. Capital is in desperately short supply due to the 5-10 fold increase in oil costs (which then correspondingly increases the cost of coal, gas, and all manufactured goods that depend on energy); yet, capital must be found to finance new alternate-energy installations. A major factor favoring gasification is that it supplies a "retrofit fuel" for existing installations, thus reducing capital expenses.

These few barriers are listed as examples of the many barriers to change because frequently they are not obvious, and lack of success may cause frustration without recognizing the hidden causes. Clearly, the continuing decrease in fossil fuel and the concomitant cost increases will force changes to alternatives no matter what the cost. Positive actions that can be taken will be discussed below.

14.3 GOVERNMENTAL AIDS TO GASIFICATION COMMERCIALIZATION

Commercialization, by its very name, is not an activity primarily assigned to government. Nevertheless, government has often had a role in aiding certain developments considered to be in the national interest; for example, the U.S. Government has been very active in developing nuclear energy in cooperation with U.S. industries, and the present close cooperation of government and industry in Japan has rapidly developed new technologies and increased foreign trade.

Because of the energy shortages that developed as a result of the OPEC oil embargo of 1973-74, the U.S. government has announced its intention to help "commercialize" various alternate energy technologies, including solar energy and biomass. With the best of intentions, however, very little has been accomplished by the government towards commercialization of biomass since the establishment of the Energy Research and Development Administration (ERDA) in 1974 and the reorganization of ERDA to form the Department of Energy (DOE) in 1977. Meanwhile, the recent rapid development of wood stoves, forest industry wood use, and gasohol was led by private groups and industries, not government.

Abstracted here are suggestions received in answer to the letters of inquiry (see Section 14.4) and others that have been gleaned from discussions with those in the field.

- Several sizes of gasifiers should be demonstrated in order to raise the public awareness of gasification as one of the most attractive alternatives to the straight combustion of biomass.
- Gasifiers should be installed at government facilities where appropriate, particularly DOE and military installations.
- Large gasifier systems, involving fuel collection, drying, and distribution as well as gasification, should be demonstrated.
- Money should be passed through to the states to support regional energy programs in whatever way the states see fit.
- There should be fuel and equipment subsidies, generally in the form of tax rebates or writeoffs, market guarantees, government purchase, etc., which aid equally all manufacturers in the field or which give potential customers the incentive to use new energy forms.
- Technical and "state-of-the-art" information, such as this survey, should be made available to all interested parties.
- Documentation should be made of the availability of feedstocks in each season to permit the manufacturer and user to assess the degree to which gasification can be implemented.
- A "strike force" of technical, business, and legal experts should be created that can visit various installations or test sites and give advice on possible development options not obvious to the individual.
- An official liaison should be established with the $150-million Canadian biomass program FIRE, instituted in 1978 to promote combustion and gasification of biomass, to learn from their successes and failures.

- Cooperation with foreign governments, which have had extensive experience in the field of gasification, should be instigated.

14.3.1 Attachment 1

From the "Bio Energy Commercialization Incentives" luncheon address by Paul F. Bente, Jr., Executive Director, BioEnergy Council, at the IGT-sponsored Conference on Energy Production from Biomass and Wastes, Orlando, Florida, January 23, 1979:

Keeping national goals and principles in mind, let us move on to several types of incentives that may be considered.

1. One type is to mandate achieving goals without specifying the means. This happened, for example, when the government told the auto industry that its cars had to reach increasingly higher mileage performance over given periods of time, without telling them what had to be done to achieve this end result. An analogy would be to mandate over a period of time the addition to gasoline of increasing amounts of alcohol fuel, regardless of origin, or perhaps even restricted to biomass origin.

2. Another approach is that of building a market by establishing economic subsidies that lower the price of a product to establish its use, much as our country now underwrites the cost of importing oil.

3. Yet another way involves offering incentives to overcome institutional barriers that are chiefly financial in nature. There are many such possibilities to consider, foremost of which are loan guarantees where bank or investor financing cannot otherwise be secured.

4. Loan guarantees have the effect of lowering the interest rate on borrowed money by about 2%. However, loan guarantees, though authorized, are not presently operative in the DOE budget. An amendment is needed to create a line-item in the budget for a loan guarantee program.

5. USDA, through its Farmers Home Adminstration, has an effective loan guarantee program. In addition, the Food and Agriculture Act of 1977 set up a $60 million loan guarantee program to guarantee loans of up to $15 million for four industrial production projects to be selected from competitive proposals.

6. About 30 requests for such assistance were received. On January 12 the Commodities Credit Corporation Board ruled on the first three firms to qualify for such assistance. A guarantee was awarded to ENERCO, Inc., of Langhorn, Pennsylvania, which has a mobile wood pyrolysis unit that can also produce hydrocarbons. The guarantee will cover about $5 million in loans for 45 mobile plants. A second guarantee was made to U.S. Sugars and Savannah Foods for a $15 million loan for facilities at Clevison, Florida to conduct acid hydrolysis of bagasse to sugars that will be fermented to make alcohol. This will be located adjacent to a sugar mill. A third guarantee is being made to Guaranty Fuels, Inc. in Independence, Kansas for $5.8 million in loans covering 2 plants to pelletize forest wastes. Sometime next month the Board will select the fourth firm to be given a loan guarantee under this program. Let us hope that the interest rates which have soared dramatically will not be so high as to stop these projects from materializing.

7. Making direct government loans may even be necessary if a loan guarantee is not a sufficient incentive for lenders, or if interest rates from conventional sources of finance are too high, even with the lower rates made possible by guarantees.

8. Utilities are vitally concerned about being able to get financing for installation of biomass facilities. Offering investor-owned utilities government loans at reduced rates may be necessary to provide a significant incentive for their using biomass as fuel.

9. Another possibility is making an outright grant of funds, possibly on the condition that it must be matched by funds from other sources. This might be necessary to

expand the resource of wood via cultivation, transportation, and energy conversion. Such a program should be applicable to public or private organizations as well as to individuals.

10. Another type of incentive is tax exemption. Under the IRS code, Economic Development Revenue Bonds of up to $1,000,000 are tax exempt if they are issued to finance the cost of some portions of "municipal solid waste facilities." It is considered legally possible to use this vehicle to finance woodfueled electric generating plants. One such case has occurred, but it is questionable if others will. When and if tested, the IRS ruling will have to classify wood residues or wastes as "municipal solid wastes." Quite possibly this may not be the case. This situation could be clarified by amending the IRS act so that it clearly qualifies wood residues or wastes for such commercialization.

11. There are other taxes, such as the inventory tax and the capital gains tax, which can discourage production, harvesting, and use of biomass for energy. Amendments to exempt biomass from these taxes could help to spur commercialization.

12. There are still other possibilities to consider, including amendment to the IRS code for allowing rapid amortization to be applied against the cost of retrofitting or converting an existing energy production unit to use of biomass as a source of energy.

13. Another example might be amending the National Energy Act to allow a 20-40% investment tax credit on the basis of capital costs incurred for converting biomass as a source of energy.

14. We have heard of the solar tax credit that just went into effect for those who install solar devices to heat water, to heat or air condition buildings, or to insulate them. Heating homes with wood, which is stored up solar energy, seems just as deserving and could have a far greater impact, for it is more readily put to use by Mr. Public. Hence, there is a possibility of increasing self-sufficiency of homeowners and reducing their use of gas and oil by amending the law to allow wood heating stoves to qualify under the solar tax credit. [However, it is necessary that wood stoves meet emission standards in high population-density areas.]

15. Another incentive that would be both controversial and complicated to administer is redirecting funds used to pay farmers to set land aside in order to reduce production. Indeed, the funds could be used to pay farmers to produce biomass for fuel. This might be a bio-energy crop to trees, corn, or other crops for conversion to fuels and possibly other valuable coproducts such as feed supplements and fertilizers.

16. Another approach to incentives might be linked to environmental regulations involving the issuing of permits, including grandfathering arrangements. Combustion of biomass materials on a large scale will no doubt require emission control devices, which are expensive. Commercialization incentives might be offered by allowing quick amortization of capital expenditures for such equipment or by providing federal subsidies via procedures such as tax exempt industrial development bonds. Another possibility is to allow an investment tax credit, or to provide Small Business Administration loans of the economic injury type. These are designed to assist small industries that cannot benefit from the other procedures because they don't yet have enough cash flow to take a tax write-off or because they aren't yet making a profit.

Our government might emulate the commercialization effort being put forth by Canada. Canadians already use wood to the extent of 3-1/2 percent of total energy consumption. Their government desires to increase this several fold and last July launched a strong commercialization program earmarking funds to get industry to use more wood. Canada launched 5 programs that commit over $300 million toward commercialization over the next 5 years.

The Forest Industry Renewable Energy (FIRE) program sets up $140 million to be used over a 5-year period to contribute up to 20% of approved capital costs of systems using wood as an energy form. A companion program, Energy from the Forest (ENFOR), provides $30 million over 5 years for a new contracted-out research program to implement large-scale use of forests to provide greater amounts of transportable fuels that will substitute for hydrocarbon fossil fuels in the late 1980s.

To spur these two programs, a series of cost-shared Federal-Provincial agreements will be set up involving a Federal contribution of $114 million allocated over the next 5 years to bring current expensive prototypes to full-scale application. The Provincial contribution will be additional; but if this has been announced, I'm not aware of it.

In addition, a loan guarantee program is being set up to encourage generation of electricity from wood and municipal waste. The first project of its kind in any province is eligible for a guarantee of 50% of loan capital for a direct generating station and 66-2/3% for a cogenerating station.

With the aid of these programs, a 10% contribution of Canada's energy supply is considered possible by the year 2000.

14.3.2 Attachment 2

Specific suggestions of Richard C. Wright:

1. Improve accuracy of media releases. There has been too much controversial and misleading publicity.

2. Differentiate between air-blown coal gas producers and biomass gasifiers. These are entirely different devices.

3. Promote recognition of forest products as equally important for renewable energy sources as for pulp and timber production.

4. Encourage refining raw biomass into a uniform high-grade fuel. This is essential for optimum fuel utilization efficiency.

5. Sponsor voluntary grade or type specifications for refined biomass products. For example, identifying specifications such as ASTM D-396 for fuel oil, or the now obsolete "Commercial Standards" such as CS-95, anthracite coal size standards, etc.

6. Avoid massive financial grants for hardware development. Too much hardware is now being reinvented at public expense.

7. U.S. Federal support for a gasifier industry should be limited. Biomass gasification is now off to a good start. If left to serious competition in private industry, it will develop on a sound basis. Scientific help from a few well-qualified institutions, i.e., Georgia Tech., U. of C. - Davis, etc., will be an advantage. Government grants to more, presently unqualified, agencies are not desirable.

14.3.3 Attachment 3

Summary of provision under Energy Tax Act of 1978 (part of NEA)—from DOE Summary:

1. Business Energy Tax Credits

   A variety of tax credits for investment by business is provided. An additional 10% investment tax credit (nonrefundable except for solar equipment) is provided for investment in:

a. Alternative Energy Property: This applies to boilers and other combustors which use coal or an alternative fuel, equipment to produce alternative fuels, pollution control equipment, equipment for handling and storage of alternate fuels, and geothermal equipment. This credit compliments and provides a major economic underpinning for the coal conversion regulatory program. The credit is not available to utilities.

14.4 RETROFIT '79 FOLLOW-UP

"Appropriate Near-Term Role of Federal Government and Other Actions to Support a U.S. Air Gasifier Industry"

ORGANIZATION Name	Type	Primary Interest	Federal Action	RECOMMENDATIONS State/Local Action	General Comments
Arkansas Power & Light Co. Little Rock, Arkansas	User/Utility	Development of: solid fuel gasifier for cogeneration; close-coupled biomass gasification system capable of switching from coal to wood. Concern: clean fuel availability.	Improve flexibility of combining technologies.		
Bio-Energy Council	Consultant	1. Fixed plant development. 2. Mobile-plant category.	1. '77 Farm Bill, Sec. 1420 pilot project loan guarantee program - possible approach. 2. Direct grants for several small-scale demos (e.g., bus/truck, auto, boat). Perhaps SERI could initiate. Note: Attachment 1 is a list of general bioenergy commercialization incentive suggestions by Dr. Paul Bente.	1. Market guarantees & major gasifier investment tax credit (e.g., EPA-California program).	Bob Kennel/Ultrasystems has concept for "strike-force," i.e., forester, economist, plant engineer on demand who make immediate recommendation on practical conversion of wood through direct burning or gasification.
Biomass Energy Institute, Inc. Winnipeg, Manitoba, Canada	Canadian Government	General			1. Need to identify location and economic statistics of any commercially viable biomass gasifiers, not just those in development stage. 2. Need closer look at shortcomings of stationary-type gasification (automation, ease of control, long-term consistency of operation). 3. Need active experimentation with rotary gasifiers, one of the most constructive activities toward technology commercialization.

Century Research, Inc. Gardena, California	Hardware developer/ manufacturer	1. Feedstock supply. 2. Importation of foreign technology. 3. Financing for air gasifier installation.	1. Government documentation of availability of feedstocks, i,e., ag/animal/wood industry wastes, low-grade lignitic deposits, etc. 2. Encourage importation of foreign technology and related research, development, and engineering experience. 3. Legislation authorizing government guarantee of special type of mortgage loan.		
Richard Wright Energy Research Associates Monroe, Wisconsin	Manufacturer	Economic growth of industry—general.		State funding/local sponsorship preferable —less cost and better able to meet local needs.	Favor "normal evolution," i.e., "hands off" by federal government. See Attachment 2 for specific recommendations.
Environmental Energy Engineering, Inc. Morgantown, West Virginia	Manufacturer	Biomass gasification development—general.	U.S. support of Sweden's already developed facility/staff. This, plus cooperation, will help move U.S. to earliest possible commercialization.		1. Get units operating on modest scale to provide visual exposure—may require subsidy for extra labor needed. Concurrently with above, develop less labor-intensive continuous units, larger units, and more effective units. 2. Establish environmental consequences associated with biogas utilization in small and large units. 3. Update Hessleman gasifier to a continuous operating and compact unit to serve as demo and operating unit for small-scale uses—demo of engine operation and firing existing gas burners of right size.

ORGANIZATION Name	Type	Primary Interest	RECOMMENDATIONS Federal Action	State/Local Action	General Comments
Gorham International, Inc. Portland, Maine	Paper Mill/ (Consultant)	Wood harvesting and distribution.	Specific: interest in DOE funding of joint demo project, with an industrial partner, involving use of downdraft gasifier chipper and dryer at or near harvesting site. (Industry partner to use chips itself or establish fuel distribution system for dry/graded chips.)		Attributes slow growth of industry to dependence of small-scale in-place gasifiers on secondary wastes (sometime negative value)—more rapid growth will require use of primary forest wastes as fuel. Need new/more economical harvesting methods, such as downdraft gasifier chipper/dryer described at left.
Halcyon Associates, Inc. E. Andover, New Hampshire	Hardware developer/ manufacturer	Economic growth of biomass gasifier industry—general.			Technology advancing slowly largely due to lack of DOE aid to smaller companies doing actual inventing/design/development. Small companies have no "in" at DOE to obtain funding for efforts to prove feasibility/practicality/economy in commercial applications. Technology design promotion—get number of units installed and operating, requires liaison with users. Small companies also hampered by terms, conditions, and guarantees required by purchasing agents and bureaucrats. In view of 'inevitable' lack of DOE or other federal support, as public attention turns to alternate fuels, small companies: (a) may form alliances with larger ones—which have "in" with DOE—to obtain funds; (b) go public to get venture capital.

Lamb/Cargate Industries, Ltd. New Westminster British Columbia	Supplier (British Columbia)	Gasifiers in energy-saving-related equipment.	Federal role should be to reduce risks undertaken by supplier & purchaser of new technology. Canada has several such programs: 1. EDP - govt. matches funds with supplier; income from sale divided equally between govt. & supplier. 2. Dept. Energy, Mines, and Resources (DEMR) - offer buyer 25% grant on cost of total energy saving system. 3. DEMR - one-time 66% loan guarantee for financing cogeneration from wood (1 per province).	DOE should reduce excessive time for processing applications.
National Center for Appropriate Technology Butte, Montana	Consultant	Air gasifiers—general.	Should install units in government facilities. DOD—largest energy user—should be prime target. (Would aid self-sufficiency of military installations and be good PR.)	Most private users taking "show me" attitude toward use of air gasifiers.
Pioneer Hi-Bred International, Inc. Des Moines, Iowa	User/Developer (large seed and grain company)	General	In general, government should stay out and let profit-oriented private industry handle. However, tax credit for private industry's investment in technology development might be helpful, although documentation to satisfy tax authorities may be difficult.	Feel their work (use of corn cobs as fuel) different from other alternate fuel projects & not practical to "wait" for government sponsorship. Time required to get govt. support too long.
Ripley & Sun Richland, Washington	?		Funding for development of portable/mobile equipment, and personnel training.	Vertical energy integration needed in agricultural, forestry, and municipal wastes (areas where sources & potential uses physically close). Ag—demo in larger agribusiness sector using available biomass resources, transport, and storage for use in air gasifiers to power farm machinery.

ORGANIZATION Name	Type	Primary Interest	RECOMMENDATIONS Federal Action	State/Local Action	General Comments
Ripley & Sun (continued)					Forestry—collect, transport, process, and transport processed fuel form to sites for use in stokers/gasifiers. Possibly use gen-gas fueled trucks for transport. Municipal wastes—similar to above.
Stanford Research Institute Menlo Park, California	Consultant	Proper design rather than just building gasifiers, which is currently the case.	Federal funding of R&D required to make air-blown gasification a commercially acceptable success. Suggestions: (1) technical and environmental evaluation of operating gasifier; (2) test varied feedstocks in commercial gasifier; (3) thermochemical modeling of data from (2) by computer; (4) cost analysis of biomass pretreatment and handling; (5) cost-effective analysis of preprocessed vs. "as-received" materials for gasification; (6) comparative cost benefit analysis of biomass gasifier vs. combustion unit for refitting gas/oil-fired industrial boilers; (7) study factors around biomass gasifier installation in terms of availability/quality/cost of feedstock, local air pollution and residue disposal regulations, tax incentives for producing syngas, and socioeconomic impact of facility.		
Texas Tech University Lubbock, Texas	Research	Effective Utilization of gasifiers—general.	Federal funding of information programs and demonstrations re: small gasifiers (for transportation and agriculture)—justifiable because small users can't make required technical/economic decisions themselves.		See rather limited utilization of air-blown biomass gasifiers. Direct combustion most effective, for new construction, for using biomass to produce steam, space heating, and electricity.

Texas Tech University (continued)			Adapt World War II gasifier data to today's technology.	See little need for govt. financial support & research in development and testing. Due to problems of high cost and fuel supply, govt. should not intervene but let marketplace determine outcome.
U.S. Forest Service, Forest Products Laboratory Madison, Wisconsin	Federal Government	General	Federal govt. could be helpful in moving gasification technology from pilot stage to commercialization. Specific suggestions: 1. Sec. 1420, '77 Farm Bill pilot project loan guarantees. 2. 1978 NEA—get clarification for manufacturers of how additional 10% tax credit for combustion units not using fossil fuels might apply to gasification units. (See Attachment 3 for summary of provision.) 3. Funding additional research to solve problems re: slag prevention & handling, tar cleanup pressurization, fuel bridging in unit, fuel handling outside unit.	
Vermont Wood Energy Corporation Stowe, Vermont	Hardware developer/ manufacturer	Home heating size gasification units.	Financial assistance for: 1. Development of small residential gasifiers (particularly where socioeconomically beneficial, as in New England). 2. Development of retail fuel distribution system, via aid to interested individuals/ groups.	
Washington State Energy Office Olympia, Washington	State Government	General	1. Conduct gasification workshops every 6-8 months to introduce & educate new prospective private industry users to gasification products. Should also include how to handle dangers of gas use, potentially a significant barrier to commercialization.	

ORGANIZATION Name	Type	Primary Interest	RECOMMENDATIONS Federal Action	State/Local Action	General Comments
Washington State Energy Office (continued)			2. Tax incentives, i.e., rapid write-off of capital investment in gasification equipment.		
Wood Energy Consultants, Charlottetown, Prince Edward Is., Canada	Consultant (Canadian)	General	Sees major problem as lack of capital. Suggests Federal and/or state assistance by: 1. Purchase, by prepayment, a number of gasifiers up to $250,000 per company. These gasifiers would be for future delivery at the stabilized production cost of the future. In the meantime, the manufacturer would have this money to finish development work and be capable of manufacturing units. 2. Loan guarantees to purchasers to buy units so that the financial risk of nonperformance is on the govt. With the massive importation of oil, the Federal govt. is spending much more money than it would lose by the failure of a few "prototype" gasifiers. These gasifiers would help replace oil that may not even be available within 30 years. These guarantees would be only to the extent of the cost of the gasifier. 3. The first installations should be in rural applications near sawmills, where the wood is readily available and the economics make most sense. After these successes, the government can take its purchased units (prepaid as in #1) and retrofit the applicable government buildings. 4. At this point, with working models and successful applications of the technology, the government could order a large enough number to help the manufacturer establish his assembly line. The units under the government guarantee program could go to normal purchasers or the excess into government buildings.		

Chapter 15

Recommendations for Future Gasification Research and Development

T. B. Reed
SERI

15.1 INTRODUCTION

It is believed that the development of biomass gasification should be at the maximum rate possible, consistent with sustainable supplies of feedstock, because biomass can supplement fuel supplies as oil and gas become increasingly costly or unavailable. Gasification can provide the gas needed for clean heat and power in our cities, and it is the basis for the synthesis of liquid fuels, SNG, and ammonia.

This survey outlines the value of gasification, the technical base for future work, and the activities now under way. The various people reading it will draw different conclusions. The conclusions on which work will be based at SERI and towards which it is recommended the national program be guided are given here. These are not immutable, and comment is invited as to their validity and completeness.

This chapter is divided into recommendations on processes and recommendations on systems using those processes.

15.2 BIOMASS AND THERMAL CONVERSION PROCESSES

15.2.1 Pyrolysis Processes

Pyrolysis processes are complementary to gasification processes, since they produce some gas, but also char and oil. Thus, they can produce gaseous fuel for continuous use, while at the same time producing storable liquid and solid fuels that can be used for peak loads or sold on the market.

Charcoal can be produced very simply in existing pyrolysis processes. An evaluation of the degree to which char and charcoal may be used in the evolving renewable energy society is recommended. Presently, charcoal has many uses and commands prices of $80 to $200/ton, depending on its quality. It is used for cooking, water purification, manufacture, chemical synthesis, etc. To what extent could the United States consume more charcoal?

Pyrolysis oils are also produced very simply and cheaply in pyrolysis processes. As produced today, they are smelly, high in oxygen, corrosive, and of uncertain value. However, crude oil was viewed similarly when it was first discovered. An integrated program to evaluate improved methods for oil production and collection, as well as laboratory work on chemical and thermal treatment to make higher-value products from the oil, is recommended.

Pyrolytic gasifiers are not as well developed as oxygen gasifiers, but the majority of the research supported by the EPA and DOE has been in this area. Continuing research and pilot work on many of these systems is recommended, because they promise higher efficiencies and lower costs than oxygen gasification in production of medium- or high-energy gas. However, it is not now clear the degree to which medium-energy gas will be distributed in the United States, and so full-scale development of pyrolytic gasifiers must wait on decisions still to be made on the gas infrastructure in the United States. These decisions hinge on the future costs and availability of natural gas versus the costs of conversion of gas to methane for distribution. One possible development would be the use of medium-energy gas from biomass in captive installations and industrial parks, combined with conversion of coal to methane for domestic distribution.

Top-priority development of flash pyrolysis processes that give a high yield of olefins and little oil or char is recommended. The olefins, in turn, can be converted directly to gasoline or alcohols. This seems to be the one truly new development in gasification since World War II. Evaluation of time-temperature and of various feedstocks and particle size options on yields at the bench level, combined with bench and engineering studies of process designs giving the very high heat transfer necessary to produce these nonequilibrium products is recommended. Evaluation of processes for reducing particle size at reasonable costs is also recommended, since this may be a necessary adjunct to flash pyrolysis. Fast pyrolysis is a major part of the biomass thermal conversion program at SERI.

Finally, a continuing effort to sort out the molecular details of pyrolysis under carefully controlled, but realistic, laboratory conditions is recommended to provide a firm foundation for understanding and improving all gasification processes. For this purpose, a molecular beam sampling apparatus is being assembled at SERI to examine the molecular details of the pyrolysis reactions. In addition, thermogravimetric techniques are being used to study the mechanisms of thermal pyrolysis.

15.2.2 Air Gasification

Air gasifiers may find a place in domestic and commercial heating; such gasifiers will certainly be used in process heating and power for the biomass industries. Although research may improve air gasification, immediate commercialization at the present level of development is recommended. A gasification reactor has been constructed at SERI to make accurate measurements of the temperatures and compositions associated with each stage of air, oxygen, and steam gasification.

An expanded support for commercialization of air gasification at the national level is recommended. Many states are already buying gasifiers in the 1-100 MBtu/h range, appropriate for process heat in small- to medium-sized industries. Evaluative technical assistance and tax incentives would accelerate this effort.

There are no air gasifiers presently available that are larger than 100 MBtu/h—yet larger sizes are needed, for instance, to retrofit the very large boilers of the paper industry, which collectively burn 1 to 2 quad of oil. A joint government/industry effort to develop very large air gasifiers suitable for retrofitting large boilers is recommended.

15.2.3 Oxygen Gasification

Development of a high-pressure oxygen gasifier capable of producing clean gas directly rather than by downstream treatment is recommended. This gas would be useful for synthesis of liquid fuels and ammonia, for limited pipeline distribution, or for operation of turbines for combined cycle-cogeneration. The present SERI program includes operation of a 100 lb/h proof-of-concept gasifier of this type. Development of oxygen gasifiers for municipal waste is also recommended, since the use of waste provides energy foı urban areas, recycles metals, and eliminates landfills.

Support is recommended for research on energy-efficient and smaller-scale methods for separation of oxygen from air.

15.2.4 New Gasification Methods

There should be continuing studies of the scientific feasibility of novel thermochemical schemes to gasify biomass to a variety of desired products (e.g., C_2H_2).

15.3 BIOMASS THERMAL CONVERSION SYSTEMS

In the past, economies of large scale have favored the use of coal as a gasification feedstock, while ease and cleanliness of gasification have favored using biomass. Now, biomass is produced in much larger quantities than previously (up to 3000 tons/day in modern paper plants and 10,000 tons/day of SMW in larger cities). Other factors that may favor the use of biomass as a gasifier feedstock will be improved methods and materials of construction, particularly new high-temperature, low-U-factor insulations; new methods of automatic sensing and control using microprocessors; and mass production of smaller units rather than individual engineering of large units.

There are a number of system studies that should also be performed as adjuncts to the biomass gasification program. We recommend that the relevance of scale to gasification plants should be studied immediately and, where appropriate, programs be initiated to overcome scale limitations. In particular, coal is likely to supply gas heat for our cities, where large plants can clean the gas sufficiently and make methane for distribution. Because biomass is much cleaner, it can be used on a smaller scale—and this is compatible with its wider distribution. If biomass residues must be processed at the 1000 ton/day level or greater to be economically viable, very little biomass will be used in this country. If it can be processed economically at the 100 ton/day level, it can be used widely.

A system study of biomass energy refineries is recommended to be used in conjunction with farming and forestry operations, taking in residues and converting them to the ammonia and fuel required to operate the farm and forest, while shipping any surplus of energy to the cities in the form of gaseous or liquid fuels.

15.4 BIOMASS BENEFICIATION

There are a number of processes that are being developed in the laboratory or commercially that alter the form of the biomass to make it more susceptible to thermal or biological processing. While not a direct part of biomass gasification, such processing can increase the ease and efficiency of conversion and would aid the integrated gasification program. They include:

- Densification—pelletizing of miscellaneous biomass forms to uniform pellets, briquettes, or logs that are easier to store and process than the natural forms
- Comminution—The pyrolysis rate depends on heat transfer to the biomass surface, followed by heat transfer through the biomass. The latter step is limiting in many cases, and use of small-particle biomass can affect both the process efficiency and the product distribution. A number of new, interesting processes for comminution are now being developed.
- Drying—Most gasification processes operate best on dry biomass, and a number of ingenious systems can be used for moisture reduction.
- Thermolysis—There are indications that some of the above processes also cause fundamental chemical changes that alter the energy content and structure of the biomass, making further thermal or biological processing more effective.

15.5 BIOMASS PRODUCTION/CONVERSION SYSTEMS

For the longer term, and for biomass conversion plants of large and small scale, economic analyses should be performed to identify suitable hybrid schemes. These include: production of methanol using a combination of biomass (low H/C ratio) and natural gas (high H/C ratio); joint electrolysis/gasification systems in which H_2 and O_2 are generated electrolytically, the oxygen is consumed in gasification, and the hydrogen increases the H/C ratio; and solar flash pyrolysis in which the high rate of heat transfer is supplied by solar collectors.

In the larger analysis, production of biomass should be an integral part of conversion processes. Therefore systems studies are recommended that include integral "energy farms," or "energy plantations," in which the central processing plant may produce fuels, chemicals, and fertilizers needed for increased production of biomass.

Finally, the production of biomass must be regarded as a steady-state activity for any continuing society. The fall of many past civilizations can be traced to an abuse of the land engendered by the pursuit of ever-increasing biomass yields. Therefore, it is recommended that the long-range ecological effects of various land-use patterns be evaluated as soon as possible. It is recommended that these studies consider biomass production for energy as an opportunity for land improvement, as well as considering its possible role in land degradation.